Topics in Atomic Physics

Charles E. Burkhardt

Department of Physics, St. Louis Community College,
St. Louis, Missouri

Jacob J. Leventhal

Department of Physics, University of Missouri—St. Louis,
St. Louis, Missouri

Topics in Atomic Physics

With 75 Illustrations

 Springer

Charles E. Burkhardt
Department of Physics
St. Louis Community College
St. Louis, MO 63121
USA

Jacob J. Leventhal
Department of Physics
University of Missouri—St. Louis
St. Louis, MO 63121
USA

ISBN 978-1-4419-2068-3   e-ISBN 978-0-387-31074-9

Printed on acid-free paper.

Printed in the United States of America.   (TB/EB)

9 8 7 6 5 4 3 2 1

springeronline.com

Helen, Charlie, Sarah, and Michelle
Bette, Steve, Andy, Dan, and Tina

Preface

The importance of the field of atomic physics to modern technology cannot be overemphasized. Atomic physics served as a major impetus to the development of the quantum theory of matter in the early part of the twentieth century and, due to the availability of the laser as a laboratory tool, it has taken us into the twenty-first century with an abundance of new and exciting phenomena to understand. Our intention in writing this book is to provide a foundation for students to begin research in modern atomic physics. As the title implies, it is not, nor was it intended to be, an all-inclusive tome covering every aspect of atomic physics.

Any specialized textbook necessarily reflects the predilection of the authors toward certain aspects of the subject. This one is no exception. It reflects our belief that a thorough understanding of the unique properties of the hydrogen atom is essential to an understanding of atomic physics. It also reflects our fascination with the distinguished position that Mother Nature has bestowed on the pure Coulomb and Newtonian potentials, and thus hydrogen atoms and Keplerian orbits. Therefore, we have devoted a large portion of this book to the hydrogen atom to emphasize this distinctiveness. We attempt to stress the uniqueness of the attractive $1/r$ potential without delving into group theory. It is our belief that, once an understanding of the hydrogen atom is achieved, the properties of multielectron atoms can be understood as departures from hydrogenic properties.

From the beginning, it was our intention to include information in this book that is not easily located elsewhere. Thus, while the book can be used as a text, it is hoped that it will also be a useful reference. To this end, we have incorporated derivations that are difficult to find in other books and, indeed, even in the literature. For example, the quantum mechanical Lenz vector operator is not often discussed in books on quantum mechanics and atomic physics. When it is discussed, it is usually stated that it commutes with the hydrogen atom Hamiltonian, but it is difficult to prove. However, this book gives this proof in some detail. In fact, one of the general features of our book is that we often include more algebraic steps than are traditionally given in textbooks. By doing this, we wish to relieve the reader of the tedium of reproducing algebra and, thus, permit concentration on the physics.

This material is intended to be suitable for a one-semester graduate or advanced undergraduate-level course in atomic physics. It is assumed that the student has had at least introductory quantum mechanics, although pertinent topics such as perturbation theory and variational techniques are briefly reviewed. Chapter 1 presents some background material which, in large part, is review. In this chapter the utility of the Bohr atom is discussed and the derivation performed as Bohr did it. This is in contrast to most modern presentations. Chapters 2 and 3 are standard reviews of angular momentum algebra with emphasis on aspects pertinent to atomic physics. Chapter 4 is a discussion of the quantum mechanical hydrogen atom and the separation of the Schrödinger equation in both spherical and parabolic coordinates. Emphasis is placed on the conditions that force quantization. This chapter also includes an attempt to clarify the difference between two commonly used definitions of Laguerre and associated Laguerre polynomials. In most treatments of the quantum mechanical hydrogen atom, no mention is made of alternate definitions of these special functions. Chapter 5 is a discussion of the classical hydrogen atom. Naturally it draws heavily on Keplerian orbits and the consequences of the additional constant of the classical motion, the Lenz vector. Chapter 6 discusses the accidental degeneracy of the hydrogen atom and its causes in the context of the quantum mechanical analog of the Lenz vector. To our knowledge, the material in Chapter 6 is not covered in any modern text. Chapter 7 discusses the breaking of the accidental degeneracy via fine structure, the Lamb shift, and hyperfine structure. The treatment is fairly standard. Chapter 8 treats the hydrogen atom in external fields. The description of the Zeeman effect is standard, but the weak field Stark effect is described in quantum mechanically and classically. The classical treatment leans heavily on the results of Chapter 5 while the quantum mechanical treatment exploits the operator formalism developed in Chapter 6.

Chapters 9 and 10 are discussions of multielectron atoms, beginning with helium in Chapter 9. The presentation is quite standard. Chapter 11 presents the quantum defect in a way that is seldom seen in texts. In keeping with the theme of this book, the quantum defect is related to classical concepts and the correspondence principle. Chapter 12 deals with multielectron atoms in external fields. Here again the Zeeman effect is treated in the standard manner, but the Stark effect is presented in a way that leans heavily on the material presented in Chapter 11. Finally, in Chapter 13, radiation is discussed at length. We emphasize how the concept of a stationary state is not at odds with the classical concept of radiation from accelerating charges. Otherwise, the presentation in this chapter is standard, but, we hope, thorough.

SI units are used except in those instances for which we believe that atomic units are considerably more convenient. For instance, the Zeeman effect is treated using SI units because the Bohr magneton times the magnetic induction field provides a convenient measure of the Zeeman energy. However, we find atomic units to be more convenient in the treatment of the Stark effect.

There are more than eighty problems listed at the ends of chapters, with selected answers given at the end of the book. Some are merely exercises, but others are more challenging. A few are derivations of results that are used later in the book. A detailed solutions manual, in many cases showing more than one way to solve a given problem, is available to qualified instructors: visit www.springeronline.com/0-387-25748-9.

A list of corrections to the book is available on the Internet at:

http://www.umsl.edu/ jjl/homepage/

and

http://www.stlcc.edu/fv/physics/CBurkhardt/

Comments or previously unreported errors can be transmitted to the authors through these Web sites.

We wish to thank our graduate students Joseph F. Baugh, Marco Ciocca, and Lucy Wenzhong He. Thanks are also due to the numerous undergraduates who have worked in the UMSL Atomic Physics Laboratory over the past 40 years. We also wish to acknowledge the contributions of our many collaborators and our UMSL colleagues, Ta-Pei Cheng, Wayne P. Garver, and Philip B. James with whom we had many valuable discussions over the years.

Thanks also to Wai-Yim Ching of UMKC for many valuable comments on the manuscript during its preparation.

Charles E. Burkhardt
Jacob J. Leventhal

Contents

Preface... vii

Chapter 1 Background................................... 1
1.1 Introduction... 1
1.2 The Bohr Model of the Atom.................................. 1
1.3 Numerical Values and the Fine Structure Constant............. 7
1.4 Atomic Dimensions—Is a_0 a Reasonable Atomic Diameter?.... 8
1.5 Localizing the Electron: Is a Point Particle Reasonable?...... 10
1.6 The Classical Radius of the Electron......................... 11
1.7 Atomic Units.. 11

Chapter 2 Angular Momentum....................... 14
2.1 Introduction.. 14
2.2 Commutators.. 18
2.3 Angular Momentum Raising and Lowering Operators.............. 20
2.4 Angular Momentum Commutation Relations with
     Vector Operators... 25
2.5 Matrix Elements of Vector Operators......................... 26
2.6 Eigenfunctions of Orbital Angular Momentum Operators......... 29
2.7 Spin.. 33
2.8 The Stern–Gerlach Experiment................................ 41

Chapter 3 Angular Momentum—Two Sources........... 46
3.1 Introduction.. 46
3.2 Two Sets of Quantum Numbers—Uncoupled and Coupled........... 46
3.3 Vector Model of Angular Momentum........................... 51
3.4 Examples of Calculation of the Clebsch–Gordan Coefficients... 55
3.5 Hyperfine Splitting in the Hydrogen Atom.................... 61

Chapter 4   The Quantum Mechanical Hydrogen Atom.... 73
4.1   The Radial Equation for a Central Potential 73
4.2   Solution of the Radial Equation in Spherical
        Coordinates—The Energy Eigenvalues 75
4.3   The Accidental Degeneracy of the Hydrogen Atom 77
4.4   Solution of the Hydrogen Atom Radial Equation in
        Spherical Coordinates—The Energy Eigenfunctions 79
4.5   The Nature of the Spherical Eigenfunctions 82
4.6   Separation of the Schrödinger Equation in
        Parabolic Coordinates 82
4.7   Solution of the Separated Equations in Parabolic
        Coordinates—The Energy Eigenvalues 85
4.8   Solution of the Separated Equations in Parabolic
        Coordinates—The Energy Eigenfunctions 87

Chapter 5   The Classical Hydrogen Atom 92
5.1   Introduction 92
5.2   The Classical Degeneracy 95
5.3   Another Constant of the Motion—The Lenz Vector 97

**Chapter 6   The Lenz Vector and the
Accidental Degeneracy 105**
6.1   The Lenz Vector in Quantum Mechanics 105
6.2   Lenz Vector Ladder Operators; Conversion of a Spherical
        Eigenfunction into Another Spherical Eigenfunction 109
6.3   Application of Lenz Vector Ladder Operators to a
        General Spherical Eigenfunction 114
6.4   A New Set of Angular Momentum Operators 116
6.5   Energy Eigenvalues 118
6.6   Relations Between the Parabolic Quantum Numbers 120
6.7   Relationship Between the Spherical and
        Parabolic Eigenfunctions 122
6.8   Additional Symmetry Considerations 123

Chapter 7   Breaking the Accidental Degeneracy 126
7.1   Introduction 126
7.2   Relativistic Correction for the Electronic Kinetic Energy 127
7.3   Spin-Orbit Correction 128
7.4   The Darwin Term 130
7.5   Evaluation of the Terms That Contribute to the Fine-Structure
        of Hydrogen 130

7.6   The Total Fine-Structure Correction.. 135
7.7   The Lamb Shift.. 137
7.8   Hyperfine Structure... 139
7.9   The Solution of the Dirac Equation... 142

Chapter 8   The Hydrogen Atom in External Fields......... 145

8.1   Introduction... 145
8.2   The Zeeman Effect—The Hydrogen Atom in a Constant
       Magnetic Field.. 146
8.3   Weak Electric Field—The Quantum Mechanical Stark Effect........ 159
8.4   Weak Electric Field—The Classical Stark Effect......................... 171

Chapter 9   The Helium Atom..................................... 178

9.1   Indistinguishable Particles.. 178
9.2   The Total Energy of the Helium Atom.. 180
9.3   Evaluation of the Ground State Energy of the Helium Atom
       Using Perturbation Theory.. 183
9.4   The Variational Method... 186
9.5   Application of the Variational Principle to the Ground
       State of Helium... 187
9.6   Excited States of Helium.. 189
9.7   Doubly Excited States of Helium: Autoionization....................... 192

Chapter 10   Multielectron Atoms............................... 196

10.1   Introduction... 196
10.2   Electron Configuration.. 196
10.3   The Designation of States—LS Coupling 198
10.4   The Designation of States—jj Coupling.................................. 207

Chapter 11   The Quantum Defect 214

11.1   Introduction... 214
11.2   Evaluation of the Quantum Defect.. 216
11.3   Classical Formulation of the Quantum Defect and the
        Correspondence Principle.. 220
11.4   The Connection Between the Quantum Defect and the
        Radial Wave Function.. 225

Chapter 12   Multielectron Atoms in External Fields 230

12.1   The Stark Effect... 230
12.2   The Zeeman Effect... 238

Chapter 13   Interaction of Atoms with Radiation **246**

13.1   Introduction 246

13.2   Time Dependence of the Wave Function 248

13.3   Interaction of an Atom with a Sinusoidal
          Electromagnetic Field 249

13.4   A Two-State System—The Rotating Wave Approximation 251

13.5   Stimulated Absorption and Stimulated Emission 254

13.6   Spontaneous Emission 260

13.7   Angular Momentum Selection Rules 266

13.8   Selection Rules for Hydrogen Atoms 267

13.9   Transitions in Multielectron Atoms 272

Answers to Selected Problems **279**

Index **285**

1
Background

1.1. Introduction

A student's introduction to atomic physics usually occurs as an undergraduate in a course entitled "Modern Physics". The subject is introduced by describing the results of some early twentieth century experiments that demonstrated the inadequacy of classical physics to describe events on the microscopic scale of atoms. Also included in these introductory courses are descriptions of theoretical efforts to explain the surprising experimental results.

Because these concepts are counter to everyday experience, it is often the case that the student does not fully appreciate their significance. Therefore, in this chapter we again present some of this introductory material from the perspective of a student who has, by now, taken a first course in quantum mechanics. This review will, it is hoped, lead the student to a fuller comprehension of some concepts that, having learned quantum mechanics, are taken for granted.

The machinery used in this book is nonrelativistic quantum mechanics, also sometimes referred to as "first quantization" or "Schrödinger wave mechanics". In this formulation, the particle energies are determined by the necessity of "fitting" deBroglie wavelengths into a "box," the dimensions of which are determined by the system. In this book, this box is almost always an atom, but, for example, nuclei and nucleons also confine particles and, in this context, may be regarded as boxes. In first quantization electromagnetic fields are treated classically and the particle motion quantized. When the electromagnetic field is also quantized, "second quantization", additional observable atomic effects are described.

1.2. The Bohr Model of the Atom

Inevitably, the study of atomic physics requires application of the principles of quantum mechanics. Although the required formalism can be quite sophisticated, the physicist's view of atoms remains a planetary system. Probability distributions notwithstanding, most physicists envision the Bohr model of the atom as a first approximation. But, the Bohr model is wrong! Atomic electrons do not execute well-defined Keplerian orbits. Why then, students ask, study the Bohr atom? There are many reasons, but three of them are:

1. Atomic parameters obtained from the Bohr model scale more or less correctly.
2. The result for the energy of the hydrogen atom, indeed, all one-electron atoms, is correct.
3. The units of length, time, electric field, and so on in a widely used system of units, atomic units, can be easily related to the atomic parameters obtained from the Bohr model.

Because of the importance of the Bohr model to concepts in atomic physics we present a derivation of the atomic parameters obtained from it. We begin by stating Bohr's two assumptions in his own words:[1]

Assumption I: That an atomic system can, and can only, exist permanently in a certain series of states corresponding to a discontinuous series of values for its energy, and that consequently any change of the energy of the system, including emission and absorption of electromagnetic radiation, must take place by a complete transition between two such states. These states will be denoted as the "stationary states" of the system.

Assumption II: That the radiation absorbed or emitted during a transition between two stationary states is "unifrequentic" and possesses a frequency f, given by the relation

$$E' - E'' = hf \tag{1.1}$$

where h is Planck's constant and where E' and E'' are the values of the energy in the two states under consideration.

A third assumption, although not stated as an assumption by Bohr, is the correspondence principle. Loosely stated, the correspondence principle asserts that as microscopic (quantum) systems become macroscopic (classical) the quantum result must go over to the classical.

Assumption I above was quite heretical because, according to classical electrodynamics, accelerating charges radiate away their energy. An electron in a circular orbit is surely accelerating. Bohr stated, however, that in these special favored "stationary states" the electron could execute circular motion without radiating away its energy. Classically, of course, as the electron loses energy by emitting radiation, it would slow down and ultimately spiral into the nucleus. Calculations show that it would take on the order of 10^{-9} seconds for the electron to spiral into the nucleus.

Assumption II was also at odds with classical concepts because, classically, the radiation given off by a periodically accelerating charge should have the same frequency as the charge. There seems to be no relation between the frequency of radiation predicted by Bohr's second assumption and the laws of classical electrodynamics.

To calculate the atomic parameters resulting from Bohr's assumptions we use classical physics together with the empirically determined Balmer formula for the wavelengths of radiation emitted by hydrogen atoms. The model assumes circular

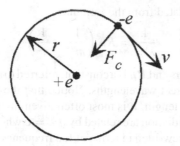

FIGURE 1.1. Parameters used in the Bohr model of the atom.

orbits in which the electron is bound to the proton by the Coulomb force. The parameters used in the derivation are shown in FIGURE 1.1.

In SI units, the Coulomb force on the electron executing a circular orbit of radius r provides the centripetal acceleration that keeps the electron in the circular orbit. We may obtain the total energy by equating the centripetal force to the Coulomb force that keeps the electron in orbit. We have

$$\frac{e^2}{4\pi\varepsilon_0} \cdot \frac{1}{r^2} = \frac{mv^2}{r} \tag{1.2}$$

and the kinetic energy is

$$T = \frac{1}{2} \cdot \left(\frac{e^2}{4\pi\varepsilon_0}\right) \cdot \frac{1}{r} \tag{1.3}$$

The total energy is then

$$\begin{aligned} E &= T + V \\ &= \frac{1}{2} \cdot \left(\frac{e^2}{4\pi\varepsilon_0}\right) \cdot \frac{1}{r} - \frac{e^2}{4\pi\varepsilon_0} \cdot \frac{1}{r} \\ &= -\frac{1}{2} \cdot \frac{e^2}{4\pi\varepsilon_0} \cdot \frac{1}{r} \end{aligned} \tag{1.4}$$

The only variable in this equation is r. Therefore, quantizing the energy must somehow involve quantizing the orbital radius of Bohr's stationary states. If a photon is emitted in a transition between states of energies E' and E'' (we assume that $E' > E''$) then the frequency f is

$$\begin{aligned} hf &= E' - E'' \\ &= \frac{1}{2}\left(\frac{e^2}{4\pi\varepsilon_0}\right)\left(\frac{1}{r'} - \frac{1}{r''}\right) \end{aligned} \tag{1.5}$$

where r' and r'' are the coordinates of the electron in the states of energy E' and E'', respectively. Now, Bohr knew of the famous Balmer formula that predicted accurately the wavelengths of radiation given off in an electrical discharge of hydrogen atoms. According to the Balmer formula, the wavelengths of the hydrogen transitions can be calculated from the formula

$$\frac{1}{\lambda_{nm}} = R_H \left(\frac{1}{m^2} - \frac{1}{n^2} \right) \tag{1.6}$$

where n and m are integers and R_H is a constant, referred to as the Rydberg constant, required to give the correct wavelengths. Notice that the Rydberg constant must have units of reciprocal length. It is most often given in cm^{-1}. But the frequency and wavelength of the radiation are related by $f\lambda = c$ where c is the speed of light so this formula may be rewritten in terms of the frequency and compared with the expression for hf derived from Bohr's second postulate. We then have

$$hf_{nm} = cR_H \left(\frac{1}{m^2} - \frac{1}{n^2} \right) \tag{1.7}$$

which has the same form as Bohr's expression for the photon energy if the orbital radii are quantized as

$$r_n = a_0 n^2 \tag{1.8}$$

where the constant a_0 has units of length and is referred to as the Bohr radius. It represents the radius of the electronic orbit when the electron is in the lowest of the stationary states.

How do we calculate a_0? The method that is most often employed in elementary texts is to assume that Bohr postulated that the angular momentum of the electron is quantized in units of $\hbar$. The electronic angular momentum in the nth Bohr orbit is then

$$m_e v_n(a_0 n^2) = n\hbar \tag{1.9}$$

where m_e is the mass of the electron and v_n is the velocity of the electron in the nth Bohr orbit and the radius r_n has been replaced by Equation (1.8). The velocity can be eliminated from Equation (1.9) using the kinetic energy

$$T = \frac{1}{2} m_e v_n^2$$
$$= \frac{1}{2} \cdot \left(\frac{e^2}{4\pi\varepsilon_0} \right) \cdot \frac{1}{(a_0 n^2)} \tag{1.10}$$

thus leading to an expression for a_0. It is, however, absurd to think that Bohr had a divine revelation that led him to postulate that the angular momentum was quantized in units of $\hbar$. Why not h, or some other multiple of h? In fact, Bohr formulated and used what is now called the correspondence principle to obtain the quantized energy. It reveals the true genius of Niels Bohr.

To find the energy as Bohr did we recall that, according to classical electro-dynamics the accelerating electron should emit radiation at the frequency of the acceleration. Bohr assumed that the frequency of the radiation as given by his Assumption II above must approach the frequency of the electron in its stationary state for adjacent highly energetic states, that is, states of large r_n. The frequency of the electron in an orbit of radius r_n is

$$f_{orbit} = \frac{\text{orbital speed}}{\text{circumference of orbit}}$$

$$= \frac{v_n}{2\pi (a_0 n^2)} \tag{1.11}$$

Substituting for v_n from Equation (1.10) and working with the square of the frequency we have

$$f_{orbit}^2 = \frac{1}{4\pi^2 n^4 a_0^2} \cdot \left[\frac{1}{m_e} \cdot \left(\frac{e^2}{4\pi\varepsilon_0} \right) \cdot \frac{1}{n^2 a_0} \right] \tag{1.12}$$

Using Bohr's second assumption we may calculate $f_{radiation}^2$ for transitions between adjacent stationary states characterized by "quantum numbers" n and $(n+1)$. It is

$$f_{radiation}^2 = \left\{ \frac{1}{2} \cdot \left(\frac{e^2}{4\pi\varepsilon_0} \right) \frac{1}{ha_0} \left[\frac{1}{n^2} - \frac{1}{(n+1)^2} \right] \right\}^2$$

$$= \left\{ \frac{1}{2} \cdot \left(\frac{e^2}{4\pi\varepsilon_0} \right) \frac{1}{ha_0} \left[\frac{(2n+1)}{n^2 (n+1)^2} \right] \right\}^2 \tag{1.13}$$

The limit of this as n becomes large is

$$\lim_{n \to \infty} f_{radiation}^2 = \left\{ \left(\frac{e^2}{4\pi\varepsilon_0} \right) \frac{1}{ha_0} \cdot \frac{1}{n^3} \right\}^2 \tag{1.14}$$

which we may equate to the expression for f_{orbit}^2 in Equation (1.12) and solve for a_0 to obtain

$$a_0 = (4\pi\varepsilon_0) \cdot \frac{\hbar^2}{m_e e^2}$$

$$= 0.529 \times 10^{-11} \text{ m} \tag{1.15}$$

It is now a simple matter to compute the energy of the electron in nth orbit as well as the Rydberg constant. The Bohr energy is

$$E_n = -\frac{1}{2} \cdot \left(\frac{e^2}{4\pi\varepsilon_0 r_n} \right)$$

$$= -\left(\frac{e^2}{4\pi\varepsilon_0} \right) \cdot \frac{1}{2n^2 a_0} \tag{1.16}$$

which turns out to be the correct energy as calculated using Schrödinger wave mechanics. This energy, of course, gives the correct wavelengths in the spectrum of atomic hydrogen.

We can obtain the Rydberg constant by comparing Equation (1.16) with Equation (1.7) which yields

$$R_H = \frac{m_e e^4}{8\varepsilon_0^2 h^3 c}$$
$$= 2.1798741 \times 10^{-18} \text{ J}$$
$$= 13.6 \text{ eV} \tag{1.17}$$

in agreement with the empirically determined value. Because the joule is a very large unit, the electron volt, abbreviated eV, is often used in atomic physics. One electron volt is defined as the kinetic energy gained by an electron that is accelerated through a potential difference of one volt. Also in wide use in atomic physics as a unit of energy is the inverse centimeter, abbreviated cm^{-1}, but read "inverse centimeters" or, more simply "wave numbers". (In fact, the cm^{-1} has an "official" name, the Kaiser, but nobody uses it.) A simple way of viewing these units is as follows. The energy interval in wave numbers between two quantized states is the reciprocal of the wavelength in centimeters of a photon that would be emitted in a transition between these two states, according to Bohr's second assumption. In terms of these units the Bohr energy is

$$E_n^{(0)} = \frac{109{,}737 \text{ cm}^{-1}}{n^2}$$
$$= \frac{R_H}{n^2} \tag{1.18}$$

In actuality, R_H in the above equation is not precise. The experimentally determined value, denoted R_∞, is $R_\infty = 109{,}737.31568525 \text{ cm}^{-1}$. The reason that R_H is not precise is that it ignores effects such as the electron-proton reduced mass (which we are ignoring here). Sometimes the energy is measured in "Rydbergs". One Rydberg is the ionization energy of hydrogen, 13.6 eV. Thus, the lowest energy state of hydrogen has an energy of -13.6 eV. This lowest energy state of any atom is referred to as the "ground state" and any other states of the atom are "excited states".

Finally, we can obtain the angular momentum of the electron in the nth Bohr orbit. From Equations (1.11) and (1.12) the velocity of the electron in the nth Bohr orbit is

$$v_n = \frac{\hbar}{m_e a_0} \cdot \frac{1}{n} \tag{1.19}$$

so the angular momentum is

$$L_n = m_e v_n r_n$$
$$= m_e \cdot \left(\frac{\hbar}{m_e a_0} \cdot \frac{1}{n} \right) \cdot (n^2 a_0)$$
$$= n\hbar \tag{1.20}$$

As noted previously though, this result is a consequence of the genius of Bohr in deducing the correspondence principle, not a magical pronouncement that angular momentum is quantized in units of $\hbar$. In fact, this "postulate" is incorrect! According to it, the angular momentum in the ground state is one unit of angular

TABLE 1.1. Bohr parameters in SI units and their dependences on the principal quantum number.

Orbital parameter	Bohr parameter in SI units for the nth orbit	Scaling
Energy	$E_n = -\dfrac{1}{2}\left(\dfrac{1}{4\pi\varepsilon_0}\right)^2\left(\dfrac{m_e e^4}{\hbar^2}\right)\cdot\dfrac{1}{n^2}$	$-R_H/n^2$
Radius	$r_n = \left(\dfrac{4\pi\varepsilon_0\hbar^2}{m_e e^2}\right)n^2$	$a_0 n^2$
Velocity	$v_n = \left(\dfrac{e^2}{4\pi\varepsilon_0\hbar}\right)\dfrac{1}{n}$	$\dfrac{v_1}{n}$
Period	$\tau_n = \left[(2\pi)\sqrt{\dfrac{(4\pi\varepsilon_0)m_e}{e^2}}\right]a_0^{3/2}n^3$	$\tau_1 n^3$
Frequency	$\omega_n = \dfrac{2\pi}{T_n}$	$\dfrac{\omega_1}{n^3}$

momentum $\hbar$. We now know that the electronic angular momentum in the ground state is actually zero. Thus, not only is the picture envisioned by the Bohr model incorrect, but it also gives *some* incorrect answers. Nonetheless, the quantities obtained from the model provide reasonable approximations to atomic parameters. TABLE 1.1 is a listing of some of these parameters in SI units.

1.3. Numerical Values and the Fine Structure Constant

Because the Bohr model of the atom provides order of magnitude estimates of atomic parameters, it is useful to cast the Bohr parameters in convenient form. This is easily done by writing them in terms of the fine structure constant α, a dimensionless quantity that is, in fact, a measure of the strength of the electromagnetic interaction. It is defined as

$$\alpha = \left(\frac{1}{4\pi\varepsilon_0}\right)\cdot\frac{e^2}{\hbar c}$$

$$\approx \frac{1}{137} \tag{1.21}$$

It is a pure number. That is, it has no units and is the same in every system of units. In terms of the fine structure constant the energy of a hydrogen atom is given by

$$E_n^{(0)} = \frac{(m_e c^2)\alpha^2}{2n^2} \tag{1.22}$$

This expression for the energy is particularly convenient because most physics students know that the rest mass of the electron $m_e c^2$ is 0.51 MeV. Because $\alpha^2 \approx 10^{-4}$ it is clear that atomic energies are in the electron volt range.

Suppose that, instead of a hydrogen atom, we have a one-electron atom with nuclear charge Z, for example, the He^+ or Li^{++} ions. What is the correct expression

TABLE I.2. Bohr parameters in terms of the fine structure constant.

Parameter	Formula	Numerical value for $Z = 1$
Rydberg constant $= R_H$	$\dfrac{1}{2}(Z\alpha)^2(m_e c^2)$	$13.6\,\text{eV} = 109{,}737.31568525\,\text{cm}^{-1}$
Bohr radius $= a_0$	$\dfrac{1}{Z}\dfrac{1}{\alpha}\left(\dfrac{\hbar}{m_e c}\right)$	$137 \cdot (3.86 \times 10^{-13}\,\text{m}) = 5.29 \times 10^{-11}\,\text{m}$
Electron velocity in the first Bohr orbit $= v_1$	$Z\alpha c$	$2.19 \times 10^6\,\text{m/s}$
Period in first Bohr orbit	$\dfrac{1}{\alpha}\left(\dfrac{2\pi a_0}{c}\right)$	$1.345 \times 10^{-16}\,\text{s}$

for the energy levels? The Coulomb force for such an atom (ion) is

$$F(r) = \frac{1}{4\pi\varepsilon_0} \cdot \frac{(Ze)(-e)}{r^2} \tag{1.23}$$

so one of the electronic charges in the expressions for the hydrogen atom must be replaced by Ze. We see that the energy of the hydrogen atom is proportional to R_H and, by Equation (1.17), $R_H \propto e^4$ which leads us to write, for a one-electron atom of nuclear charge Z,

$$E_n^{(0)} = \frac{Z^2(m_e c^2)\alpha^2}{2n^2} \tag{1.24}$$

Some Bohr parameters for one-electron atoms of nuclear charge Z are listed in TABLE 1.2 in terms of the fine structure constant.

Because the meter is an inconvenient unit to use for atomic dimensions a_0 is often given in nanometers (nm): 10^{-9} nm $= 1$ m. Although not an SI unit, the Angström (Å) is also in wide use: 1 Å $= 0.1$ nm. Thus, according to the Bohr theory, the diameter of an atom is ~ 1 Å. Is this value reasonable?

The expression for the orbital velocity of the electron in terms of the fine structure constant makes it clear that, for light atoms, this is a nonrelativistic problem. That is, the highest orbital velocity that can occur in hydrogen (under the assumptions of the Bohr model) is the speed of light divided by 137.

1.4. Atomic Dimensions—Is a_0 a Reasonable Atomic Diameter?

The most unforgiving of all principles to which any physical parameter must conform is the uncertainty principle. According to this principle, the product of the uncertainty in position Δx and uncertainty in momentum Δp must be greater than $\hbar/2$. Thus, for an electron bound to a proton we may assume that the uncertainty in the momentum is equal to the momentum itself, p, and the uncertainty in the position is the radius of the "orbit". With these loose assumptions on the uncertainties the relation between momentum and position that is dictated by the uncertainty principle is

$$p = \frac{\hbar}{2r} \tag{1.25}$$

The total energy is then

$$E = \frac{p^2}{2m_e} - \left(\frac{e^2}{4\pi\varepsilon_0}\right) \cdot \frac{1}{r}$$

$$= \frac{\hbar^2}{8m_e r^2} - \left(\frac{e^2}{4\pi\varepsilon_0}\right) \cdot \frac{1}{r} \tag{1.26}$$

Now, if the electronic motion is to be stable the energy must be a minimum. Differentiating with respect to r and setting the derivative equal to zero, we obtain the electronic radius for stability r_s to be

$$r_s = 2 \cdot \left(\frac{4\pi\varepsilon_0}{e^2}\right) \cdot \frac{\hbar^2}{m_e}$$

$$= 2a_0 \tag{1.27}$$

Thus, for an electron bound to a proton the uncertainty principle dictates that the dimensions of the system are of the order of the Bohr radius. That is, the electron cannot be localized to a volume that is smaller than $\sim \text{Å}^3$. The factor of two in the above expression is inconsequential. We are only interested in orders of magnitude. We conclude therefore that the electronic radii obtained from the Bohr model of the atom are of the correct order of magnitude.

How about the energy? Of course, if we put a_0 in the expression for the total energy we will obtain an energy of order eV. That this is a reasonable energy for atoms can be confirmed using the simplest of all potentials for which one solves the Schrödinger equation, the one-dimensional infinite potential well of width L. The energy levels are given by

$$E_n = \frac{n^2 \pi^2 \hbar^2}{2mL^2} \tag{1.28}$$

This expression gives the energy levels for a system consisting of a particle of mass m confined to an infinite potential well of width L. The n-dependence is unimportant. What is important is the relation between the parameters m and L to the constants that provides an estimate of the energy levels for a system consisting of a particle of mass m confined to a "box" of length L. We may cast E_n in the form

$$E_n = \frac{n^2 \pi^2 (\hbar c)^2}{2(m_e c^2) L^2}$$

$$= n^2 \frac{\pi^2 (1973\, eV \cdot \text{Å})^2}{2M(5 \times 10^5)(1\, \text{Å})^2}$$

$$= \frac{38n^2}{ML^2}\, eV \tag{1.29}$$

where M is the mass in terms of the mass of the electron and L is in Å. Thus, for an "atom" ($M = 1$ and $L \approx 1$ Å) the energy level spacings are on the order of eV. This is somewhat larger than most atomic energies, but not an order of magnitude. Remember, we are using this as an estimate. Notice that this expression is good for estimating other energies as well. For example, for a nucleon confined to a box the size of a nucleus ($M \approx 2000$ and $L \approx 10^{-5}$ Å), the energy level spacing is roughly seven orders of magnitude greater than for the atom. Although this is hardly an exact formula we can use the simplest of all potential energy functions to estimate energy levels for real systems.

1.5. Localizing the Electron: Is a Point Particle Reasonable?

We have seen that although the Bohr model is in conflict with the principles of quantum mechanics, it does produce the correct answer for the energy levels of the hydrogen atom. It also predicts, more or less correctly, the dimensions of atoms, ~ 1 Å. The fact that Bohr assumed that the electron is a point particle is not at variance with nonrelativistic quantum mechanics (also sometimes referred to as "first quantization" or "Schrödinger wave mechanics"). Recall that in Schrödinger wave mechanics the particles are assumed to be point particles; it is the probability density that does the waving. Thus, the probability density for the stationary states of hydrogen that is obtained from the Schrödinger equation is smeared over the roughly 1 Å diameter of the atom. According to the uncertainty principle, however, the electron cannot be localized to a region smaller than roughly its deBroglie wavelength.

In order to confine an electron in a "box" the dimensions of the box must be at a minimum the deBroglie wavelength. For an electron traveling with a speed v this wavelength is given by

$$\lambda_D = \frac{h}{p}$$

$$= \frac{h}{m_e v} \tag{1.30}$$

To make this wavelength as small as possible we replace v with c, the speed of light. Of course, the electron will not be traveling with the speed of light, but this replacement gives a lower limit on the value of λ_D. In this limit, $\lambda_D \rightarrow \lambda_C$ where λ_C is called the Compton wavelength because it appears in the correct description of the Compton effect. Thus, we conclude that λ_C is the smallest length to within which the electron can be localized. We may rewrite λ_C as

$$\lambda_C = \frac{h}{m_e c} \cdot \left(\frac{4\pi \varepsilon_0 \hbar^2}{m_e e^2} \right) \frac{1}{4\pi \varepsilon_0} \frac{m_e e^2}{\hbar^2}$$

$$= 2\pi a_0 \alpha \tag{1.31}$$

which shows that the "fuzziness" of the electron is much smaller than atomic dimensions so the assumption that it is a point particle in "orbit" is justified.

1.6. The Classical Radius of the Electron

Although we have concluded that it is permissible to treat the electron (and, indeed, the proton) as point particles, there is another parameter that is often used to obtain a sense of scale. If one assumes that the electron is actually a charged sphere it is possible to calculate the "classical radius of the electron" by equating the electronic rest energy $m_e c^2$ to the electrostatic energy. The exact electrostatic energy of the sphere depends upon the assumed distribution of charge within it. For example, if a sphere of radius R is uniformly charged the energy is given by

$$E_{electrostatic} = \left(\frac{3}{5}\right) \frac{e^2}{4\pi\varepsilon_0 R} \tag{1.32}$$

Because only orders of magnitude are important, the fraction is usually dropped and the classical radius of the electron r_e is calculated from

$$m_e c^2 = \frac{e^2}{4\pi\varepsilon_0 r_e} \tag{1.33}$$

From this we obtain r_e which, as with the Compton wavelength, we write in terms of the Bohr radius and the fine structure constant.

$$
\begin{aligned}
r_e &= \frac{e^2}{4\pi\varepsilon_0} \cdot \frac{1}{m_e c^2} \\
&= \frac{e^2}{4\pi\varepsilon_0} \cdot \frac{1}{m_e c^2} \left(\frac{4\pi\varepsilon_0 \hbar^2}{m_e e^2}\right) \frac{1}{4\pi\varepsilon_0} \frac{m_e e^2}{\hbar^2} \\
&= \alpha^2 a_0 \tag{1.34}
\end{aligned}
$$

We see then that the particle "dimensions" are roughly $1/137$ times smaller than the fuzziness of the particle as dictated by the uncertainty principle.

1.7. Atomic Units

Although SI units are often preferred, in atomic physics it is useful to employ atomic units (a.u.). The convenience is that many of the constants are set equal to unity so that atomic parameters are expressed in terms of Bohr atom parameters. Atomic units are defined so that

$$1/(4\pi\varepsilon_0) \equiv 1; \quad e \equiv 1; \quad \hbar \equiv 1; \quad m_e \equiv 1 \tag{1.35}$$

TABLE 1.3 is a listing of some parameters in atomic units. Note that each of the entries in the second column is unity in atomic units.

It is interesting to note that in atomic units the speed of light is 137 atomic units of length (Bohr radii) per atomic unit of time. This is evident from the definition of the fine structure constant, Equation (1.21), because $\alpha = 1/137$ is independent of the system of units. In addition, the atomic unit of length is a_0, the radius of the first Bohr orbit, which is frequently referred to as one Bohr. The unit of energy in

TABLE 1.3. Atomic units.

Physical quantity	Unit	Physical significance	Value (SI)
Mass	m_e	Electron mass	9.109534×10^{-31} kg
Charge	e	Absolute value of the electron charge	$1.6021892 \times 10^{-19}$ C
Angular momentum	$\hbar$		$1.05465887 \times 10^{-34}$ J ?s
Length	$\dfrac{4\pi\varepsilon_0\hbar^2}{m_e e^2}$	Bohr radius of atomic hydrogen	$5.2917706 \times 10^{-11}$ m
Velocity	$v_0 = e^2/(4\pi\varepsilon_0\hbar)$ $= \alpha c$	Velocity in first Bohr orbit	2.19769×10^6 m/s
Time	$\dfrac{a_0}{v_0}$	Period of the first Bohr orbit divided by 2π	2.41889×10^{-17} s
Frequency	$\dfrac{v_0}{a_0}$	Orbital frequency (cycles/s) times 2π	6.57968×10^{-15} s^{-1}
Energy	$\dfrac{e^2}{4\pi\varepsilon_0 a_0}$	Twice the ionization potential of atomic hydrogen	$4.35974417 \times 10^{-18}$ J 27.2113845 eV
Potential	$\dfrac{e}{4\pi\varepsilon_0 a_0}$	Potential at the first Bohr orbit	27.2113845 V
Electric field	$\dfrac{e}{4\pi\varepsilon_0 a_0^2}$	Electric field at the first Bohr orbit	5.14225×10^{11} V/m

atomic units, often referred to as one hartree, is twice the ionization potential of hydrogen, that is, two Rydbergs or 27.2 eV.

So far we have introduced four different units of energy. It is typical of atomic physics that different units of energy are used to describe certain energy intervals. This is because the range of atomic energy intervals is roughly seven orders of magnitude. It is also typical of atomic physics that energy units are often expressed in terms of the units measured in the experimental techniques that are used to study them. For example, the cm^{-1} is a convenient unit in spectroscopy where the wavelength is measured. Another unit that is often used for small energy intervals is frequency measured in Hz or MHz (10^6 Hz). The usefulness of frequency as a measure of energy is that many experiments employ microwave techniques. Of course, the energy measured in Hz is simply the energy divided by Planck's constant h in consistent units. For convenience, some of the frequently used energy conversion factors are listed in TABLE 1.4.

TABLE 1.4. Conversion factors between energy units used in atomic physics.

1 hartree = 27.2113845 eV	1 eV = $3.67493245 \times 10^{-2}$ hartree
1 eV = $1.60217653 \times 10^{-19}$ J	1 J = $6.24150947 \times 10^{18}$ eV
1 eV = 8065.54445 cm^{-1}	1 cm^{-1} = $1.23984191 \times 10^{-4}$ eV
1 eV = 2.417989×10^8 MHz	1 MHz = $4.13566743 \times 10^{-9}$ eV
1 cm^{-1} = 2.997924×10^4 MHz	1 MHz = 3.335640×10^{-5} cm^{-1}

Problems

1.1. Bohr's theory of the atom pre-dated deBroglie's hypothesis of what we now call the deBroglie wavelength λ_D. Suppose, however, that Bohr was aware of this hypothesis and, rather than using the correspondence principle stated:

Assumption III: The orbits of the stationary states are defined by the condition that only integral multiples of $\lambda_D = h/p$ "fit" into the orbital circumference.

Show that this is identical to the "assumption" that the angular momentum is quantized in units of $\hbar$, and, thus, it would lead to the correct energies and the same atomic parameters that he derived using the correspondence principle.

1.2. Positronium is a bound state of a positron (anti-electron) and an electron. It is essentially a hydrogen atom with the proton replaced by a positron. What is the ground state energy of positronium?

1.3. Obtain the expression for the energy levels of a particle in an infinite square well of length L, Equation (1.29), by fitting the deBroglie wave in the box. Note that in this special case the deBroglie wavelength for each level is constant throughout the box because there is no potential energy; that is, the kinetic energy (and hence the momentum) is constant for each level.

1.4. What is the ionization energy of ground state He^+? In other words, what is the ground state energy of He^+?

1.5. A diatomic molecule for which the reduced mass of the nuclei is μ undergoes vibrational motion.

(a) Use the uncertainty principle to minimize the energy to show that an approximate value for the amplitude of vibration A is $A = \sqrt{\hbar/2\mu\omega}$.

(b) Show that, using A from part (a), the minimum energy is the correct zero point energy of a harmonic oscillator.

(c) The spacing between the quantized vibrational energy levels of a diatomic hydrogen molecule is roughly 0.4 eV. Find the approximate amplitude of vibration for this molecule.

References

1. N. Bohr, in *Sources of Quantum Mechanics*, edited by B.L.v.d. Waerden (Dover, New York, 1918), p. 95.
2. H. Goldstein, *Classical Mechanics* (Addison-Wesley, Reading, MA, 1980).
3. L.I. Schiff, *Quantum Mechanics* (McGraw-Hill, New York, 1968).

2
Angular Momentum

2.1. Introduction

An understanding of angular momentum algebra is crucial for proper description of the properties and interactions of atoms. We therefore devote this chapter to a review of angular momentum. We begin with a quantum mechanical description of a particle of mass M confined to move on a plane circular ring of radius R. Such a description leads to the quantum mechanically allowable values of the energy and, of course, the angular momentum. In addition, this problem, which is often referred to as the rigid rotor, provides an introduction to a very important concept for the study of atomic structure—degeneracy.

We set up the coordinate system shown in FIGURE 2.1. We wish to solve the eigenvalue problems for energy and angular momentum. The energy equation is the familiar time independent Schrödinger equation in plane polar coordinates ρ and ϕ

$$\hat{H}\psi(\rho, \phi) = E\psi(\rho, \phi) \tag{2.1}$$

where $\hat{H}$ is the Hamiltonian operator and E is the energy eigenvalue. We use the "hat" over a quantity to signify that it is a quantum mechanical operator although occasionally the hat is omitted when it is unnecessary, for example, for coordinates such as x, y, or z. The Hamiltonian for the system is given by

$$\hat{H} = \frac{1}{2M}\left(\hat{p}_x^2 + \hat{p}_y^2\right)$$

$$= -\frac{\hbar^2}{2M}\left(\frac{\partial^2}{\partial x^2} + \frac{\partial^2}{\partial y^2}\right) \tag{2.2}$$

because there is no potential energy. Because of the symmetry of the problem cylindrical coordinates are ideal. The equations for the transformation from Cartesian to cylindrical coordinates are

$$x = \rho \cos \phi \qquad y = \rho \sin \phi$$

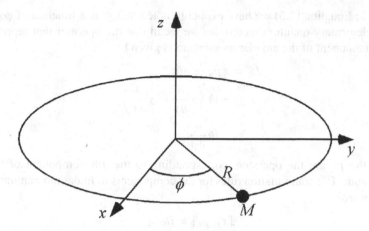

FIGURE 2.1. Coordinates for the rigid rotor problem.

and the derivatives transform according to

$$\frac{\partial}{\partial \rho} = \frac{\partial x}{\partial \rho} \cdot \frac{\partial}{\partial x} + \frac{\partial y}{\partial \rho} \cdot \frac{\partial}{\partial y} \quad \& \quad \frac{\partial}{\partial \phi} = \frac{\partial x}{\partial \phi} \cdot \frac{\partial}{\partial x} + \frac{\partial y}{\partial \phi} \cdot \frac{\partial}{\partial y}$$

These transformations lead to the Laplacian operator in polar coordinates:

$$\frac{\partial^2}{\partial x^2} + \frac{\partial^2}{\partial y^2} = \frac{1}{\rho} \frac{\partial}{\partial \rho} \left(\rho \frac{\partial}{\partial \rho} \right) + \frac{1}{\rho^2} \frac{\partial^2}{\partial \phi^2}$$

For the particle confined to the ring the radial distance remains constant; that is, $\rho = R$. The time-independent Schrödinger wave equation is therefore

$$\frac{d^2 \psi}{d\phi^2} + \frac{2IE}{\hbar^2} \psi = 0 \tag{2.3}$$

where $I = MR^2$, the moment of inertia of the particle about the z-axis. In this last equation the partial derivative has been replaced by a total derivative because for $\rho = R$, ψ is a function of ϕ only.

Equation (2.3) could have been deduced by noting that the energy of the rotating particle is given by

$$E = \frac{(L_z)^2}{2I} \tag{2.4}$$

where L_z is the angular momentum. Thus, from Equation (2.4) the quantum mechanical operator $\hat{H} = \hat{L}_z^2/2I$ where $\hat{L}_z$ is the operator corresponding to the z-component of angular momentum. Because $\hat{L}_z$ commutes with $\hat{H}$ they will have simultaneous eigenfunctions and the Schrödinger equation becomes

$$\frac{\hat{L}_z^2}{2I} \psi(\phi) = E\psi(\phi) \tag{2.5}$$

where, in Equation (2.5) we have explicitly noted that ψ is a function of ϕ only. From elementary quantum mechanics we recall that the operator that represents the z-component of the angular momentum is given by

$$\hat{L}_z = x\hat{p}_y - y\hat{p}_x$$
$$= -i\hbar \left(x\frac{\partial}{\partial y} - y\frac{\partial}{\partial x} \right)$$
$$= -i\hbar \frac{\partial}{\partial \phi} \tag{2.6}$$

where the $\hat{p}_j$ are the operators corresponding to the jth component of linear momentum. The commutation rules for the components of linear momentum with position are

$$[x_j, \hat{p}_k] = i\hbar\delta_{jk} \tag{2.7}$$

Equations (2.5) and (2.6) lead to the time-independent Schrödinger equation for the rigid rotor

$$\frac{d^2\psi}{d\phi^2} + \omega^2\psi = 0 \tag{2.8}$$

where $\omega^2 = (2IE)/\hbar^2$.

Equation (2.8) is perhaps the most familiar differential equation in all of physics, first encountered as the description of the motion of an undamped simple harmonic oscillator. The solutions are

$$\psi(\phi) = Ae^{\pm i\omega\phi} \tag{2.9}$$

where A is a constant. Because $\psi(\phi)$ must be single-valued, ω must be an integer. That is, we must have

$$\psi(\phi) = \psi(\phi + 2\pi) \tag{2.10}$$

which can be satisfied only if

$$e^{2\pi i\omega} = 1 \tag{2.11}$$

Therefore,

$$Ae^{i\omega\phi} = Ae^{i\omega\phi}[e^{2\pi i\omega}] \tag{2.12}$$

Normalization of the wave function shows that $A = 1/\sqrt{2\pi}$.

To satisfy Equation (2.12) we require $\omega = 0, \pm 1, \pm 2, \ldots$. Because ω is an integer it is usually designated by the letter m. Consequently, the energy is given by

$$E_m = \frac{\hbar^2}{2I}m^2 \quad \text{where} \quad m = 0, \pm 1, \pm 2, \ldots \tag{2.13}$$

It is clear from Equation (2.13) that the energy is quantized. Moreover, because the eigenvalues depend on the square of m, two values of m, positive and negative, correspond to the same energy; that is, the system is degenerate.

Now, because $\hat{H}$ and $\hat{L}_z$ commute they must have simultaneous eigenfunctions. Thus, we examine the effect of $\hat{L}_z$ operating on these energy eigenfunctions.

$$\hat{L}_z \psi(\phi) = \frac{\hbar}{i} \frac{\partial}{\partial \phi} \psi(\phi)$$

$$= \frac{\hbar}{i} \frac{\partial}{\partial \phi} A e^{im\phi}$$

$$= m\hbar \psi(\phi) \tag{2.14}$$

Therefore, as expected, $\psi(\phi)$ is also an eigenfunction of $\hat{L}_z$ with eigenvalue $m\hbar$. In this case, however, the sign of m determines the sign of the eigenvalue of the angular momentum. In fact, it is obvious that, classically, the positive eigenvalue corresponds to L_z pointing in the direction of the z-axis and the negative eigenvalue corresponds to the angular momentum vector pointing in the direction of the negative z-axis. Because, however, the energy is proportional to m^2 either direction of the angular momentum leads to the same energy and the system is degenerate.

Clearly it is the symmetry of the system that is responsible for this degeneracy. In general, degeneracies are associated with symmetries. Mathematically, the operators that represent such symmetries commute with the Hamiltonian and therefore have simultaneous eigenfunctions. In the case of the particle confined to a ring, L_z (the z-component of the angular momentum) is that operator.

To summarize, we have:

Eigenvalues of $\hat{H}$:

$$E_m = \frac{\hbar^2}{2I} m^2 \quad \text{where} \quad m = 0, \pm 1, \pm 2, \ldots \tag{2.15}$$

Eigenvalues of $\hat{L}_z$:

$$L_z = m\hbar \quad \text{where} \quad m = 0, \pm 1, \pm 2, \ldots \tag{2.16}$$

The degeneracy corresponds to different directions of rotation for which $|L_z|$ is the same. This "symmetry" is, as usual, responsible for the degeneracy. Notice that the Schrödinger equation can also be separated in Cartesian coordinates. Frequently symmetries manifest themselves by making the Schrödinger equation separable in more than one coordinate system. A familiar example is the three-dimensional isotropic harmonic oscillator. If a Hamiltonian can be written as the sum of terms, each of which contains only a single variable, for example,

$$\hat{H}(x, y, z) = \hat{H}_x(x) + \hat{H}_y(y) + \hat{H}_z(z) \tag{2.17}$$

then the eigenfunctions and eigenvalues are given by

$$\psi(x, y, z) = \psi_x(x)\psi_y(y)\psi_z(z)$$

$$E = E_x + E_y + E_z \tag{2.18}$$

where the ψ_i and E_i are solutions of the equations

$$\hat{H}_i \psi_i = E_i \psi_i \tag{2.19}$$

Because the Hamiltonian for the isotropic oscillator is given by

$$\hat{H}(x, y, z) = \frac{1}{2m} \left(p_x^2 + p_y^2 + p_z^2 \right) + \frac{1}{2} k(x^2 + y^2 + z^2)$$
$$= \hat{H}_x(x) + \hat{H}_y(y) + \hat{H}_z(z) \qquad (2.20)$$

it is clear that the Schrödinger equation can indeed be separated in Cartesian coordinates. Moreover, by virtue of the central nature of this potential $V(r) = (1/2)kr^2$, the Schrödinger equation can also be separated in spherical coordinates. Accordingly, the isotropic oscillator possesses a degree of symmetry that is higher than the symmetry associated with the isotropic nature of *any* central potential. This is discussed in more detail later in this book.

2.2. Commutators

Classsically, the angular momentum is a vector in three dimensions

$$L = L_x i + L_y j + L_z k \qquad (2.21)$$

Throughout this book the symbols i, j, and k in boldface designate the unit vectors in Cartesian coordinates. The customary "hat" is omitted to avoid confusion with quantum mechanical operators. In quantum mechanics a more generalized approach to angular momentum is taken than to consider angular momentum according to the classical definition

$$L = r \times p \qquad (2.22)$$

An *operator* $\hat{J}$

$$\hat{J} = \hat{J}_x i + \hat{J}_y j + \hat{J}_z k \qquad (2.23)$$

is defined to be an angular momentum if the commutation relations between its Cartesian components are given by

$$[\hat{J}_i, \hat{J}_j] = i\hbar \hat{J}_k \, \varepsilon_{ijk} \qquad (2.24)$$

where ε_{ijk} is the Levi–Cevita symbol for which even permutations of i, j, k yield $+1$ and odd permutations yield -1; $\varepsilon_{ijk} = 0$ if any two of the indices are the same. For example,

$$[\hat{J}_y, \hat{J}_x] = -i\hbar \hat{J}_z \qquad (2.25)$$

but

$$[\hat{J}_x, \hat{J}_x] = 0 \qquad (2.26)$$

Note that $\hat{J}$ is "an angular momentum" whether or not there is a classical counterpart of the operator that represents angular motion if the commutation rules of Equation (2.24) are obeyed. In spite of this generalized nature of angular momentum, we visualize such a motion and refer to rotations when discussing angular momentum operators. Later, we encounter quantum mechanical operators for which the above commutation relations apply, but that cannot be visualized classically. Of course, the most familiar of these is the "intrinsic" angular momentum—spin.

The commutation rules of Equation (2.24) indicate that an angular momentum can have only one of its components specified precisely. In keeping with accepted tradition, we choose $\hat{J}_z$ as this component. Then $\hat{J}_x$ and $\hat{J}_y$ cannot have sharply defined values. If we know $\hat{J}_z$ we wish to find out if we can, *simultaneously*, know the magnitude of $\boldsymbol{J}$.

To answer this question we examine $\hat{J}^2$, the square of the *magnitude* of the total angular momentum.

$$\hat{J}^2 = \hat{J}_x^2 + \hat{J}_y^2 + \hat{J}_z^2 \tag{2.27}$$

If $\hat{J}_z$ commutes with $\hat{J}^2$ then the eigenvalues of these operators can be specified simultaneously. This commutator is

$$[\hat{J}^2, \hat{J}_z] = [\hat{J}_x^2, \hat{J}_z] + [\hat{J}_y^2, \hat{J}_z] + [\hat{J}_z^2, \hat{J}_z] \tag{2.28}$$

where the last term is obviously zero. Using the easily derived identity

$$[\hat{A}\hat{B}, \hat{C}] = [\hat{A}, \hat{C}]\hat{B} + \hat{A}[\hat{B}, \hat{C}] \tag{2.29}$$

the first term is

$$[\hat{J}_x^2, \hat{J}_z] = \hat{J}_x\hat{J}_x\hat{J}_z - \hat{J}_z\hat{J}_x\hat{J}_x \tag{2.30}$$

Now subtract and add $\hat{J}_x\hat{J}_z\hat{J}_x$ to obtain

$$\begin{aligned}
[\hat{J}_x^2, \hat{J}_z] &= \hat{J}_x\hat{J}_x\hat{J}_z + \{-\hat{J}_x\hat{J}_z\hat{J}_x + \hat{J}_x\hat{J}_z\hat{J}_x\} - \hat{J}_z\hat{J}_x\hat{J}_x \\
&= \hat{J}_x[\hat{J}_x, \hat{J}_z] + [\hat{J}_x, \hat{J}_z]\hat{J}_x \\
&= \hat{J}_x(-i\hbar\hat{J}_y) + (-i\hbar\hat{J}_y)\hat{J}_x \\
&= -i\hbar\{\hat{J}_x\hat{J}_y + \hat{J}_y\hat{J}_x\}
\end{aligned} \tag{2.31}$$

The same procedure yields

$$[\hat{J}_y^2, \hat{J}_z] = +i\hbar\{\hat{J}_x\hat{J}_y + \hat{J}_y\hat{J}_x\} \tag{2.32}$$

so that

$$\begin{aligned}
[\hat{J}^2, \hat{J}_z] &= [(\hat{J}_x^2 + \hat{J}_y^2 + \hat{J}_z^2), \hat{J}_z] \\
&= [\hat{J}_x^2, \hat{J}_z] + [\hat{J}_y^2, \hat{J}_z] + [\hat{J}_z^2, \hat{J}_z] \\
&= -i\hbar\{\hat{J}_x\hat{J}_y + \hat{J}_y\hat{J}_x\} + i\hbar\{\hat{J}_x\hat{J}_y + \hat{J}_y\hat{J}_x\} + 0 \\
&= 0
\end{aligned} \tag{2.33}$$

By symmetry we have

$$[\hat{J}^2, \hat{J}_x] = 0 = [\hat{J}^2, \hat{J}_y]$$

which shows that $|\hat{J}|$ can be specified together with any one of its components. Because, however, the components do not themselves commute, only one of them can be specified. Customarily this component is chosen be $\hat{J}_z$.

2.3. Angular Momentum Raising and Lowering Operators

Manipulation of angular momentum operators is facilitated by defining two linear combinations of $\hat{J}_x$ and $\hat{J}_y$ which are usually referred to as raising and lowering operators or, simply, ladder operators. They are defined as

$$\hat{J}_{\pm} = \hat{J}_x \pm i\hat{J}_y \qquad (2.34)$$

Note that $\hat{J}_+$ and $\hat{J}_-$ are Hermitian conjugates of each other.

Using Equation (2.25) we obtain the following commutation relations.

$$
\begin{aligned}
[\hat{J}_{\pm}, \hat{J}_z] &= [\hat{J}_x, \hat{J}_z] \pm i[\hat{J}_y, \hat{J}_z] \\
&= \mp i\hbar \hat{J}_y \mp \hbar \hat{J}_x \\
&= \mp \hbar(\hat{J}_x + i\hat{J}_y) \\
&= \mp \hbar \hat{J}_{\pm} \qquad (2.35)
\end{aligned}
$$

Similarly,

$$[\hat{J}_+, \hat{J}_-] = 2\hbar \hat{J}_z \qquad (2.36)$$

Note also that because $\hat{J}^2$ commutes with all components of $\hat{J}$, it commutes with $\hat{J}_+$ and $\hat{J}_-$.

Now, we use $\hat{J}_+$ and $\hat{J}_-$ to find the eigenvalues and eigenfunctions of the angular momentum. Because $\hat{J}^2$ and $\hat{J}_z$ commute, we find simultaneous eigenfunctions of these two operators. Suppose that the eigenstates are distinguished by two quantum numbers, j and m, and denote the eigenfunctions by the ket $|jm\rangle$. If m is the quantum number associated with $\hat{J}_z$ we have

$$\hat{J}_z |jm\rangle = m\hbar |jm\rangle \qquad (2.37)$$

We insert the $\hbar$ for future convenience because we know that angular momentum is quantized in units of $\hbar$.

Now, m must be real because $\hat{J}_z$ is a Hermitian operator (it represents an observable). We do not, however, know that it is an integer; it might be a continuous function. Nor do we know the *range* of m. Also, operation on $|jm\rangle$ by $\hat{J}^2$ must give

$$\hat{J}^2 |jm\rangle = \hbar^2 f(j, m) |jm\rangle \qquad (2.38)$$

where $f(j, m)$ is a dimensionless function (because $\hbar$ has units of angular momentum). We have assumed that the eigenvalues of $\hat{J}^2$ depend upon both j and m.

We first investigate restrictions on the relative magnitudes of $f(j, m)$ and the quantum numbers. The difference in the expectation values of $\hat{J}^2$ and $\hat{J}_z$ is

$$
\begin{aligned}
\langle \hat{J}^2 - \hat{J}_z^2 \rangle &= \langle jm| \hat{J}^2 - \hat{J}_z^2 |jm\rangle \\
&= \{f(j, m) - m^2\}\hbar^2 \qquad (2.39)
\end{aligned}
$$

but it is also given by

$$\langle jm | \hat{J}^2 - \hat{J}_z^2 | jm \rangle = \langle \hat{J}_x^2 \rangle + \langle \hat{J}_y^2 \rangle \tag{2.40}$$

which is manifestly positive. Therefore,

$$f(j, m) \geq m^2 \tag{2.41}$$

Now apply the raising and lowering operators recalling that they commute with $\hat{J}^2$ and cannot affect the magnitude of the angular momentum. We show this as follows.

Operating with $\hat{J}_+$ on $\hat{J}^2 | jm \rangle = \hbar^2 f(j, m) | jm \rangle$ we obtain

$$\hat{J}_+ \{ \hat{J}^2 | jm \rangle \} = \hbar^2 f(j, m) \{ \hat{J}_+ | jm \rangle \}$$

which may be rewritten as

$$\hat{J}^2 \{ \hat{J}_+ | jm \rangle \} = \hbar^2 f(j, m) \{ \hat{J}_+ | jm \rangle \} \tag{2.42}$$

Thus, $\{ \hat{J}_+ | jm \rangle \}$ is an eigenstate of $\hat{J}^2$ with the same eigenvalue as $| jm \rangle$, $\hbar^2 f(j, m)$. Because $\hat{J}^2$ represents the magnitude of the angular momentum $\{ \hat{J}_+ | jm \rangle \}$ has the same magnitude as $| jm \rangle$. The same argument applies to $\hat{J}_-$.

The same proof does not work for $\hat{J}_z$ because $\hat{J}_+$ and $\hat{J}_-$ do not commute with $\hat{J}_z$. Therefore, application of $\hat{J}_+$ or $\hat{J}_-$ to $| jm \rangle$ can shift the value of m, but it must leave the magnitude unchanged. To find the eigenvalue of $\hat{J}_z$ corresponding to the eigenstate $\{ \hat{J}_+ | jm \rangle \}$ we apply $\hat{J}_z \{ \hat{J}_+ | jm \rangle \}$.

$$\hat{J}_z \{ \hat{J}_+ | jm \rangle \} = (\hat{J}_z \hat{J}_+ + \hat{J}_+ \hat{J}_z - \hat{J}_+ \hat{J}_z) | jm \rangle$$
$$= ([\hat{J}_z, \hat{J}_+] + \hat{J}_+ \hat{J}_z) | jm \rangle \tag{2.43}$$

where the second and third terms were added and subtracted to put the commutator on the right-hand side. Using $[\hat{J}_+, \hat{J}_z] = -\hbar \hat{J}_+$ we have

$$\hat{J}_z \{ \hat{J}_+ | jm \rangle \} = (\hbar \hat{J}_+ + \hat{J}_+ \hat{J}_z) | jm \rangle$$
$$= (m + 1) \hbar \{ \hat{J}_+ | jm \rangle \}$$

from which we see that $\{ \hat{J}_+ | jm \rangle \}$ is an eigenstate of $\hat{J}_z$ with eigenvalue $(m + 1)\hbar$. $\hat{J}_+$ therefore *raises* the eigenvalue by one unit of angular momentum $\hbar$. The action of $\hat{J}_+$ on $| jm \rangle$ produces an eigenstate of $\hat{J}^2$ and $\hat{J}_z$ that is proportional to $| j (m + 1) \rangle$. That is,

$$\hat{J}_+ | jm \rangle = \hbar C_{nm}^+ | j (m + 1) \rangle \tag{2.44}$$

Using an identical procedure we find that

$$\hat{J}_- | jm \rangle = \hbar C_{nm}^- | j (m - 1) \rangle \tag{2.45}$$

Now, $\hat{J}_+$ and $\hat{J}_-$ do not change j, only m. Also, because $f(j, m) \geq m^2$, we cannot apply $\hat{J}_+$ and $\hat{J}_-$ indefinitely. There must be maximum and minimum values such that

$$\hat{J}_+ | jm_{\text{max}} \rangle = 0 \tag{2.46}$$

Applying $\hat{J}_-$ to Equation (2.46) we have

$$\hat{J}_-\hat{J}_+|jm_{\max}\rangle = 0 \qquad (2.47)$$

But,

$$
\begin{aligned}
\hat{J}_-\hat{J}_+ &= (\hat{J}_x - i\hat{J}_y)(\hat{J}_x + i\hat{J}_y) \\
&= \hat{J}_x^2 + \hat{J}_y^2 + i\hat{J}_x\hat{J}_y - i\hat{J}_y\hat{J}_x \\
&= \hat{J}^2 - \hat{J}_z^2 + i[\hat{J}_x, \hat{J}_y] \\
&= \hat{J}^2 - \hat{J}_z^2 - \hbar\hat{J}_z
\end{aligned} \qquad (2.48)
$$

so that

$$
\begin{aligned}
\hat{J}_-\hat{J}_+|jm_{\max}\rangle &= \left(\hat{J}^2 - \hat{J}_z^2 - \hbar\hat{J}_z\right)|jm_{\max}\rangle \\
&= 0
\end{aligned} \qquad (2.49)
$$

which leads to

$$
\begin{aligned}
\hat{J}^2|jm_{\max}\rangle &= \left(\hat{J}_z^2 + \hbar\hat{J}_z\right)|jm_{\max}\rangle \\
&= \{(m_{\max})^2\hbar^2 + \hbar(m_{\max}\hbar)\}|jm_{\max}\rangle \\
&= m_{\max}(m_{\max} + 1)\hbar^2|jm_{\max}\rangle
\end{aligned} \qquad (2.50)
$$

We have therefore solved for $f(j, m_{\max})$. We have

$$f(j, m_{\max}) = m_{\max}(m_{\max} + 1) \qquad (2.51)$$

We see then that the magnitude of the angular momentum, the square root of the eigenvalue of $\hat{J}^2$, is determined by $m_{\max}$ which we replace by j so that the eigenvalue equation becomes

$$\hat{J}^2|jm\rangle = j(j + 1)\hbar^2|jm\rangle \qquad (2.52)$$

We must still determine the nature of j and m and the lower bound on m. We begin by finding the matrix elements of $\hat{J}_-\hat{J}_+$.

$$
\begin{aligned}
\hat{J}_-\hat{J}_+|jm\rangle &= (\hat{J}^2 - \hat{J}_z^2 - \hbar\hat{J}_z)|jm\rangle \\
&= (\hbar^2 j(j + 1) - m^2\hbar^2 - m\hbar^2)|jm\rangle \\
&= \hbar^2(j(j + 1) - m(m + 1))|jm\rangle
\end{aligned}
$$

Another expression for this matrix element may be obtained by recalling that

$$\hat{J}_+|jm\rangle = C_{jm}^+|j(m + 1)\rangle \text{ and } \hat{J}_-|jm\rangle = C_{jm}^-|j(m - 1)\rangle$$

so that

$$
\begin{aligned}
\langle jm|\hat{J}_-\hat{J}_+|jm\rangle &= \langle jm|\hat{J}_-\hbar C_{jm}^+|j(m + 1)\rangle \\
&= \hbar^2 C_{jm}^+ C_{jm+1}^- \langle jm||jm\rangle \\
&= \hbar^2 C_{jm}^+ C_{jm+1}^-
\end{aligned}
$$

Comparing the two expressions for the matrix element we find that

$$C_{jm}^{+} C_{jm+1}^{-} = j(j+1) - m(m+1) \tag{2.53}$$

Now to find the relation between C_{jm}^{+} and C_{jm}^{-} we use

$$\langle jm|\hat{J}_{-}|j(m+1)\rangle = \langle j(m+1)|\hat{J}_{+}|jm\rangle^{*} \tag{2.54}$$

which is true because $\hat{J}_{+}$ and $\hat{J}_{-}$ are Hermitian conjugates. Because

$$\langle j(m+1)|\hat{J}_{+}|jm\rangle = \hbar C_{jm}^{+} \tag{2.55}$$

and

$$\langle jm|\hat{J}_{-}|j(m+1)\rangle = \hbar C_{jm+1}^{-} \tag{2.56}$$

we have

$$C_{jm+1}^{-} = C_{jm}^{+} \tag{2.57}$$

which we insert in Equation (2.53) and arrive at

$$C_{jm}^{+} \left(C_{jm}^{+} \right)^{*} = j(j+1) - m(m+1) \tag{2.58}$$

By convention, we choose C_{jm}^{+} to be real and positive which leads to

$$\begin{aligned} C_{jm}^{+} &= \sqrt{j(j+1) - m(m+1)} \\ &= \sqrt{(j-m)(j+m+1)} \end{aligned} \tag{2.59}$$

and

$$\begin{aligned} C_{jm}^{-} &= \left(C_{jm-1}^{+} \right)^{*} = \sqrt{j(j+1) - m(m-1)} \\ &= \sqrt{(j+m)(j-m+1)} \end{aligned} \tag{2.60}$$

so the actions of the ladder operators on the ket $|jm\rangle$ are

$$\begin{aligned} \hat{J}_{+}|jm\rangle &= \hbar\sqrt{j(j+1) - m(m+1)}|j(m+1)\rangle \\ &= \hbar\sqrt{(j-m)(j+m+1)}|j(m+1)\rangle \end{aligned} \tag{2.61}$$

and

$$\begin{aligned} \hat{J}_{-}|jm\rangle &= \hbar\sqrt{j(j+1) - m(m-1)}|j(m-1)\rangle \\ &= \hbar\sqrt{(j+m)(j-m+1)}|j(m-1)\rangle \end{aligned} \tag{2.62}$$

We may learn more about the nature of the quantum numbers by considering the effect of $\hat{J}_{-}$ on the state with the lowest possible value of m; call it $m_{\min}$. Clearly,

$$\hat{J}_{-}|jm_{\min}\rangle = 0 \tag{2.63}$$

and

$$\langle j(m_{\min}-1)|\hat{J}_{-}|jm_{\min}\rangle = 0 \tag{2.64}$$

Furthermore, because

$$\hat{J}_-|jm\rangle = \hbar C_{jm}^-|j\,(m-1)\rangle \tag{2.65}$$

the above matrix element is also given by

$$\langle j\,(m_{min}-1)|\hat{J}_-|jm_{min}\rangle = \hbar C_{jm}^-\langle j\,(m_{min}-1)||j\,(m_{min}-1)\rangle$$
$$= \hbar C_{jm}^- \tag{2.66}$$

Because

$$C_{jm}^- = \sqrt{j(j+1)-m(m-1)} \tag{2.67}$$

we have

$$C_{jm_{min}}^- = 0$$
$$= \sqrt{j(j+1)-m_{min}(m_{min}-1)} \tag{2.68}$$

so that $m_{min} = -j$ and we have $-j \le m \le +j$.

The symmetry of this relationship imposes two restrictions on j: it must be either an integer or a half-integer. For example,

$$j = 2 \quad \Rightarrow \quad m = -2, -1, 0, +1, +2$$

and

$$j = 3/2 \quad \Rightarrow \quad m = -3/2, -1/2, +1/2, +3/2$$

We show that when boundary conditions are applied to spatial wave functions then j will have to take on integral values. Half-integral angular momenta do, however, exist. Electron spin is such an angular momentum, an "internal" or "intrinsic" angular momentum.

By convention, integral values of angular momentum are orbital angular momenta and are designated by ℓ and m_ℓ, that is, the quantum numbers $j \to \ell$ and $m \to m_\ell$. For spin angular momentum the convention is that $j \to s$ and $m \to m_s$. Also, it is customary to designate orbital angular momenta by letters that seem to make no sense. These letters are relics of the early days of spectroscopy when the origin of atomic emissions and absorptions was unknown. Nonetheless, the designations persist so we list them in TABLE 2.1.

TABLE 2.1. Letter designations for orbital angulart momentum.

ℓ	Letter designation
0	s
1	p
2	d
3	f
4 and higher	Alphabetically $g, h, \dots$

The lowercase letters used in this notation refer to a single electron. When there is more than one electron the script ℓ is replaced by a capital L and the letter designations are also replaced by capital letters.

2.4. Angular Momentum Commutation Relations with Vector Operators

There are a number of commutation relations between angular momentum operators and a specific class of vector operators that are useful in atomic physics. If a vector operator obeys certain commutation rules it is often referred to as a class $\hat{T}$ operator.[1,2] We simply refer to it as a "vector operator". These commutation rules are, in fact, the same commutation rules that are obeyed by the components of $\hat{J}$. Thus, if the components of an operator $\hat{V}$ obey

$$[\hat{J}_i, \hat{V}_j] = i\hbar \hat{V}_k \varepsilon_{ijk} \tag{2.69}$$

it is a vector operator. Obviously, $\hat{J}$ itself is a vector operator. Others are the position vector r and the linear momentum $\hat{p}$.

A number of relations can be derived from this definition. Among them is the important commutation relation

$$[\hat{J}, (\hat{V}_1 \cdot \hat{V}_2)] = 0 \tag{2.70}$$

If we let $\hat{V}_1 = \hat{J}$ and $\hat{V}_2 = \hat{V}$ we obtain a relation that will be useful later; *viz.*

$$[\hat{J}, (\hat{J} \cdot \hat{V})] = 0 \tag{2.71}$$

It is often useful to define combinations of the components of a vector operator as

$$\hat{V}_\pm = \hat{V}_x \pm i\hat{V}_y \tag{2.72}$$

Using the defining commutation rules together with these definitions we obtain several commutator relations which are summarized in TABLE 2.2.

TABLE 2.2. Some useful commutator relations.

$$[\hat{J}_i, \hat{V}_j] = i\hbar \hat{V}_k \varepsilon_{ijk}$$
$$[\hat{V}_\pm, \hat{J}_z] = \mp \hbar \hat{V}_\pm$$
$$[\hat{V}_\pm, \hat{J}_x] = \pm \hbar \hat{V}_z$$
$$[\hat{V}_\pm, \hat{J}_y] = i\hbar \hat{V}_z$$
$$[\hat{V}_\pm, \hat{J}_\pm] = 0$$
$$[\hat{V}_\pm, \hat{J}_\mp] = \pm 2\hbar \hat{V}_z$$
$$[\hat{V}_z, \hat{J}_\pm] = \pm \hbar \hat{V}_\pm$$

These are often cast in terms of the "spherical components" of the operators defined as

$$\hat{V}_0 = \hat{V}_z$$

$$\hat{V}_{\pm 1} = \mp \frac{1}{\sqrt{2}} \hat{V}_{\pm} \tag{2.73}$$

2.5. Matrix Elements of Vector Operators

It is often necessary to compute the matrix elements of vector operators between two angular momentum eigenstates. This is especially useful when finding the selection rules for radiative transitions between two atomic states.

We consider systems for which the operators $\hat{J}^2$ and its z-component $\hat{J}_z$ commute with the Hamiltonian and stress that $\hat{\boldsymbol{J}}$ can be *any* angular momentum. To derive the form of the matrix elements of the total angular momentum we make use of a rather specialized commutator relation[2]

$$[\hat{J}^2, [\hat{J}^2, \hat{V}]] = 2\hbar^2 (\hat{J}^2 \hat{V} + \hat{V} \hat{J}^2) - 4\hbar^2 \hat{\boldsymbol{J}} (\hat{\boldsymbol{J}} \cdot \hat{V}) \tag{2.74}$$

We form the matrix element of these operators with eigenstates of $\hat{J}^2$ and $\hat{J}_z$ which we designate $|jm\rangle$ and $|j'm'\rangle$. These eigenstates are also eigenstates of the Hamiltonian so there can be other quantum numbers such as those corresponding to the energy, but inasmuch as these matrix elements depend upon angular momentum we simplify the notation by omitting them. Our goal in evaluating the matrix elements of the commutator relation in Equation (2.74) is to obtain an expression involving the matrix element $\langle jm|\hat{V}|j'm'\rangle$.

Taking the matrix element of the operators in Equation (2.74) between the states $\langle jm|$ and $|j'm'\rangle$ we have

$$\langle jm| \hat{J}^4 \hat{V} - 2\hat{J}^2 \hat{V} \hat{J}^2 + \hat{V} \hat{J}^4 - 2\hbar^2 (\hat{J}^2 \hat{V} + \hat{V} \hat{J}^2)$$
$$+ 4\hbar^2 \hat{\boldsymbol{J}} (\hat{\boldsymbol{J}} \cdot \hat{V}) |j'm'\rangle = 0 \tag{2.75}$$

To implement our goal we seek relations between the initial and final quantum numbers for which the matrix element $\langle jm|\hat{V}|j'm'\rangle$ does not (necessarily) vanish. We first examine the properties of the operator in the last term of Equation (2.75), $\hat{\boldsymbol{J}}(\hat{\boldsymbol{J}} \cdot \hat{V})$.

From Equation (2.71) we know that $[\hat{\boldsymbol{J}}, (\hat{\boldsymbol{J}} \cdot \hat{V})] = 0$ so $\hat{\boldsymbol{J}}$ and $(\hat{\boldsymbol{J}} \cdot \hat{V})$ have simultaneous eigenfunctions. Moreover, $\hat{\boldsymbol{J}}$ and $(\hat{\boldsymbol{J}} \cdot \hat{V})$ commute with $\hat{J}^2$. *Any* operator that commutes with $\hat{J}^2$ cannot have nonzero matrix elements between states of different j. This can be seen as follows.

Suppose $\hat{O}$ is an operator that commutes with $\hat{J}^2$. We may form the matrix element of their commutator.

$$0 = \langle jm|[\hat{O}, \hat{J}^2]|j'm'\rangle$$
$$= \{j(j+1) - j'(j'+1)\}\langle jm|\hat{O}|j'm'\rangle \tag{2.76}$$

Thus, unless $j = j'$ the matrix element vanishes. In our case, the term containing $\hat{J}(\hat{J} \cdot \hat{V})$ in Equation (2.75) vanishes unless $j = j'$.

First we consider the case for which $j \neq j'$ so the matrix element of the last term in Equation (2.75) vanishes. Equation (2.75) becomes

$$[j(j+1)]^2 - 2j(j+1)j'(j'+1) + [j'(j'+1)]^2$$
$$- 2[j(j+1) + j'(j'+1)]\langle jm|\hat{V}|j'm'\rangle = 0 \qquad (2.77)$$

or

$$[(j+j'+1)^2 - 1][(j-j')^2 - 1]\langle jm|\hat{V}|j'm'\rangle = 0 \qquad (2.78)$$

The coefficient of $\langle jm|\hat{V}|j'm'\rangle$ in Equation (2.78) vanishes under the following circumstances.

1. $j + j' = -2$.
2. $j = j' = 0$.
3. $j + j' = 0$.
4. $j - j' = \pm 1$.

The first of these is impossible because $j \geq 0$. The second is irrelevant because $j = j'$ is (temporarily) excluded. The third is excluded for the same reasons as the first two. The fourth of these conditions is indeed possible. If $j - j' = \pm 1$ the first factor can vanish and $\langle jm|\hat{V}|j'm'\rangle$ can be nonzero.

For the case of $j = j'$ we must actually evaluate the matrix element of $\hat{J}(\hat{J} \cdot \hat{V})$. To do so we need an important general relation for vector operators that holds only for $j = j'$. It is called the Landé formula.[3] It states

$$\langle jm|\hat{V}|jm'\rangle = \frac{\langle jm|\hat{V} \cdot \hat{J}|jm\rangle}{j(j+1)}\langle jm|\hat{J}|jm'\rangle \qquad (2.79)$$

Letting $\hat{V} \rightarrow \hat{J}(\hat{J} \cdot \hat{V})$ we have

$$\langle jm|\hat{J}(\hat{J} \cdot \hat{V})|jm'\rangle = \frac{\langle jm|\hat{J}(\hat{J} \cdot \hat{V}) \cdot \hat{J}|jm\rangle}{j(j+1)}\langle jm|\hat{J}|jm'\rangle$$

$$= \frac{\langle jm|(\hat{J} \cdot \hat{V})\hat{J}^2|jm\rangle}{j(j+1)}\langle jm|\hat{J}|jm'\rangle$$

$$= \langle jm|(\hat{J} \cdot \hat{V})|jm\rangle\langle jm|\hat{J}|jm'\rangle \qquad (2.80)$$

where we have used the fact that $\hat{J}$ and $(\hat{J} \cdot \hat{V})$ commute. Solving Equation (2.79) for the term $\langle jm|\hat{J}|jm'\rangle$ we obtain

$$\langle jm|\hat{J}|jm'\rangle = j(j+1)\frac{\langle jm|\hat{V}|jm\rangle}{\langle jm|\hat{V} \cdot \hat{J}|jm'\rangle} \qquad (2.81)$$

which we insert in Equation (2.80) and arrive at

$$\langle jm|\hat{J}(\hat{J} \cdot \hat{V})|jm'\rangle = j(j+1)\langle jm|\hat{V}|jm'\rangle \qquad (2.82)$$

which is precisely what we seek, an expression for $\langle jm|\hat{J}(\hat{J} \cdot \hat{V})|jm'\rangle$ in terms of the matrix element of $\hat{V}$.

To find the matrix element $\langle jm|\hat{V}|j'm'\rangle$ for $j = j'$ we include the term given in Equation (2.82) in Equation (2.75). This amounts to adding $[4j(j+1)]$ to the first factor in Equation (2.78). We obtain

$$\{[(2j+1)^2 - 1](-1) + [4j(j+1)]\}\langle jm|\hat{V}|j'm'\rangle = 0 \qquad (2.83)$$

Because the factor that multiplies the matrix element is identically zero $\langle jm|\hat{V}|j'm'\rangle$ can be nonzero for $j = j'$.

There is, however, an important exception to this last conclusion. It is the case for which $j = j' = 0$. The matrix element for $j = j' = 0$ is

$$\langle 00|\hat{V}|00\rangle = \langle 00|\hat{V}_x|00\rangle i + \langle 00|\hat{V}_j|00\rangle j + \langle 00|\hat{V}_z|00\rangle k \qquad (2.84)$$

so we must evaluate the matrix element of each of the components of $\hat{V}$. This can be done most easily by using the commutation relations in TABLE 2.2.

The matrix element of the z-component is

$$\langle 00|\hat{V}_z|00\rangle = -\frac{1}{2\hbar}\langle 00|(\hat{J}_-\hat{V}_+ - \hat{V}_+\hat{J}_-)|00\rangle$$
$$= 0 \qquad (2.85)$$

because $\hat{J}_\pm|00\rangle \equiv 0$. Because $\hat{V}_\pm = \hat{V}_x \pm i\hat{V}_y$ we may write

$$\langle 00|\hat{V}_x|00\rangle = \frac{1}{2}\{\langle 00|\hat{V}_+|00\rangle + \langle 00|\hat{V}_-|00\rangle\}$$
$$= \frac{1}{\hbar}\{\langle 00|(\hat{V}_z\hat{J}_+ - \hat{J}_+\hat{V}_z)|00\rangle - \langle 00|(\hat{V}_z\hat{J}_- - \hat{J}_-\hat{V}_z)|00\rangle\}$$
$$= 0 \qquad (2.86)$$

Clearly $\langle 00|\hat{V}_y|00\rangle = 0$ as well thus establishing that the matrix element $\langle jm|\hat{V}|j'm'\rangle$ vanishes identically if $j = j' = 0$.

We may also deduce selection rules on the m quantum number using operator techniques. First, because $[\hat{J}_z, \hat{V}_z] = 0$ we have

$$\langle jm|[\hat{J}_z, \hat{V}_z]|j'm'\rangle = 0$$
$$= \langle jm|(\hat{J}_z\hat{V}_z - \hat{V}_z\hat{J}_z)|j'm'\rangle$$
$$= m\langle jm|\hat{V}_z|j'm'\rangle - m'\langle jm|\hat{V}_z|j'm'\rangle$$
$$= (m - m')\langle jm|\hat{V}_z|j'm'\rangle \qquad (2.87)$$

which shows that the matrix element $\langle jm|\hat{V}_z|j'm'\rangle$ vanishes unless $m = m'$. This is, however, not the entire story for m. We must examine the other components of $\hat{V}$ as well. For this purpose we work with $\hat{V}_\pm$ using a commutation relation from TABLE 2.2,

$$[\hat{J}_z, \hat{V}_\pm] = \pm\hbar\hat{V}_\pm \qquad (2.88)$$

from which we have

$$\langle jm|\{[\hat{J}_z, \hat{V}_\pm] \mp \hbar \hat{V}_\pm\}|j'm'\rangle = 0$$
$$= \langle jm|(\hat{J}_z \hat{V}_\pm - \hat{V}_\pm \hat{J}_z \mp \hbar \hat{V}_\pm)|j'm'\rangle$$
$$= \hbar(m - m' \mp 1)\langle jm|\hat{V}_\pm|j'm'\rangle \qquad (2.89)$$

which leads to two equations for the two unknowns $\langle jm|\hat{V}_x|j'm'\rangle$ and $\langle jm|\hat{V}_y|j'm'\rangle$.

$$(m - m' - 1)\langle jm|\hat{V}_x|j'm'\rangle + i(m - m' - 1)\langle jm|\hat{V}_y|j'm'\rangle = 0$$
$$(m - m' + 1)\langle jm|\hat{V}_x|j'm'\rangle - i(m - m' + 1)\langle jm|\hat{V}_y|j'm'\rangle = 0 \qquad (2.90)$$

These homogeneous equations have a nontrivial solution only if the determinant of the coefficients vanishes, so for the matrix elements to be nonzero we must have

$$(m - m' - 1)(m - m' + 1) = 0$$

or

$$(m - m') = \pm 1 \qquad (2.91)$$

The results of this section may be summarized as follows. The matrix element $\langle jm|\hat{V}|j'm'\rangle$ vanishes unless

$$j - j' = 0, \pm 1 \quad \text{but } j = j' = 0 \text{ is not allowed}$$
$$m - m' = 0, \pm 1 \qquad (2.92)$$

2.6. Eigenfunctions of Orbital Angular Momentum Operators

For orbital angular momenta it is customary to let $\hat{J} = \hat{L}$ and $j \to \ell$, a positive integer. We wish to find the explicit functions $|\ell\, m\rangle$ in a specific coordinate system. We choose spherical coordinates r, θ, ϕ because we will be dealing with central potentials and for *any* central potential the Schrödinger equation is separable in spherical coordinates. The work that we have performed in obtaining the properties of the shift operators means that we need only find the eigenfunction $|\ell\ell\rangle$ and successively lower it with $\hat{L}_-$ to generate all of the eigenfunctions.

We find $|\ell\ell\rangle$ by solving the equation

$$\hat{L}_+|\ell\,\ell\rangle = 0 \qquad (2.93)$$

We must at this point choose a representation in which to work. We can no longer use the abstract representation of operators. From the expressions for the Cartesian components of the vector operator for linear momentum

$$\hat{p}_{x_j} \to \frac{\hbar}{i}\frac{\partial}{\partial x_j} \qquad (2.94)$$

where x_j represents any of x, y, or z, Equation (2.22), and the transformation equations from Cartesian to spherical coordinates we have

$$\hat{L}_x = -\frac{\hbar}{i} \left\{ \sin\phi \frac{\partial}{\partial\theta} + \cot\theta \cos\phi \frac{\partial}{\partial\phi} \right\}$$

$$\hat{L}_y = \frac{\hbar}{i} \left\{ \cos\phi \frac{\partial}{\partial\theta} - \cot\theta \sin\phi \frac{\partial}{\partial\phi} \right\}$$

$$\hat{L}_z = -\frac{\hbar}{i} \frac{\partial}{\partial\phi} \tag{2.95}$$

We may find the shift operators from their definitions, and the above equations.

$$\hat{L}_+ = \hat{L}_x + i\hat{L}_y$$

$$= -\frac{\hbar}{i} \left\{ \sin\phi \frac{\partial}{\partial\theta} + \cot\theta \cos\phi \frac{\partial}{\partial\phi} \right\} + i\frac{\hbar}{i} \left\{ \cos\phi \frac{\partial}{\partial\theta} - \cot\theta \sin\phi \frac{\partial}{\partial\phi} \right\}$$

$$= -\frac{\hbar}{i} \sin\phi \frac{\partial}{\partial\theta} - \frac{\hbar}{i} \cot\theta \cos\phi \frac{\partial}{\partial\phi} + \hbar \cos\phi \frac{\partial}{\partial\theta} - \hbar \cot\theta \sin\phi \frac{\partial}{\partial\phi}$$

$$= \hbar(\cos\phi + i\sin\phi)\frac{\partial}{\partial\theta} + \hbar \cot\theta(\cos\phi + i\sin\phi)\frac{\partial}{\partial\phi}$$

$$= \hbar e^{i\phi} \left\{ \frac{\partial}{\partial\theta} + i\cot\theta \frac{\partial}{\partial\phi} \right\} \tag{2.96}$$

where we have used $\frac{1}{i} = -i$ and $e^{\pm i\phi} = \cos\phi \pm i\sin\phi$. Similarly, we find that

$$\hat{L}_- = -\hbar e^{-i\phi} \left\{ \frac{\partial}{\partial\theta} - i\cot\theta \frac{\partial}{\partial\phi} \right\} \tag{2.97}$$

Because $\hat{L}_+|\ell\,\ell\rangle = 0$ we have

$$\hbar e^{-i\phi} \left\{ \frac{\partial}{\partial\theta} + i\cot\frac{\partial}{\partial\phi} \right\} \psi_{\ell\ell}(\theta,\phi) = 0 \tag{2.98}$$

where

$$\psi_{\ell\ell}(\theta,\phi) = |\ell\,\ell\rangle \tag{2.99}$$

Now, as usual, we try separation of variables. Let

$$\psi_{\ell\ell}(\theta,\phi) = \Theta(\theta)\Phi(\phi) \tag{2.100}$$

which leads to

$$\frac{\tan\theta}{\Theta} \cdot \frac{d\Theta}{d\theta} = -i\frac{1}{\Phi} \cdot \frac{d\Phi}{d\phi} \tag{2.101}$$

There are two important aspects of Equation (2.101):

1. The derivatives are *total* derivatives.
2. The left side contains only θ and the right side only ϕ.

Therefore each side must equal a constant; call it κ. The ϕ equation integrates trivially to

$$\Phi(\phi) \propto e^{i\kappa\phi} \tag{2.102}$$

The θ equation is only slightly more difficult. The solution is

$$\Theta(\theta) \propto \sin^\kappa \theta \qquad (2.103)$$

which is easily verified by substitution. Therefore,

$$\psi_{\ell\ell}(\theta, \phi) \propto \sin^\kappa \theta \cdot e^{i\kappa\phi} \qquad (2.104)$$

We normalize this wave function later.

Now we must find κ. We do this by requiring

$$\hat{L}_z \psi_{\ell\ell}(\theta, \phi) = \ell\hbar \psi_{\ell\ell}(\theta, \phi) \qquad (2.105)$$

and, using Equation (2.95), $\hat{L}_z \psi_{\ell\ell}(\theta, \phi)$ can also be written

$$\hat{L}_z \psi_{\ell\ell}(\theta, \phi) = \frac{\hbar}{i} \frac{\partial}{\partial \phi} \psi_{\ell\ell}(\theta, \phi)$$

$$= \frac{\hbar}{i}(i\kappa)\psi_{\ell\ell}(\theta, \phi) \qquad (2.106)$$

Comparison of the last two equations shows that $\kappa = \ell$ so that, to within the normalization constant A,

$$\psi_{\ell\ell}(\theta, \phi) = A \sin^\ell \theta \cdot e^{i\ell\phi} \qquad (2.107)$$

Because we know $\psi_{\ell\ell}(\theta, \phi)$ we can generate the remaining eigenfunctions. We begin by symbolically applying the lowering operator to $|\ell\,\ell\rangle$ to obtain $|\ell(\ell - 1)\rangle$.

$$\hat{L}_- |\ell\,\ell\rangle = C_{\ell\ell}^- \hbar |\ell\,(\ell - 1)\rangle$$

$$= \sqrt{\ell(\ell + 1) - \ell(\ell - 1)}\,\hbar |\ell\,(\ell - 1)\rangle$$

$$= \sqrt{2\ell}\,\hbar |\ell\,(\ell - 1)\rangle \qquad (2.108)$$

Now actually operate on $\psi_{\ell\ell}(\theta, \phi)$ with $\hat{L}_-$ and compare the two results.

$$\hat{L}_- \psi_{\ell\ell}(\theta, \phi) = -\hbar e^{-i\phi} \left\{ \frac{\partial}{\partial \theta} - i \cot\theta \frac{\partial}{\partial \phi} \right\} A \sin^\ell \theta \cdot e^{i\ell\phi}$$

$$= -A\hbar e^{-i\phi} \{\ell[\sin^{\ell-1}\theta]\cos\theta - i(i\ell)\cot\theta \sin^\ell \theta\} e^{i\ell\phi}$$

$$= -2A\hbar\ell \sin^{\ell-1}\theta \cos\theta \cdot e^{i(\ell-1)\phi} \qquad (2.109)$$

Comparing the two expressions for $\hat{L}_- |\ell\,\ell\rangle$ we see that

$$\sqrt{2\ell}\,\hbar |\ell\,(\ell - 1)\rangle = -2A\hbar\ell \sin^{\ell-1}\theta \cos\theta \cdot e^{i(\ell-1)\phi} \qquad (2.110)$$

so that

$$|\ell\,(\ell - 1)\rangle = \psi_{\ell(\ell-1)} = -\sqrt{2\ell}A \sin^{\ell-1}\theta \cos\theta \cdot e^{i(\ell-1)\phi} \qquad (2.111)$$

Now, the expressions for $\psi_{\ell\ell}$ and $\psi_{\ell(\ell-1)}$ are the familiar spherical harmonics that are traditionally designated $Y_{\ell m}(\theta, \phi)$. In fact, we have actually generated $Y_{\ell\ell}(\theta, \phi)$ and $Y_{\ell(\ell-1)}(\theta, \phi)$, although they have not been normalized.

TABLE 2.3. Spherical harmonics in spherical and Cartesian coordinates.

Spherical harmonic	Spherical coordinates	Cartesian coordinates
Y_{00}	$\dfrac{1}{\sqrt{4\pi}}$	$\dfrac{1}{\sqrt{4\pi}}$
Y_{10}	$\sqrt{\dfrac{3}{4\pi}}\cos\theta$	$\sqrt{\dfrac{3}{4\pi}}\left(\dfrac{z}{r}\right)$
$Y_{1\pm1}$	$\mp\sqrt{\dfrac{3}{8\pi}}\sin\theta\, e^{\pm i\phi}$	$\mp\sqrt{\dfrac{3}{8\pi}}\left(\dfrac{x\pm iy}{r}\right)$
Y_{20}	$\sqrt{\dfrac{5}{16\pi}}(3\cos^2\theta-1)$	$\sqrt{\dfrac{5}{16\pi}}\left(\dfrac{3z^2-r^2}{r^2}\right)$
$Y_{2\pm1}$	$\mp\sqrt{\dfrac{15}{8\pi}}\cos\theta\sin\theta\, e^{\pm i\phi}$	$\mp\sqrt{\dfrac{15}{8\pi}}\left[\dfrac{(x\pm iy)z}{r^2}\right]$
$Y_{2\pm2}$	$\sqrt{\dfrac{15}{32\pi}}\sin^2\theta\, e^{\pm 2i\phi}$	$\sqrt{\dfrac{15}{32\pi}}\left[\dfrac{(x\pm iy)^2}{r^2}\right]$

Spherical harmonics[4] are products of $e^{im\phi}$ and associated Legendre functions. The associated Legendre functions are defined as

$$P_\ell^m(\mu)=\frac{(1-\mu^2)^{m/2}}{2^\ell\cdot\ell!}\frac{d^{\ell+m}}{d\mu^{\ell+m}}(\mu^2-1)^\ell \qquad (2.112)$$

Properly normalized they are given by

$$Y_{\ell m}(\theta,\phi)=(-)^m\sqrt{\frac{(2\ell+1)}{4\pi}\cdot\frac{(\ell-m)!}{(\ell+m)!}}\,e^{im\varphi}P_\ell^m(\cos\theta) \qquad (2.113)$$

The orthogonalality relation is

$$\int_0^{2\pi}d\varphi\int_0^{\pi}Y_{\ell m}(\theta,\phi)[Y_{\ell'm'}(\theta,\phi)]^*\sin\theta\,d\theta=\delta_{\ell\ell'}\delta_{mm'} \qquad (2.114)$$

For convenience the spherical harmonics listed in TABLE 2.3 are shown in both spherical and Cartesian coordinates.

We also quote a few other important results.

$$Y_{\ell m}(\theta,\phi)=(-)^m[Y_{\ell,-m}(\theta,\phi)]^* \qquad (2.115)$$

$$\cos\theta Y_{\ell m}(\theta,\phi)=\sqrt{\frac{(\ell+m+1)(\ell-m+1)}{(2\ell+1)(2\ell+3)}}Y_{\ell+1,m}(\theta,\phi)$$
$$+\sqrt{\frac{(\ell+m)(\ell-m)}{(2\ell+1)(2\ell-1)}}Y_{\ell-1,m}(\theta,\phi) \qquad (2.116)$$

$$\sin\theta e^{\pm i\phi}Y_{\ell m}(\theta,\phi)=\mp\sqrt{\frac{(\ell\pm m+1)(\ell\pm m+2)}{(2\ell+1)(2\ell+3)}}Y_{\ell+1,m\pm1}(\theta,\phi)$$
$$\pm\sqrt{\frac{(\ell\mp m)(\ell\mp m-1)}{(2\ell+1)(2\ell-1)}}Y_{\ell-1,m\pm1}(\theta,\phi) \qquad (2.117)$$

There is an interesting relation between the spherical harmonics and the unit vector in the r direction which we designate a_r. This unit vector is given by[5]

$$a_r = \sin\theta\cos\phi i + \sin\theta\sin\phi j + \cos\theta k \qquad (2.118)$$

According to the definition of the spherical components of a vector, Equation (2.73), the spherical components of a_r are

$$(a_r)_0 = \cos\theta$$
$$(a_r)_{\pm 1} = \mp\frac{1}{\sqrt{2}}(\sin\theta\cos\phi \pm i\sin\theta\sin\phi) \qquad (2.119)$$

Using TABLE 2.3, however, Equation (2.119) may be rewritten in terms of the spherical harmonics.

$$(a_r)_0 = \sqrt{\frac{4\pi}{3}}Y_{10}(\theta,\phi)$$
$$(a_r)_{\pm 1} = \mp\frac{1}{\sqrt{2}}\sin\theta(\cos\phi \pm i\sin\phi)$$
$$= \mp\frac{1}{\sqrt{2}}\sin\theta e^{\pm i\phi}$$
$$= \mp\sqrt{\frac{4\pi}{3}}Y_{1\pm 1}(\theta,\phi) \qquad (2.120)$$

Thus, the spherical harmonics are the spherical components of the unit vector a_r which may be written as

$$a_r = \sqrt{\frac{4\pi}{3}}\left\{\left[\frac{(-i+ij)}{\sqrt{2}}\right]Y_{11}(\theta,\phi)\right.$$
$$\left. + \left[\frac{(i+ij)}{\sqrt{2}}\right]Y_{1-1}(\theta,\phi) + Y_{10}(\theta,\phi)k\right\} \qquad (2.121)$$

2.7. Spin

Before discussing spin angular momentum we discuss the magnetic dipole moment associated with a Bohr atom resulting from the electric current of the orbiting electron. This simple classical picture permits calculation of the magnetic moment (yet another virtue of the Bohr atom). We assume that the electron of charge e and mass m_e executes a circular orbit of radius r with velocity v about the proton as illustrated in FIGURE 2.2.

The circulating electron causes the system to mimic a bar magnet. The magnetic moment of this bar magnet is given by

$$\mu_\ell = i \cdot A$$

where i is the current of the circulating electron. The current i is simply the electronic charge e divided by the period of the electronic motion $2\pi r/v$.

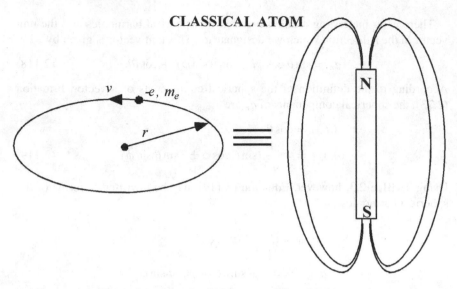

FIGURE 2.2. Illustration of the classical concept of the orbital magnetic moment.

Thus,

$$\mu_\ell = \frac{ev}{2\pi r} \cdot \pi r^2$$

$$= \frac{evr}{2}$$

But, for this circular orbit, the orbital angular momentum is $L = mvr$ so

$$\mu_\ell = -\frac{e}{2m_e} \cdot L \tag{2.122}$$

The minus sign arises because of the negative charge on the electron; that is, the magnetic dipole moment and the angular momentum point in opposite directions. This equation shows the direct relationship between the orbital angular momentum and the orbital magnetic moment. Multiplying and dividing this expression by $\hbar$ we define the Bohr magneton

$$\mu_B = \frac{e\hbar}{2m_e} \tag{2.123}$$

as the orbital magnetic moment of the electron in the first Bohr orbit. It is convenient to measure magnetic moments in terms of the Bohr magneton so we write the magnetic moment as

$$\mu_\ell = -\frac{g_\ell \mu_B}{\hbar} \cdot L \tag{2.124}$$

where $g_\ell = 1$ is known as the orbital g-factor. Although it is equal to unity, the inclusion of g_ℓ in Equation (2.124) is convenient because the g-factors of other magnetic moments have other values.

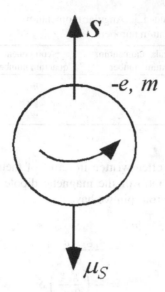

FIGURE 2.3. Conceptualization of the electron as a spinning ball of charge that produces a magnetic dipole field.

The term "spin" represents the magnetic moment associated with a "spinning" sphere of charge, the electron. Note that in nonrelativistic quantum mechanics the electron is still regarded as a point particle. A point particle cannot spin, but the concept is carried over from classical notions. Because the magnetic moment is associated with the electron itself, and has nothing to do with its orbital motion, the magnetic dipole moment associated with the spin is an *intrinsic* magnetic moment. It is built into the electron, even if the electron is isolated in space. One may think of the electron as being equivalent to a bar magnet as in FIGURE 2.3.

By analogy with the orbital magnetic moment, the spin magnetic moment μ_S is assumed proportional to an angular momentum so there is an angular momentum associated with the electron irrespective of its orbital motion, an intrinsic angular momentum. In this context the word "intrinsic" means that the electron possesses a magnetic moment whether or not it is an atomic electron. Indeed, the electron behaves as a bar magnet even if it is isolated in space.

For the spin magnetic moment μ_S we may write

$$\mu_S = -\frac{g_e \mu_B}{\hbar} \cdot S \tag{2.125}$$

where g_e is the electron spin g-factor. It is equal to $(2 + \varepsilon)$ where ε is a small number. For an electron it is found that the angular momentum quantum number $j \to s = 1/2$. The quantum number s is *always* equal to one-half. Furthermore, the spin is not included in the Schrödinger equation so it must be included "by hand." Spin only appears in relativistic treatments so it is clear that the intrinsic angular momentum is a relativistic property.

TABLE 2.4. Angular momentum
quantum numbers.

Angular momentum quantum number	z-component quantum number
j	m_j
ℓ	m_ℓ
s	m_s

Magnetic moments are often written in terms of their gyromagnetic ratios γ which are defined as the ratio of the magnetic dipole moment to the angular momentum. Thus, for electron spin we have

$$\mu_S = -\gamma_e S$$
$$= -\frac{g_e \mu_B}{\hbar} \cdot S$$
$$= -\left(\frac{g_e e}{2m_e}\right) S \tag{2.126}$$

from which we see that the gyromagnetic ratio is

$$\gamma_e = \frac{g_e e}{2m_e} \tag{2.127}$$

This general definition, suitably modified, is true for all magnetic moments, for example, the intrinsic magnetic moment of the proton.

We must now modify our notation to distinguish between values of m that correspond to different types of angular momenta. The conventional designations are shown in TABLE 2.4.

For an electron $s = 1/2$. Therefore $m_s = \pm 1/2$. These are the only possibilities! Let $|\chi_{s\,m_s}\rangle$ be the simultaneous eigenfunctions of the square of the spin angular momentum operator $\hat{S}^2$ and its z-component $\hat{S}_z$. Then, because $\hat{S}$ is an angular momentum

$$\hat{S}^2|\chi_{s\,m_s}\rangle = s(s+1)\hbar^2|\chi_{s\,m_s}\rangle \quad \text{and} \quad \hat{S}_z|\chi_{s\,m_s}\rangle = m_s\hbar|\chi_{s\,m_s}\rangle$$

There are a number of ways of condensing the notation for $|\chi_{s\,m_s}\rangle$ A few of the common designations are

$$|+\rangle = |\uparrow\rangle = |\alpha\rangle = \alpha = |\chi_{1/2\,1/2}\rangle = \text{"spin up"}$$
$$|-\rangle = |\downarrow\rangle = |\beta\rangle = \beta = |\chi_{1/2\,-1/2}\rangle = \text{"spin down"}$$

We use the $|\alpha\rangle$ and $|\beta\rangle$. Because the total spin operator $\hat{S}$ corresponds to a physical observable, it is Hermitian and $|\alpha\rangle$ and $|\beta\rangle$ are orthogonal. It is assumed that they are also normalized so that

$$\langle\alpha\,|\,\alpha\rangle = \langle\beta\,|\,\beta\rangle = 1 \quad \text{and} \quad \langle\alpha\,|\,\beta\rangle = 0 \tag{2.128}$$

A complete basis set may be represented by the kets $|q, \alpha\rangle$ and $|q, \beta\rangle$ where q represents all orbital quantum numbers, for example, n, ℓ, m if we are considering an atom. An arbitrary wave function may be expanded on this basis set

$$|\psi\rangle = \sum_q K_q^\alpha |q, \alpha\rangle + \sum_q K_q^\beta |q, \beta\rangle \qquad (2.129)$$

so that $|\psi\rangle$ contains both spin and orbital coordinates. Let $|r, \alpha\rangle / |r, \beta\rangle$ represent the state in which the electron is localized at r and has spin up/down. We can form two wave functions in the position representation

$$|\psi_\alpha(r)\rangle = \langle r, \alpha \mid \psi\rangle$$
$$= \sum_q K_q^\alpha \langle r, \alpha \mid q, \alpha\rangle$$
$$= \text{probability amplitude for finding the electron at } r \text{ with spin up}$$

with an analogous expression for $|\psi_\beta(r)\rangle$, the probability amplitude for finding the electron at r with spin down.

We can combine these two amplitudes into a single two-component object called a spinor,

$$\begin{pmatrix} \psi_\alpha(r) \\ \psi_\beta(r) \end{pmatrix} = |\psi_\alpha(r)\rangle \begin{pmatrix} 1 \\ 0 \end{pmatrix} + |\psi_\beta(r)\rangle \begin{pmatrix} 0 \\ 1 \end{pmatrix} \qquad (2.130)$$

The unit "vectors" $\begin{pmatrix} 1 \\ 0 \end{pmatrix}$ and $\begin{pmatrix} 0 \\ 1 \end{pmatrix}$ are representations of $|\alpha\rangle$ and $|\beta\rangle$, respectively. If we use this representation, then operators such as $\hat{S}^2$ and $\hat{S}_z$ must be represented by 2×2 matrices. Because $\begin{pmatrix} 1 \\ 0 \end{pmatrix}$ and $\begin{pmatrix} 0 \\ 1 \end{pmatrix}$ are, by definition, eigenfunctions of $\hat{S}^2$ and $\hat{S}_z$, their matrices are diagonal with the eigenvalues along the diagonal. We have then

$$\hat{S}_z = \frac{1}{2}\hbar \begin{pmatrix} 1 & 0 \\ 0 & -1 \end{pmatrix} \qquad (2.131)$$

and

$$\hat{S}^2 = \left(\frac{1}{2}\right)\left(\frac{1}{2} + 1\right)\hbar^2 \begin{pmatrix} 1 & 0 \\ 0 & 1 \end{pmatrix}$$
$$= \frac{3}{4}\hbar^2 \begin{pmatrix} 1 & 0 \\ 0 & 1 \end{pmatrix} \qquad (2.132)$$

We can use $\hat{S}_+$ and $\hat{S}_-$ to find $\hat{S}_x$ and $\hat{S}_y$. Obviously

$$\hat{S}_+ |\alpha\rangle = 0 \quad \text{and} \quad \hat{S}_- |\beta\rangle = 0 \qquad (2.133)$$

Also

$$\hat{S}_+|\beta\rangle = C_{jm}^+|\alpha\rangle$$
$$= \sqrt{j(j+1) - m(m+1)}\hbar|\alpha\rangle$$
$$= \sqrt{s(s+1) - m_s(m_s+1)}\hbar|\alpha\rangle$$
$$= \sqrt{\left(\frac{1}{2}\right)\left(\frac{3}{2}\right) - \left(-\frac{1}{2}\right)\left(\frac{1}{2}\right)}\hbar|\alpha\rangle$$
$$= \sqrt{\frac{3}{4} + \frac{1}{4}}\hbar|\alpha\rangle$$
$$= \hbar|\alpha\rangle \tag{2.134}$$

Similarly, we find that $\hat{S}_-|\alpha\rangle = \hbar|\beta\rangle$. Combining the two expressions with $|\alpha\rangle$ on the left side we get

$$\hat{S}_+|\alpha\rangle + \hat{S}_-|\alpha\rangle = 2\hat{S}_x|\alpha\rangle$$
$$= \hbar|\beta\rangle$$
$$= 0 + \hbar|\beta\rangle$$

from which

$$\hat{S}_x|\alpha\rangle = \frac{1}{2}\hbar|\beta\rangle \tag{2.135}$$

The same procedure for the two expressions with $|\beta\rangle$ on the left gives

$$\hat{S}_x|\beta\rangle = \frac{1}{2}\hbar|\alpha\rangle \tag{2.136}$$

We note that the analogous expressions for $\hat{S}_y$ operating on $|\alpha\rangle$ and $|\beta\rangle$ are

$$\hat{S}_y|\alpha\rangle = \frac{i}{2}\hbar|\beta\rangle \quad \text{and} \quad \hat{S}_y|\beta\rangle = -\frac{i}{2}\hbar|\alpha\rangle \tag{2.137}$$

Then, using $\hat{S}_x|\alpha\rangle = \frac{1}{2}\hbar|\beta\rangle$ we have

$$\begin{pmatrix} (\hat{S}_x)_{11} & (\hat{S}_x)_{12} \\ (\hat{S}_x)_{21} & (\hat{S}_x)_{22} \end{pmatrix} \begin{pmatrix} 1 \\ 0 \end{pmatrix} = \frac{1}{2}\hbar \begin{pmatrix} 0 \\ 1 \end{pmatrix} \tag{2.138}$$

or

$$\begin{pmatrix} (\hat{S}_x)_{11} \\ (\hat{S}_x)_{21} \end{pmatrix} = \frac{1}{2}\hbar \begin{pmatrix} 0 \\ 1 \end{pmatrix} \tag{2.139}$$

so that

$$(\hat{S}_x)_{11} = 0 \quad \text{and} \quad (\hat{S}_x)_{21} = \frac{1}{2}\hbar$$

Using Equation (2.136) we have

$$\begin{pmatrix} (\hat{S}_x)_{11} & (\hat{S}_x)_{12} \\ (\hat{S}_x)_{21} & (\hat{S}_x)_{22} \end{pmatrix} \begin{pmatrix} 0 \\ 1 \end{pmatrix} = \frac{1}{2}\hbar \begin{pmatrix} 1 \\ 0 \end{pmatrix} \tag{2.140}$$

or

$$\begin{pmatrix} (\hat{S}_x)_{12} \\ (\hat{S}_x)_{22} \end{pmatrix} = \frac{1}{2}\hbar \begin{pmatrix} 0 \\ 1 \end{pmatrix} \tag{2.141}$$

from which

$$(\hat{S}_x)_{12} = \frac{1}{2}\hbar \quad \text{and} \quad (\hat{S}_x)_{22} = 0 \tag{2.142}$$

so that

$$\hat{S}_x = \frac{1}{2}\hbar \begin{pmatrix} 0 & 1 \\ 1 & 0 \end{pmatrix} \tag{2.143}$$

Applying the same technique to $\hat{S}_y$ we obtain

$$\hat{S}_y = \frac{1}{2}\hbar \begin{pmatrix} 0 & -i \\ i & 0 \end{pmatrix} \tag{2.144}$$

To avoid having to continually write $(1/2)\hbar$ the Pauli spin matrices are often used. They are defined as

$$\hat{\sigma}_x = \begin{pmatrix} 0 & 1 \\ 1 & 0 \end{pmatrix}; \quad \hat{\sigma}_y = \begin{pmatrix} 0 & -i \\ i & 0 \end{pmatrix}; \quad \hat{\sigma}_z = \begin{pmatrix} 1 & 0 \\ 0 & -1 \end{pmatrix} \tag{2.145}$$

so that

$$\hat{S} = \frac{1}{2}\hbar\hat{\sigma} \quad \text{where} \quad \hat{\sigma} = \hat{\sigma}_x i + \hat{\sigma}_y j + \hat{\sigma}_x k \tag{2.146}$$

and i, j, k are the unit vectors in Cartesian coordinates. Also, the actions of the Pauli spin matrices on the eigenkets $|\alpha\rangle$ and $|\beta\rangle$ are

$$\hat{\sigma}_x|\alpha\rangle = |\beta\rangle \ : \ \hat{\sigma}_x|\beta\rangle = |\alpha\rangle$$
$$\hat{\sigma}_y|\alpha\rangle = i|\beta\rangle \ ; \ \hat{\sigma}_y|\beta\rangle = -i|\alpha\rangle$$
$$\hat{\sigma}_z|\alpha\rangle = |\alpha\rangle \ ; \ \hat{\sigma}_z|\beta\rangle = -|\beta\rangle \tag{2.147}$$

We now find the eigenstates of $\hat{S}_x$ and $\hat{S}_y$. To do this we use the matrix representation to determine the eigenstates of the operator

$$\hat{S}_n = \hat{S}\cdot n \quad \text{where} \quad n = \cos\phi\, i + \sin\phi\, j$$

The choice $\phi = 0$ ($\phi = \pi/2$) will yield the eigenstates of $\hat{S}_x$ ($\hat{S}_y$). The eigenvalue equation is then

$$\hat{S}_n|\mu\rangle = \mu \left(\frac{\hbar}{2}\right)|\mu\rangle \tag{2.148}$$

where $|\mu\rangle$ is the eigenvector and μ is the corresponding eigenvalue. The factor $\hbar/2$ has been inserted for convenience. Note that we know that putting the $\hbar/2$ factor in the eigenvalue equation will be convenient because the eigenvalues of $\hat{S}_z$ are $\pm\hbar/2$. Because our choice of z-axis is arbitrary, $\pm\hbar/2$ must also be the eigenvalues of $\hat{S}_n$.

In matrix form

$$
\begin{aligned}
\hat{S}_n &= \hat{S} \cdot n \\
&= \left(\frac{\hbar}{2}\right) \cos\phi \cdot \hat{\sigma}_x + \left(\frac{\hbar}{2}\right) \sin\phi \cdot \hat{\sigma}_y + 0 \cdot \left(\frac{\hbar}{2}\right) \hat{\sigma}_z \\
&= \left(\frac{\hbar}{2}\right) \left[\cos\phi \begin{pmatrix} 0 & 1 \\ 1 & 0 \end{pmatrix} + \sin\phi \begin{pmatrix} 0 & -i \\ i & 0 \end{pmatrix} \right]
\end{aligned}
\tag{2.149}
$$

We wish to write the eigenvector as a linear combination of $\hat{S}_z$ eigenkets $|\alpha\rangle$ and $|\beta\rangle$.

$$
|\mu\rangle = a|\alpha\rangle + b|\beta\rangle
$$

where

$$
a = \langle\alpha \mid \mu\rangle \quad \text{and} \quad b = \langle\beta \mid \mu\rangle
$$

In matrix form we have

$$
\left(\frac{\hbar}{2}\right) \left[\cos\phi \begin{pmatrix} 0 & 1 \\ 1 & 0 \end{pmatrix} + \sin\phi \begin{pmatrix} 0 & -i \\ i & 0 \end{pmatrix} \right] = \mu \left(\frac{\hbar}{2}\right) \begin{pmatrix} a \\ b \end{pmatrix}
\tag{2.150}
$$

Collecting, we have

$$
\begin{pmatrix} -\mu & e^{-i\phi} \\ e^{i\phi} & -\mu \end{pmatrix} \begin{pmatrix} a \\ b \end{pmatrix} = 0 \cdot \begin{pmatrix} a \\ b \end{pmatrix}
\tag{2.151}
$$

Multiplying produces two homogeneous equations in a and b. A nontrivial solution exists only if the determinant of the coefficients vanishes

$$
\begin{vmatrix} -\mu & e^{-i\phi} \\ e^{i\phi} & -\mu \end{vmatrix} = 0
\tag{2.152}
$$

or

$$
\mu^2 - 1 = 0 \quad \Rightarrow \quad \mu = \pm 1
\tag{2.153}
$$

Thus, the eigenvalues are $\pm(\hbar/2)$, which, of course, we already knew.

We require the eigenkets that correspond to each of these eigenvalues. For $m = +1$ call the eigenket $|m_+\rangle$. We have

$$
\begin{pmatrix} -1 & e^{-i\phi} \\ e^{i\phi} & -1 \end{pmatrix} \begin{pmatrix} a_+ \\ b_+ \end{pmatrix} = 0
\tag{2.154}
$$

from which

$$
-a_+ + b_+ e^{-i\phi} = 0
\tag{2.155}
$$

and

$$a_+ e^{i\phi} - b_+ = 0 \tag{2.156}$$

which are the same equation. From this equation we have

$$b_+ = e^{i\phi} a_+ \tag{2.157}$$

But, we also have the normalization condition

$$|a_+|^2 + |b_+|^2 = 1 \tag{2.158}$$

Equations (2.156) and (2.158) lead to

$$|a_+|^2 = |b_+|^2 = \frac{1}{2} \tag{2.159}$$

so that we arrive at one eigenket of $\hat{S}_n$

$$|\mu_+\rangle = \frac{1}{\sqrt{2}} |\alpha\rangle + \frac{e^{i\phi}}{\sqrt{2}} |\beta\rangle \tag{2.160}$$

Similarly, we get

$$|\mu_-\rangle = \frac{1}{\sqrt{2}} |\alpha\rangle - \frac{e^{i\phi}}{\sqrt{2}} |\beta\rangle \tag{2.161}$$

Now, we may choose $\phi = 0$ to obtain the eigenkets of $\hat{S}_x$ which we denote by $|\alpha\rangle_x$ and $|\beta\rangle_x$.

$$|\alpha\rangle_x = \frac{1}{\sqrt{2}} |\alpha\rangle + \frac{1}{\sqrt{2}} |\beta\rangle$$

$$|\beta\rangle_x = \frac{1}{\sqrt{2}} |\alpha\rangle - \frac{1}{\sqrt{2}} |\beta\rangle \tag{2.162}$$

Because we have chosen $\hat{S}_z$ to commute with $\hat{S}^2$ thus making it special, we suppress the subscript z from the eigenfunctions of $\hat{S}_z$; that is, $|\alpha\rangle = |\alpha\rangle_z$ and $|\beta\rangle = |\beta\rangle_z$.

Finally, to obtain the eigenkets of $\hat{S}_y$ we choose $\phi = \pi/2$ and obtain

$$|\alpha\rangle_y = \frac{1}{\sqrt{2}} |\alpha\rangle + \frac{i}{\sqrt{2}} |\beta\rangle$$

$$|\beta\rangle_y = \frac{1}{\sqrt{2}} |\alpha\rangle - \frac{i}{\sqrt{2}} |\beta\rangle \tag{2.163}$$

2.8. The Stern–Gerlach Experiment

The Stern–Gerlach experiment had monumental consequences for the development of the modern quantum theory. It showed definitively that angular momentum was quantized, and, in particular, that the intrinsic angular momentum of the electron, that is, spin, was quantized. A schematic diagram of the apparatus and the observed result are shown in FIGURE 2.4.

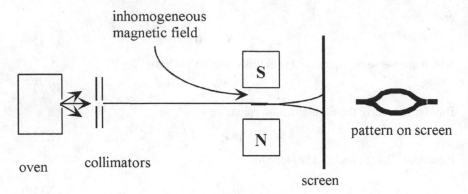

FIGURE 2.4. Schematic diagram of the Stern–Gerlach apparatus showing the observed pattern.

The original experiment was performed with Ag atoms (total spin 1/2). We may imagine that the experiment is being performed with electrons and ignore the effects of the magnetic field on the charged electrons.

If the electrons emerging from the oven are "unpolarized," that is, if the spins are randomly oriented, then we may expand the wave function on the complete set of $\hat{S}_z$ eigenfunctions, $|\alpha\rangle$ and $|\beta\rangle$. Because, however, they are randomly oriented the wave function must be

$$|\psi\rangle = \frac{1}{\sqrt{2}}|\alpha\rangle + \frac{1}{\sqrt{2}}|\beta\rangle \qquad (2.164)$$

Therefore, a Stern–Gerlach apparatus will simply split the beam into the spin-up and spin-down components as shown schematically in FIGURE 2.5. SGz represents a Stern–Gerlach apparatus with the inhomogeneous magnetic field oriented in the z-direction.

Each of the emerging beams is polarized: one spin-up, $|\alpha\rangle$ and the other spin-down, $|\beta\rangle$.

If either of the beams emerging from the SGz apparatus were passed through a second SGz apparatus only that beam would emerge because it is already in an eigenstate of $\hat{S}_z$ so "operating" on it again simply reproduces the eigenvector. In other words, the particles that constitute the remaining beam, those in the $|\alpha\rangle$ eigenstate, remain in the $|\alpha\rangle$ eigenstate.

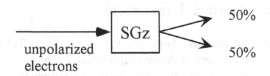

FIGURE 2.5. Schematic diagram of the results of passing an unpolarized beam through a Stern–Gerlach apparatus with the inhomogeneous magnetic field in the z-direction.

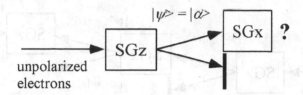

FIGURE 2.6. Schematic diagram showing a beam that is polarized in the +z-direction entering a Stern–Gerlach apparatus with the inhomogeneous magnetic field in the x-direction.

Suppose, however, that instead of passing the output beam of the first SGz apparatus into another SGz apparatus we pass it through an SGx apparatus, a Stern–Gerlach apparatus having the magnetic field oriented at right angles to the field in the first apparatus as shown schematically in FIGURE 2.6.

To determine the fate of the pure $|\alpha\rangle$ beam when it passes through the SGx device we must express $|\alpha\rangle$ in terms of $|\alpha\rangle_x$ and $|\beta\rangle_x$. We have already found $|\alpha\rangle_x$ and $|\beta\rangle_x$ in terms of $|\alpha\rangle$ and $|\beta\rangle$, Equation (2.162).

$$|\alpha\rangle_x = \frac{1}{\sqrt{2}}|\alpha\rangle + \frac{1}{\sqrt{2}}|\beta\rangle$$

$$|\beta\rangle_x = \frac{1}{\sqrt{2}}|\alpha\rangle - \frac{1}{\sqrt{2}}|\beta\rangle \qquad (2.165)$$

Adding these equations we eliminate $|\beta\rangle$ and solve for $|\alpha\rangle$.

$$|\alpha\rangle = \frac{1}{\sqrt{2}}|\alpha\rangle_x + \frac{1}{\sqrt{2}}|\beta\rangle_x \qquad (2.166)$$

Thus, the SGx device sorts the particles in the beam according to their x-components of spin. The $|\alpha\rangle$ beam is split into two equal beams as shown in FIGURE 2.7.

Now, what happens if we block the $|\beta\rangle_x$ beam and pass the $|\alpha\rangle_x$ beam through a SGz device? Because, from Equation (2.165)

$$|\alpha\rangle_x = \frac{1}{\sqrt{2}}|\alpha\rangle + \frac{1}{\sqrt{2}}|\beta\rangle$$

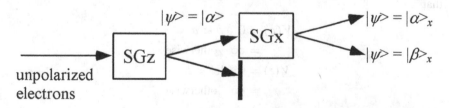

FIGURE 2.7. Schematic diagram of the results of passing a beam polarized in the +z-direction through a Stern–Gerlach apparatus with the inhomogeneous magnetic field in the x-direction.

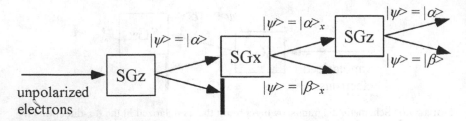

FIGURE 2.8. Schematic diagram of the results of passing a beam polarized in the $+x$-direction through a Stern–Gerlach apparatus with the inhomogeneous magnetic field in the z-direction.

the beam will be split into two equal parts again. The situation is represented graphically in FIGURE 2.8.

Problems

2.1. Consider a particle of mass μ subjected to a two-dimensional harmonic oscillator potential in which the force constants for the x- and y-components of motion are $k_x = k_y = k$. Show that the Schrödinger equation is separable in Cartesian coordinates. How many quantum numbers are there? Find an expression for the energy of this oscillator and discuss degeneracy. Do you think that the problem is separable in any other coordinate system? If so, which one? Why?

2.2. Consider a particle of mass μ confined to a two-dimensional circular "box" of radius a such that

$$V(\rho) = 0 \quad \rho < a$$
$$= \infty \quad \text{otherwise}$$

Show that the Schrödinger equation is separable in polar coordinates ρ, ϕ. Find and solve the equations for the radial and the angular motion. Indicate how to obtain the energy eigenvalues.

2.3. Discuss modifications of the energies and wave functions obtained in Problem 2.2 if the particle were in a three-dimensional cylindrical box of length L such that

$$V(\rho) = 0 \quad \rho < a$$
$$= \infty \quad \text{otherwise}$$
$$V(z) = 0 \quad |z| < L$$
$$= \infty \quad \text{otherwise}$$

2.4. Verify that $\lfloor \hat{\boldsymbol{J}}^2, \hat{J}_x \rfloor = 0 = \lfloor \hat{\boldsymbol{J}}^2, \hat{J}_y \rfloor$.

2.5. Prove that the operators $\hat{L}^2$, $\hat{S}^2$, $\hat{J}^2$, $\hat{J}_z$ form a set of commuting operators, and that $\lfloor (\hat{\boldsymbol{L}} \cdot \hat{\boldsymbol{S}}), J_z \rfloor = 0$.

2.6. A particle moving in a central potential is described (in spherical coordinates) by the wave function

$$\psi(r, \theta, \phi) = r^2 e^{-\beta r^2} [\cos\theta + e^{i\phi} \sin\theta (1 + \cos\theta)]$$

where β is a real positive constant. What is the probability that measurements of $\hat{L}^2$ and $\hat{L}_z$ yield the results $6\hbar^2$ and $\hbar$ respectively?

2.7. A system is in the following coherent superposition of angular momentum states $|\ell m>$.

$$|\psi\rangle = A|11\rangle + B|10\rangle + C|1-1\rangle$$

where A, B, and C are complex constants. Calculate the expectation value of $\hat{L}_x$.

2.8. (a) Find the eigenvectors of the operator $\hat{S}_y$ in terms of $|\alpha\rangle$ and $|\beta\rangle$, the eigenvectors of $\hat{S}_z$. Express them as spinors.

(b) Suppose that an electron is in the spin state

$$\frac{1}{\sqrt{5}} \begin{pmatrix} 2 \\ -1 \end{pmatrix}$$

with the $\hat{S}_z$ eigenvectors as the basis. If we measure the y-component of the spin, what is the probability for finding a value of $+(1/2)\hbar$?

2.9. Use the spin 1 $\hat{S}_z$ eigenstates as a basis to form the matrix representations of the angular momentum operators.

Answer:

$$\hat{S}_x = \frac{\hbar}{\sqrt{2}} \begin{pmatrix} 0 & 1 & 0 \\ 1 & 0 & 1 \\ 0 & 1 & 0 \end{pmatrix}; \; \hat{S}_y = \frac{\hbar}{\sqrt{2}} \begin{pmatrix} 0 & -i & 0 \\ i & 0 & -i \\ 0 & i & 0 \end{pmatrix}; \; \hat{S}_z = \frac{\hbar}{\sqrt{2}} \begin{pmatrix} 1 & 0 & 0 \\ 0 & 0 & 0 \\ 0 & 0 & -1 \end{pmatrix}$$

2.10. An unpolarized beam of neutral spin-1 particles is passed through a Stern–Gerlach device with magnetic field in the y-direction, an SGy device. The beam of emerging particles having $S_y = \hbar$ is then passed through an SGz device. What fraction of the particles with $S_y = \hbar$ will be found to have $S_z = \hbar$?

2.11. Show that the operators $\hat{L}_z$ and $\hat{H} = (\hat{p}^2/2m) + V(r)$ commute for a spherically symmetric potential.

References

1. R.H. Dicke and J.P. Wittke, *Introduction to Quantum Mechanics* (Addison-Wesley, Reading, MA, 1960).
2. E.U. Condon and G.H. Shortley, *The Theory of Atomic Spectra* (Cambridge, London, 1970).
3. M. Weissbluth, *Atoms and Molecules* (Academic Press, San Diego, 1978).
4. E. Merzbacher, *Quantum Mechanics* (John Wiley, New York, 1998).
5. G.B. Arfken and H.J. Weber, *Mathematical Methods for Physicists* (Harcourt, New York, 2001).

3
Angular Momentum—Two Sources

3.1. Introduction

So far we have discussed a single source of angular momentum, that is, a single angular momentum vector. Suppose, however, that there are two (or more) angular momenta. As we have seen, even a single particle can have associated with it two angular momenta, for example, spin and orbital angular momenta. Thus, there are a number of quantized variables that describe the system. For two angular momenta corresponding to the two operators $\hat{\boldsymbol{J}}_1$ and $\hat{\boldsymbol{J}}_2$, there are four quantum numbers that might be used to describe the system (j_1, m_{j_1}) and (j_2, m_{j_2}). The *total* angular momentum is represented by the operators

$$\hat{\boldsymbol{J}} = \hat{\boldsymbol{J}}_1 + \hat{\boldsymbol{J}}_2 \quad and \quad \hat{J}_z = \hat{J}_{z_1} + \hat{J}_{z_2} \qquad (3.1)$$

which have quantum numbers (j, m_j). It is necessary to find sets of commuting operators that will provide a suitable description of the system.

   We show that there are two sets of mutually commuting operators that lead to "good" quantum numbers that describe a given system. Good quantum numbers are quantum numbers that are eigenvalues of the mutually commuting operators. They usually represent quantities that are conserved classically. Each set of operators leads to a different set of angular momentum eigenkets and eigenvalues. Each set of eigenkets constitutes a basis set upon which a wave function may be expanded. The conditions of the problem determine which of these basis sets is most convenient. Moreover, we would like to know how to convert from one basis set to the other.

3.2. Two Sets of Quantum Numbers—Uncoupled and Coupled

Although a single particle can have more than one angular momentum associated with it, for simplicity, we consider two different particles, each having a single angular mometum. The state of a particle, say particle 1, is fully specified by knowledge of its quantum numbers j_1 and m_{j1} and likewise for particle 2. To see if it is possible to specify the entire state in terms of the *individual* angular momentum quantum numbers j_1, m_{j1}, j_2, and m_{j2} we must check the commutation rules. Specifically, we wish to know if $\hat{J}_1^2$ and $\hat{J}_{1z}$ commute with $\hat{J}_2^2$ and $\hat{J}_{2z}$. Clearly they do because they operate on independent coordinates. For example, spin operators act on internal coordinates of a particle whereas the orbital operators act on the space coordinates. Therefore, all components of $\hat{\boldsymbol{J}}_1$ commute with all components of $\hat{\boldsymbol{J}}_2$. Furthermore, $\hat{J}_1^2$ and $\hat{J}_2^2$ also commute. We therefore conclude that the four operators $\hat{J}_1^2, \hat{J}_{1z}, \hat{J}_2^2$, and $\hat{J}_{2z}$ constitute a mutually commuting set of operators and that j_1, m_{j1}, j_2, and m_{j2} are good quantum numbers. We write the ket as

$$|j_1, m_{j1}; j_2, m_{j2}\rangle$$

Now, how about the *total* angular momentum which we provisionally define as $\hat{\boldsymbol{J}} = \hat{\boldsymbol{J}}_1 + \hat{\boldsymbol{J}}_2$? The definition is provisional because we must first prove that $\hat{\boldsymbol{J}}$ as defined above is indeed an angular momentum. To do this we examine the commutators of the components of $\hat{\boldsymbol{J}}$ where

$$\hat{J}_i = \hat{J}_{1i} + \hat{J}_{2i} \qquad i = x, y, z \tag{3.2}$$

$\hat{\boldsymbol{J}}$ is an angular momentum if

$$[\hat{J}_i, \hat{J}_j] = i\hbar\varepsilon_{ijk}\hat{J}_k \tag{3.3}$$

Evaluating $\lfloor \hat{J}_x, \hat{J}_y \rfloor$ will be sufficient.

$$\begin{aligned}
[\hat{J}_x, \hat{J}_y] &= [(\hat{J}_{1x} + \hat{J}_{2x}), (\hat{J}_{1y} + \hat{J}_{2y})] \\
&= [\hat{J}_{1x}, \hat{J}_{1y}] + [\hat{J}_{1x}, \hat{J}_{2y}] + [\hat{J}_{2x}, \hat{J}_{1y}] + [\hat{J}_{2x}, \hat{J}_{2y}] \\
&= i\hbar\hat{J}_{1z} + 0 + 0 + i\hbar\hat{J}_{2z} \\
&= i\hbar(\hat{J}_{1z} + \hat{J}_{2z}) \\
&= i\hbar\hat{J}_z \tag{3.4}
\end{aligned}$$

Therefore, $\hat{\boldsymbol{J}}$ is indeed an angular momentum and we know immediately that the eigenvalues of the magnitude of the $\hat{\boldsymbol{J}}$ are

$$\sqrt{j(j+1)}\hbar \quad \text{where } j = 0, \frac{1}{2}, 1, \frac{3}{2}, 2, \frac{5}{2}, 3, \dots.$$

and the eigenvalues of $\hat{J}_z$ are $m_j\hbar$ where $m_j = -j, -(j-1), \dots, (j-1), j$.

It is possible to specify simultaneously the values of j, j_1, and j_2 because $\hat{J}_1^2$, $\hat{J}_2^2$, and $\hat{J}^2$ are mutually commuting operators. That is,

$$[\hat{J}^2, \hat{J}_1^2] = 0 = [\hat{J}^2, \hat{J}_2^2]$$

and, indeed, the eigenvalues of $\hat{J}^2, \hat{J}_1^2$, and $\hat{J}_2^2$ can be specified simultaneously. Moreover, m_j can also be specified because $\hat{J}_z$ commutes with $\hat{J}_1^2, \hat{J}_2^2$, and $\hat{J}^2$. We may therefore find simultaneous eigenkets of these four operators which we designate as

$$|j_1, j_2; j, m_j\rangle$$

On the other hand, because neither $\hat{J}_{1z}$ nor $\hat{J}_{2z}$ commutes with $\hat{J}^2$, m_{j1} and m_{j2} are not good quantum numbers. This may be seen by evaluating the commutators $\lfloor \hat{J}_{1z}, \hat{J}^2 \rfloor$ and $\lfloor \hat{J}_{2z}, \hat{J}^2 \rfloor$. (Of course, one will suffice.)

$$[\hat{J}_{1z}, \hat{J}^2] = [\hat{J}_{1z}, \hat{J}_x^2] + [\hat{J}_{1z}, \hat{J}_y^2] + [\hat{J}_{1z}, \hat{J}_z^2]$$
$$= [\hat{J}_{1z}, (\hat{J}_{1x} + \hat{J}_{2x})^2] + [\hat{J}_{1z}, (\hat{J}_{1y} + \hat{J}_{2y})^2] + [\hat{J}_{1z}, (\hat{J}_{1z} + \hat{J}_{2z})^2]$$

$$(3.5)$$

Ignoring the terms that are obviously zero, we have

$$[\hat{J}_{1z}, \hat{J}^2] = \left[\hat{J}_{1z}, \left(\hat{J}_{1x}^2 + 2\hat{J}_{1x}\hat{J}_{2x}\right)\right] + \left[\hat{J}_{1z}, \left(\hat{J}_{1y}^2 + 2\hat{J}_{1y}\hat{J}_{2y}\right)\right]$$
$$= \left[\hat{J}_{1z}, \left(\hat{J}_{1x}^2 + \hat{J}_{1y}^2\right)\right] + 2[\hat{J}_{1z}, \hat{J}_{1x}]\hat{J}_{2x} + 2[\hat{J}_{1z}, \hat{J}_{1y}]\hat{J}_{2y}$$
$$= \left[\hat{J}_{1z}, \left(\hat{J}_1^2 - \hat{J}_{1z}^2\right)\right] + 2i\hbar\hat{J}_{1y}\hat{J}_{2x} - 2i\hbar\hat{J}_{1x}\hat{J}_{2y}$$
$$= 2i\hbar(\hat{J}_{1y}\hat{J}_{2x} - \hat{J}_{1x}\hat{J}_{2y})$$
$$\neq 0$$

$$(3.6)$$

We conclude that there are two different sets of quantum numbers that describe a system that consists of two independent angular momenta $\hat{J}_1$ and $\hat{J}_2$. In one set, referred to as the uncoupled set, the quantum numbers are j_1, m_{j1}, j_2, and m_{j2} corresponding to the mutually commuting operators $\hat{J}_1^2$, $\hat{J}_{1z}$, $\hat{J}_2^2$, and $\hat{J}_{1z}$ and we know the magnitude of the individual angular momenta and their z-components. In this representation we have no information about the relative orientations of the two angular momenta.

In the other representation, the coupled representation, the good quantum numbers are j_1, j_2, j, m_j and the corresponding ket is designated $|j_1, j_2; j, m_j\rangle$. In this representation the magnitudes of the individual angular momenta are known as is the magnitude of the total angular momentum. We do not, however, know the z-components of the individual angular momenta, only the z-component of the total angular momentum.

These different representations are merely alternate ways of describing a system consisting of two angular momenta. It is part of our art to decide which set is most convenient for a given problem. To use the coupled representation we must find the allowed values of j and m_j. If we operate with $\hat{J}_z$ on an uncoupled ket we find that

$$\hat{J}_z|j_1, m_{j1}; j_2, m_{j2}\rangle = (\hat{J}_{1z} + \hat{J}_{2z})|j_1, m_{j1}; j_2, m_{j2}\rangle$$
$$= (m_{j1} + m_{j2})\hbar|j_1, m_{j1}; j_2, m_{j2}\rangle \qquad (3.7)$$

But the eigenvalues of $\hat{J}_z$ are $m_j\hbar$ so we must have

$$m_j = m_{j1} + m_{j2} \qquad (3.8)$$

Also, because

$$(m_j)_{max} = (m_{j1})_{max} + (m_{j2})_{max}$$
$$= j_1 + j_2 \qquad (3.9)$$

TABLE 3.1. Quantum numbers and number of states for two angular momenta.

j	m_j	Number of states
$j_1 + j_2$	$(j_1 + j_2), (j_1 + j_2 - 1), \ldots - (j_1 - j_2 + 1), -(j_1 - j_2)$	$2(j_1 + j_2) + 1$
$j_1 + j_2 - 1$	$(j_1 + j_2 - 1), (j_1 + j_2 - 2), \ldots - (j_1 - j_2 + 2), -(j_1 - j_2 + 1)$	$2(j_1 + j_2 - 1) + 1$
$\ldots$	$\ldots$	$\ldots$
$j_1 - j_2$	$(j_1 - j_2), (j_1 - j_2 - 1), \ldots - (j_1 - j_2 + 1), -(j_1 - j_2)$	$2(j_1 - j_2) + 1$

and

$$(m_j)_{\max} = j \Rightarrow j = j_1 + j_2 \tag{3.10}$$

We therefore have a system in which $j = j_1 + j_2$ and m_j takes on the values $-(j_1 + j_2) \le m_j \le +(j_1 + j_2)$. There are, however, other possible combinations. For example, the state having $m_j = j_1 + j_2 - 1$ can be formed from $m_{j1} = j_1$ and $m_{j2} = j_2 - 1$ or by $m_{j1} = j_1 - 1$ and $m_{j2} = j_2$. Although one of these states "belongs" to the $j = j_2 + j_2$ set, the other is associated with $j = j_2 + j_2 - 1$. We have then a grouping of states as shown in TABLE 3.1. For specificity, we assume that $j_1 > j_2$.

It is clear from examination of the uncoupled representation that the total number of states is

$$N = (2j_1 + 1) \times (2j_2 + 1)$$

That is, N is simply the product of the total number of m_{j1} states and the total number of m_{j2} states. To total the states using the coupled representation we add all states for a given pair of j_1 and j_2. From TABLE 3.1 we see that

$$N = \{2(j_1 + j_2) + 1\} + \{2(j_1 + j_2 - 1) + 1\} + \{2(j_1 + j_2 - 2) + 1\} + \cdots$$
$$\{2(j_1 - j_2 + 2) + 1\} + \{2(j_1 - j_2 + 1) + 1\} + \{2(j_1 - j_2) + 1\}$$
$$= \sum_{n=0}^{2j_2} \{2(j_1 + j_2 - n) + 1\} \tag{3.11}$$

where, as above, it is assumed that $j_1 > j_2$. That the summation in Equation (3.11) is correct can be seen by noting that there are $(2j_2 + 1)$ terms in the sum, j_2 added to j_1, j_2 subtracted from j_1, and one term in between. Writing out the terms, especially in the region in which the transition from $(j_1 + j_2)$ to $(j_1 - j_2)$ occurs reveals that the summation is indeed correct. The sum is easily evaluated with the aid of "Gauss' trick"[1]

$$\sum_{0}^{M} n = \frac{M(M + 1)}{2} \tag{3.12}$$

and noting that there are $(2j_2 + 1)$ terms in the summation from $n = 0$ to $2j_2$. We have

$$N = \sum_{n=0}^{2j_2} \{2(j_1 + j_2 - n) + 1\}$$

$$= 2j_1 \sum_{n=0}^{2j_2} 1 + 2j_2 \sum_{n=0}^{2j_2} 1 - 2 \sum_{n=0}^{2j_2} n + \sum_{n=0}^{2j_2} 1$$

$$= 2j_1(2j_2 + 1) + 2j_2(2j_2 + 1) - 2\frac{2j_2(2j_2 + 1)}{2} + (2j_2 + 1)$$

$$= (2j_2 + 1)(2j_1 + 1) \tag{3.13}$$

thus demonstrating that the total number of states is the same in either representation.

As a simple example of these two representations we consider a single atomic electron having orbital angular momentum one and spin angular momentum one-half. We may thus designate $j_1 = \ell = 1$ and $j_2 = s = 1/2$. There are two possible values of the total angular momentum j because there are two possible orientations of the spin with respect to the orbital angular momentum as can be seen using a Bohr picture of the hydrogen atom as shown in FIGURE 3.1.

The total angular momentum j can then take on the values

$$j = \left(1 + \frac{1}{2}\right) \quad \text{or} \quad j = \left(1 + \frac{1}{2} - 1\right)$$

$$= \frac{3}{2}, \frac{1}{2} \tag{3.14}$$

Note that we must stop subtracting unity at $1/2$ because the value of j cannot be negative. Of course, the orbiting electron produces a magnetic moment, independent of the intrinsic magnetic moment of the electron, so the system may be imagined as two bar magnets having two different possible orientations.

The energy of orientation of a magnetic dipole in a magnetic induction field B is

$$E = -\mu \cdot B$$

$$\propto \mu_S \cdot \mu_l$$

$$\propto S \cdot L \tag{3.15}$$

so that the energy of this atomic electron will depend on the relative orientations of the angular momenta vectors S and L. There will be two different

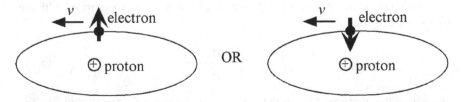

FIGURE 3.1. The two orientations of spin and orbital magnetic moments.

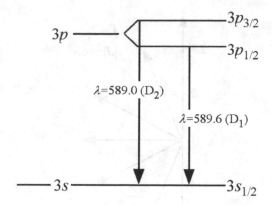

FIGURE 3.2. State that led to the Na D-line emissions.

energies corresponding to parallel ($j = 3/2$) and anti-parallel ($j = 1/2$) angular momenta.

These two energies are observed in many atomic spectra. A simple example is the spectrum of the Na atom. When Na emits a photon in undergoing a transition from the first excited state ($n = 3$, $\ell = 1$) to the lowest state ($n = 3$, $\ell = 0$), it emits photons in the yellow region of the spectrum, wavelength ~590 nm. Closer examination reveals that there are actually two different wavelengths of yellow light emitted. These emissions are referred to as the sodium D-lines, actually D_1 and D_2. This occurs because the upper state is split into two different states depending upon the value of j, either $j = 3/2$ or $j = 1/2$. These two upper states are designated $3p_{1/2}$ and $3p_{3/2}$ and the lower state $3s_{1/2}$ as discussed in Chapter 2. The energy level diagram together with the atomic emission lines is shown in FIGURE 3.2. Why is the ground state, the $3s$-state, not similarly split?

It is interesting to note that the designation of these yellow lines as "D-lines" is in no way related to angular momentum. In 1814 Fraunhofer examined the solar spectrum using a spectrograph to separate the constituent wavelengths. He observed a number of black lines against the continuum blackbody radiation from the sun. These black lines were due to absorption by atoms and molecules in the solar atmosphere at the characteristic wavelengths of these atoms and molecules. He did not know the origin of these black lines, but he labeled the strongest of them alphabetically beginning at the red end of the spectrum. Absorption by sodium atoms in the solar atmosphere at around 589 nm was the fourth of the strong lines. Thus, it is designated as the D-line. Often emissions and absorptions that originate from analogous states of other alkali metal atoms are incorrectly referred to as D-lines. There is, in fact, no such thing as the potassium D-line.

3.3. Vector Model of Angular Momentum

Because only discrete values of m_j are allowed, the z-component of angular momentum is quantized. Similarly, only discrete values of j are allowed so that

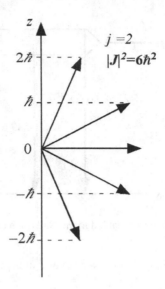

FIGURE 3.3. Two-dimensional view of the vector model of the atom for $j = 2$.

the *magnitude* of the total angular momentum is quantized. This is sometimes referred to as "space quantization". The maximum value of the z-component is $(m_j)_{max}\hbar = j\hbar$ and the magnitude of the angular momentum is $\sqrt{j(j+1)}\hbar$. Therefore, the total angular momentum vector J can never be along the z-axis; that is, it cannot coincide with its own z-component. (Recall that the choice of z-axis is arbitrary.) This can conveniently be illustrated using a vector model for angular momentum. The case of $j = 2$ illustrates the model. The length of the total angular momentum vector is

$$|J| = \sqrt{j(j+1)}\hbar$$
$$= \sqrt{2 \cdot 3}\hbar \qquad\qquad (3.16)$$

and the maximum value of the z-component is

$$(J_z)_{max} = 2\hbar \qquad\qquad (3.17)$$

This leads us to a picture of the total angular momentum and its z-component as shown in FIGURE 3.3 for $j = 2$.

The quantum numbers j and m_j, however, give no information about the values of the x- and y-components of J. This is a feature of the quantum mechanics of angular momentum and is, of course, a consequence of the commutation relation between the components of angular momentum. Thus, rather than the above two-dimensional diagram, we should more properly depict the angular momentum vector as shown in FIGURE 3.4.

We should think of the angular momentum as lying on one of the cones that is defined by the five vectors of length $\sqrt{6}\hbar$. The particular cone upon which it lies is determined by the value of m_j. We should *not* think of the angular momentum vector as sweeping out of one of these cones. Rather, it lies at a particular orientation

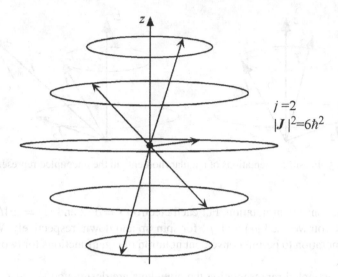

FIGURE 3.4. Three-dimensional view of the vector model of the atom for $j = 2$.

on the cone. Quantum mechanics simply refuses to let us know *where* it lies on that cone if we already know the value of the z-component. This is because the components of $\hat{J}$ do not commute.

The vector model is also useful for two angular momenta. The individual angular momentum vectors $\hat{J}_1$ and $\hat{J}_2$ have lengths

$$\sqrt{j_1(j_1 + 1)}\hbar \quad \text{and} \quad \sqrt{j_2(j_2 + 1)}\hbar$$

where j_1 and j_2 are the individual quantum numbers. The length of the total angular momentum vector is $\sqrt{j(j + 1)}\hbar$ where j is one of the permitted values of the total angular momentum $J = J_1 + J_2$; that is,

$$j = (j_1 + j_2), (j_1 + j_2 - 1), (j_1 + j_2 - 2), \ldots |j_1 - j_2| \qquad (3.18)$$

In the uncoupled representation, in addition to the individual angular momentum quantum numbers j_1 and j_2, we also know their individual z-components m_{j1} and m_{j2}. In this representation there is no information about the relative orientation of J_1 and J_2 so the magnitude of J, that is, the quantum number j, is not known. FIGURE 3.5 is an illustration of how two different orientations of individual angular momenta produce different total angular momentum vectors.

In the coupled representation we know j and its z-component m_j as well as the magnitudes of the individual angular momenta j_1 and j_2. The quantum numbers representing the individual z-components m_{j1} and m_{j2} are not known. We do, however, know the sum $m_j = m_{j1} + m_{j2}$. FIGURE 3.6 is an illustration of how two different orientations of individual angular momenta can produce identical total angular momentum vectors.

Suppose we have two particles, each with spin 1/2 such as two electrons. This is the situation for the He atom and the alkaline earth atoms. First, we must

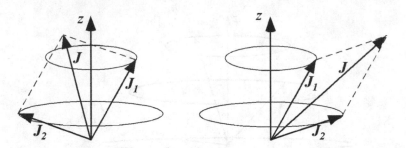

FIGURE 3.5. Possible orientations of angular momenta in the uncoupled representation.

establish a convenient notation. For each electron $s = 1/2$ and $m_s = \pm 1/2$. For a single electron we used $|\alpha\rangle$ and $|\beta\rangle$ for spin up and down, respectively. We must adapt the notation to permit convenient notation of eigenfunctions for two angular momenta.

In the uncoupled representation the eigenkets are designated $|s_1, m_{s1}; s_2, m_{s2}\rangle$ but, because s_1 and s_2 are always 1/2, we may simplify the notation as follows

$$\left|\frac{1}{2}, \frac{1}{2}; \frac{1}{2}, \frac{1}{2}\right\rangle = |\alpha_1 \alpha_2\rangle$$

$$\left|\frac{1}{2}, \frac{1}{2}; \frac{1}{2}, -\frac{1}{2}\right\rangle = |\alpha_1 \beta_2\rangle$$

$$\left|\frac{1}{2}, -\frac{1}{2}; \frac{1}{2}, \frac{1}{2}\right\rangle = |\beta_1 \alpha_2\rangle$$

$$\left|\frac{1}{2}, -\frac{1}{2}; \frac{1}{2}, -\frac{1}{2}\right\rangle = |\beta_1 \beta_2\rangle$$

In the coupled representation the eigenkets are designated

$$|s_1, s_2; S, M_S\rangle$$

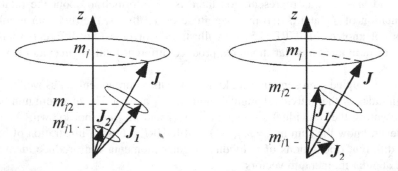

FIGURE 3.6. Possible orientations of angular momenta in the coupled representation.

but again $s_1 = s_2 = 1/2$, so we may eliminate them from the ket and designate a coupled ket as $|S, M_S\rangle$ where, for two spin 1/2 particles $S = 1$ or 0 and $M_S = m_{s1} + m_{s2}$. For $S = 0$ there is only one possible value of M_S, zero.

$$M_S = 0 \quad \Rightarrow \quad |0, 0\rangle$$

This is called a "singlet" state because there is only one value of M_S and thus only one state having $S = 0$. For $S = 1$, the "triplet" state, we have

$$M_S = -1, 0, +1 \quad \Rightarrow \quad |1, -1\rangle; |1, 0\rangle; |1, +1\rangle$$

The length of S is $\sqrt{2}\hbar$ and the length of the constituent spin vectors is $(1/2) \cdot \sqrt{(1/2) + 1}\hbar = (1/2) \cdot \sqrt{3}\hbar$. Therefore, the individual vectors cannot lie at arbitrary *relative* orientations on the cones. Their relative positions must be such that their vector sum is a vector of length $\sqrt{2}\hbar$. If one is at some orientation on its cone, the other has a definite angle on its cone. The absolute location of the vectors remains unknown.

The coupled and uncoupled sets are each complete sets so it is possible to express any coupled state as a linear combination of the uncoupled states and, of course, vice versa. Expanding a single uncoupled ket we have

$$|j_1, j_2; j, m_j\rangle = \sum_{m_{j1}} \sum_{m_{j2}} C_{m_{j1}m_{j2}} |j_1, m_{j1}; j_2, m_{j2}\rangle \qquad (3.19)$$

where the $C_{m_{j1}m_{j2}}$ are the Clebsch–Gordan coefficients.

3.4. Examples of Calculation of the Clebsch–Gordan Coefficients

Example 1

To illustrate the procedure for obtaining the Clebsch–Gordan expansion we express the coupled kets for two spin-1/2 particles in terms of the uncoupled kets. It is clear that each of the coupled kets $|S, M_S\rangle = |1, +1\rangle$ and $|S, M_S\rangle = |1, -1\rangle$ can have only one "component"; that is, each is identical to a particular uncoupled ket. We can make the independence of the two uncoupled spin 1/2 kets more obvious by letting

$$|\alpha_j \alpha_k\rangle = |\alpha_j\rangle|\alpha_k\rangle; \ |\beta_j \beta_k\rangle = |\beta_j\rangle|\beta_k\rangle; \ |\alpha_j \beta_k\rangle = |\alpha_j\rangle|\beta_k\rangle$$

Returning now to the two coupled kets $|S, M_S\rangle = |1, +1\rangle$ and $|S, M_S\rangle = |1, -1\rangle$ we see that the only uncoupled kets for which the individual z-components of angular momentum combine to give $M = \pm 1$ are $|\alpha_1\rangle|\alpha_2\rangle$ and $|\beta_1\rangle|\beta_2\rangle$ for $M = \pm 1$, respectively. Any other combination leads to $M = 0$. Therefore,

$$|1, +1\rangle = |\alpha_1\rangle|\alpha_2\rangle \quad \text{and} \quad |1, -1\rangle = |\beta_1\rangle|\beta_2\rangle \qquad (3.20)$$

The remaining coupled kets are $|S, M\rangle = |1, 0\rangle$ and $|S, M\rangle = |0, 0\rangle$. The expansion for these coupled states must contain both of the uncoupled states $|\alpha_1\rangle|\beta_2\rangle$

TABLE 3.2. Table of Clebsch–Gordan coefficients for $j_2 = 1/2$.

	$\langle j_1, \frac{1}{2}; m_{j1}, m_{j2} | j m_j \rangle$	
j	$m_{j2} = +1/2 \Leftrightarrow |\alpha_2\rangle$	$m_{j2} = -1/2 \Leftrightarrow |\beta_2\rangle$
$j_1 + 1/2$	$\sqrt{(j_1 + m_j + 1/2)/(2j_1 + 1)}$	$\sqrt{(j_1 - m_j + 1/2)/(2j_1 + 1)}$
$j_1 - 1/2$	$-\sqrt{(j_1 - m_j + 1/2)/(2j_1 + 1)}$	$\sqrt{(j_1 + m_j + 1/2)/(2j_1 + 1)}$

and $|\alpha_2\rangle|\beta_1\rangle$. Of course, it cannot contain either $|\alpha_1\rangle|\alpha_2\rangle$ or $|\beta_1\rangle|\beta_2\rangle$. We have then

$$|SM\rangle = |1, 0\rangle = C_{\frac{1}{2}, -\frac{1}{2}} |\alpha_1\rangle|\beta_2\rangle + C_{-\frac{1}{2}, \frac{1}{2}} |\beta_1\rangle|\alpha_2\rangle \tag{3.21}$$

We must obtain the two expansion coefficients for each of the coupled kets having $M_J = 0$, $|1, 0\rangle$ and $|0, 0\rangle$. The common method is to use standard tabulations which we employ first. The case at hand is particularly simple. TABLE 3.2 is a listing of Clebsch–Gordan coefficients for an arbitrary angular momentum j_1, but with j_2 specified to be $1/2$.

If, however, both angular momenta are spin so that $j_1 = j_2 = 1/2$ then there is considerable simplification as can be seen in TABLE 3.3.

To express the coupled ket $|00\rangle$ in terms of the uncoupled kets $|\alpha_1\rangle|\beta_2\rangle$ and $|\beta_1\rangle|\alpha_2\rangle$ we read the Clebsch–Gordan coefficients from the bottom row of the table (for which $S = 0$) and obtain:

SINGLET:

$$|SM\rangle = |00\rangle = -\frac{1}{\sqrt{2}} |\beta_1\rangle|\alpha_2\rangle + \frac{1}{\sqrt{2}} |\alpha_1\rangle|\beta_2\rangle, \tag{3.22}$$

This is the singlet state.

There are three coupled states $|1\ M\rangle$ corresponding to the values $M = 0, \pm 1$, the triplet states. Reading from the row of the table for which $S = 1$ we obtain

TRIPLET:

$$|SM\rangle = |11\rangle = |\alpha_1\rangle|\alpha_2\rangle + 0$$

$$= |10\rangle = \frac{1}{\sqrt{2}} |\alpha_2\rangle|\beta_1\rangle + \frac{1}{\sqrt{2}} |\alpha_1\rangle|\beta_2\rangle$$

$$= |1 - 1\rangle = 0 + |\beta_1\rangle|\beta_2\rangle \tag{3.23}$$

Now, a specific set of eigenvectors must be orthogonal, and, indeed, $\langle 00|10\rangle = 0 = \langle 00|1 - 1\rangle$ because $\langle 00|$ contains neither $|\alpha_1\rangle|\alpha_2\rangle$ nor $|\beta_1\rangle|\beta_2\rangle$. Inspection of $\langle 00|$ and $|10\rangle$ reveals that the difference in the signs of the coefficients of $|\alpha_2\rangle|\beta_1\rangle$

TABLE 3.3. Clebsch–Gordan coefficients for $j_1 = j_2 = 1/2$.

S	$|\alpha_2\rangle$	$|\beta_2\rangle$
1	$\sqrt{(1 + M)/2}$	$\sqrt{(1 - M)/2}$
0	$-1/\sqrt{2}$	$1/\sqrt{2}$

in these two coupled states leads to

$$\langle 11|00\rangle = \left[\frac{1}{\sqrt{2}}\langle\beta_1|\langle\alpha_2| + \frac{1}{\sqrt{2}}\langle\alpha_1|\langle\beta_2|\right]\left[-\frac{1}{\sqrt{2}}|\beta_1\rangle|\alpha_2\rangle + \frac{1}{\sqrt{2}}|\alpha_1\rangle|\beta_2\rangle\right]$$

$$= \frac{1}{2}(-1 + 0 + 0 + 1)$$

$$= 0 \tag{3.24}$$

so that these two states are also orthogonal. The difference in the signs of $|00\rangle$ and $|10\rangle$ can be understood in terms of the vector diagram. In the triplet, the individual angular momenta have the same phase

$$|10\rangle = \frac{1}{\sqrt{2}}[|\beta_1\rangle|\alpha_2\rangle + e^{i\delta}|\alpha_1\rangle|\beta_2\rangle] \tag{3.25}$$

where $\delta = 0$.

In the singlet

$$|00\rangle = -\frac{1}{\sqrt{2}}[|\beta_1\rangle|\alpha_2\rangle + e^{i\delta}|\alpha_1\rangle|\beta_2\rangle] \tag{3.26}$$

where $\delta = \pi$.

To illustrate the method by which the tables may be constructed we again consider two spin 1/2 particles. We begin with a coupled state for which we already know the representation in terms of the uncoupled states, that is, one for which there is only one uncoupled "component". We know that $|11\rangle = |\alpha_1\rangle|\alpha_2\rangle$ because there is no other way to get $M = +1$. (We could also start with $|1 - 1\rangle = |\beta_1\rangle|\beta_2\rangle$.)

Applying the lowering operator to $|11\rangle$ we have

$$\hat{S}_-|SM\rangle = \sqrt{S(S+1) - M(M-1)}\hbar|S(M-1)\rangle \tag{3.27}$$

which leads to

$$\hat{S}_-|11\rangle = \sqrt{2}\hbar|10\rangle \tag{3.28}$$

But, we may also compute $\hat{S}_-|11\rangle$ by applying $\hat{S}_-$ in terms of the individual spin angular momentum operators; that is, we apply $\hat{S}_- = (\hat{S}_{1-} + \hat{S}_{2-})$ to $|\alpha_1\rangle|\alpha_2\rangle$. In general

$$\hat{S}_{k-}|\alpha_k\rangle = \sqrt{\frac{1}{2}\left(\frac{1}{2}+1\right) - \frac{1}{2}\left(\frac{1}{2}-1\right)}\hbar|\beta_k\rangle$$

$$= \hbar|\beta_k\rangle \tag{3.29}$$

so that, operating on $|11\rangle$ with $\hat{S}_-$ we have

$$\hat{S}_-|11\rangle = (\hat{S}_{1-} + \hat{S}_{2-})|\alpha_1\rangle|\alpha_2\rangle$$

$$= \hbar|\beta_1\rangle|\alpha_2\rangle + \hbar|\alpha_1\rangle|\beta_2\rangle$$

$$= \hbar(|\beta_1\rangle|\alpha_2\rangle + |\alpha_1\rangle|\beta_2\rangle) \tag{3.30}$$

Now, we have two different expressions for the action of $\hat{S}_-$ on $|11\rangle$, Equations (3.28) and (3.30). Equating these expressions, we obtain

$$\sqrt{2}\hbar|10\rangle = \hbar(|\beta_1\rangle|\alpha_2\rangle + |\alpha_1\rangle|\beta_2\rangle)$$

or

$$|10\rangle = \frac{1}{\sqrt{2}}(|\beta_1\rangle|\alpha_2\rangle + |\alpha_1\rangle|\beta_2\rangle) \tag{3.31}$$

as we found using the table. We have therefore derived the Clebsch–Gordan coefficients which are, in this case, both $+(1/\sqrt{2})$.

To obtain the third of the triplet states we repeat the procedure (although we already know that the answer is $|1 - 1\rangle = |\beta_1\rangle|\beta_2\rangle$). Applying the lowering operator to the ket that we just obtained we have

$$\hat{S}_-|10\rangle = \sqrt{2}\hbar|1 - 1\rangle \tag{3.32}$$

Also

$$\hat{S}_-|10\rangle = (\hat{S}_{1-} + \hat{S}_{2-})|10\rangle$$

$$= \frac{1}{\sqrt{2}}(|\beta_1\rangle|\alpha_2\rangle + |\alpha_1\rangle|\beta_2\rangle)$$

$$= \frac{1}{\sqrt{2}}\hbar(|\beta_1\rangle|\beta_2\rangle + 0 + 0 + |\beta_1\rangle|\beta_2\rangle)$$

$$= \frac{2}{\sqrt{2}}\hbar|\beta_1\rangle|\beta_2\rangle \tag{3.33}$$

Again equating the two expressions, this time for $\hat{S}_-|10\rangle$, Equations (3.32) and (3.33), we obtain $|1 - 1\rangle = |\beta_1\rangle|\beta_2\rangle$ as expected. Clearly, we could have begun with $|1 - 1\rangle = |\beta_1\rangle|\beta_2\rangle$ and applied the raising operator twice to obtain the other two of the triplet states.

The singlet state must be constructed from $|\alpha_1\rangle|\beta_2\rangle$ and $|\alpha_2\rangle|\beta_1\rangle$ so that we cannot apply a shift operator to one of the sets. It must, however, be orthogonal to all of the triplet states because $\hat{S}^2$ is a Hermitian operator. Therefore, let

$$|00\rangle = A|\alpha_1\rangle|\beta_2\rangle + B|\alpha_2\rangle|\beta_1\rangle \tag{3.34}$$

so that

$$\langle10|00\rangle = \left[\frac{1}{\sqrt{2}}\langle\beta_1|\langle\alpha_2| + \frac{1}{\sqrt{2}}\langle\alpha_1|\langle\beta_2|\right][A|\beta_1\rangle|\alpha_2\rangle + B|\alpha_1\rangle|\beta_2\rangle]$$

$$= \frac{1}{\sqrt{2}}(A + B)$$

Because $\langle10|00\rangle = 0$ we must have $A + B = 0$. Also, the normalization condition is

$$|A|^2 + |B|^2 = 1$$

so that

$$2|A|^2 = 1 \Rightarrow A = \pm 1/\sqrt{2} \quad \text{and} \quad B = \pm 1/\sqrt{2}$$

We have then

$$|00\rangle = -\frac{1}{\sqrt{2}}|\alpha_1\rangle|\beta_2\rangle + \frac{1}{\sqrt{2}}|\alpha_2\rangle|\beta_1\rangle \tag{3.35}$$

Notice that A was chosen to be negative, but, no matter which sign was chosen, B must have the opposite sign for orthogonality.

Example 2

Consider now two p-electrons in an atom and ignore their spin. In this case $j_1 = 1 = j_2$. Because these are orbital angular momenta it is customary to let $j_1 = \ell_1$ and $j_2 = \ell_2$ and designate the coupled angular momentum J by L. Now, each of these ℓ-states has three substates so there will be a total of nine states. Of course, we may also count the states from the point of view of the coupled representation, but the number of states must be the same. Standard notation for the L-states is

$$L = 0, 1, 2, 3, 4, 5 \ldots$$
$$= S, P, D, F, G, H \ldots$$

Note that here S is *not* the spin operator; it refers to $L = 0$. Also, after F (for $L = 3$) the letters proceed alphabetically. The coupled kets are designated

$$|\ell_1, \ell_2; L, M_L\rangle \equiv |LM_L\rangle$$

and the uncoupled kets are

$$|\ell_1, m_{\ell 1}; \ell_2, m_{\ell 2}\rangle \equiv |m_{\ell 1}, m_{\ell 2}\rangle$$

We know that $|D2\rangle = |11\rangle$, where D stands for $L = 2$, is the highest value of M_L; that is, $M_L = 2$. Note that even though the coupled and uncoupled kets each contain only two elements in our simplified notation there should be no confusion between them because the coupled ket will always contain one capital letter and one number whereas the uncoupled kets contain two numbers. Now apply the lowering operator in two different forms to $|D2\rangle = |11\rangle$.

$$\hat{L}_-|D2\rangle = \sqrt{2(2+1) - 2(2-1)}\,\hbar|D1\rangle$$
$$= 2\hbar|D1\rangle \tag{3.36}$$

and

$$(\hat{L}_{1-} + \hat{L}_{2-})|11\rangle = \sqrt{1(1+1) - 1(1-1)}\,\hbar|01\rangle + \sqrt{1(1+1) - 1(1-1)}\,\hbar|10\rangle$$
$$= \sqrt{2}\hbar(|01\rangle + |10\rangle) \tag{3.37}$$

TABLE 3.4. Clebsch–Gordan coefficients for $j_1 = 1 = j_2$.

$j_1=1$	$j_2=1$	$j=2$					$j=1$			$j=0$
m_1	m_2	$m=2$	$m=1$	$m=0$	$m=-1$	$m=-2$	$m=1$	$m=0$	$m=-1$	$m=0$
1	1	1								
1	0		$\sqrt{1/2}$				$\sqrt{1/2}$			
1	−1			$\sqrt{1/6}$				$\sqrt{1/2}$		$\sqrt{1/3}$
0	1		$\sqrt{1/2}$				$-\sqrt{1/2}$			
0	0			$\sqrt{2/3}$				0		$-\sqrt{1/3}$
0	−1				$\sqrt{1/2}$				$\sqrt{1/2}$	
−1	1			$\sqrt{1/6}$				$-\sqrt{1/2}$		$\sqrt{1/3}$
−1	0				$\sqrt{1/2}$				$-\sqrt{1/2}$	
−1	−1					1				

Equating the two expressions, Equations (3.36) and (3.37), we have

$$|D1\rangle = \frac{1}{\sqrt{2}}(|01\rangle + |10\rangle) \tag{3.38}$$

The remaining three kets for $L = D$ may be generated by successive application of $\hat{L}_-$. For the $L = 1$, the P states, we start with $|P1\rangle$ which must be composed of only $|10\rangle$ and $|01\rangle$. Moreover, by symmetry, these uncoupled kets must be present in equal amounts. We do not, however, know the sign (phase) so we have

$$|P1\rangle = \frac{1}{\sqrt{2}}(|10\rangle \pm |01\rangle) \tag{3.39}$$

But we know that it must be orthogonal to all other eigenkets such as $|D1\rangle$. Now, we know that $\langle D1|P1\rangle = 0$ because they must be orthogonal. Evaluating this inner product using Equations (3.38) and (3.39) we have

$$\langle D1|P1\rangle = \frac{1}{\sqrt{2}}(\langle 01| + \langle 10|)\frac{1}{\sqrt{2}}(|10\rangle \pm |01\rangle)$$

$$= \frac{1}{2}(\pm 1 + 1) \tag{3.40}$$

We see that the only way $\langle D1|P1\rangle$ can vanish is if we choose the minus sign so

$$|P1\rangle = \frac{1}{\sqrt{2}}(|10\rangle - |01\rangle) \tag{3.41}$$

The remaining P-states may be generated by again applying $\hat{L}_-$. Continued, this procedure may be used to construct the entire table of Clebsch–Gordan coefficients for the case $j_1 = 1 = j_2$ as shown in TABLE 3.4.

Note that we obtained the coupled states $|\ell_1, \ell_2; L, M_L\rangle \equiv |LM_L\rangle$ in terms of the uncoupled states $|\ell_1, m_{\ell 1}; \ell_2, m_{\ell 2}\rangle \equiv |m_{\ell 1}, m_{\ell 2}\rangle$. To use the table we would simply read down the appropriate column. For example, $|D1\rangle = \frac{1}{\sqrt{2}}(|10\rangle + |01\rangle)$. To obtain the uncoupled states in terms of the coupled states we read across a row.

For example, if we subtract Equation (3.41) from Equation (3.38) we obtain

$$|01\rangle = \frac{1}{\sqrt{2}}(|D1\rangle - |P1\rangle)$$

which is also obtained reading across the row of Table 3.4 having $m_1 = 0$ and $m_2 = 1$.

3.5. Hyperfine Splitting in the Hydrogen Atom

Both the electron and the proton possess intrinsic magnetic dipole moments by virtue of their half-integral spins. Because there are different energies associated with the two different orientations of these magnetic moments as shown in FIGURE 3.7, the energy of each state will be split by this interaction. Solving the eigenvalue problem for this energy is a good exercise in angular momentum algebra. We concentrate on the ground state of the hydrogen atom for which the orbital magnetic moment is zero so we deal only with the spin–spin interaction.

This spin–spin interaction energy is rather small compared with other effects in the hydrogen atom so it is given the name hyperfine structure. In Chapter 2 it was seen that the magnetic moment of the electron is given by

$$\mu_S = -\frac{g_e \mu_B}{\hbar} \cdot S$$

$$= -\frac{2}{\hbar}\left(\frac{e\hbar}{2m_e}\right)S \quad (3.42)$$

An approximation to the magnetic moment of the proton that is analogous to Equation (3.42) for the electron is obtained by replacing m_e by M_P, the rest mass of the proton, and g_e by $g_p \approx 5.6$. Thus, the magnetic moment of the proton is smaller than that of the electron by a factor of ~700.

The term in the Hamiltonian, referred to as the hyperfine interaction, can be shown to be proportional to the dot product of the two spins which we

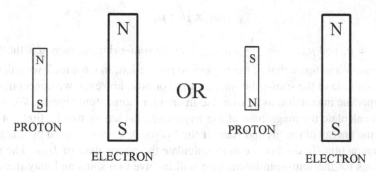

FIGURE 3.7. Two possible orientations of the proton spin with respect to the electron spin.

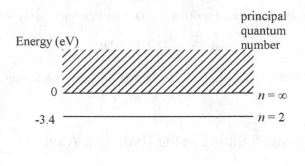

FIGURE 3.8. Energy levels of the hydrogen atom.

write as

$$\hat{H}_{HF} = K\hat{S}_1 \cdot \hat{S}_2 \tag{3.43}$$

where K is a constant, and $\hat{S}_1$ and $\hat{S}_2$ are the spin angular momentum operators. (We need not concern ourselves with which is the electron and which is the proton.) Also, it can be shown that $K > 0$.

FIGURE 3.8 shows the gross energy level structure of the hydrogen atom, that is, the Bohr energy levels.

The hyperfine interaction is very small because the magnetic moment of the proton is small. To see this we can compare the hyperfine interaction energy to that of the spin–orbit interaction for an excited state of hydrogen, that is, the interaction between μ_s and the magnetic moment that results from the electron's "orbit" around the proton. (It has to be an excited state because there is no orbital magnetic moment for the $\ell = 0$ ground state.) This spin–orbit interaction energy is a component of the splitting due to several factors and collectively known as fine-structure splitting and is discussed in Chapter 7.

The spin–orbit Hamiltonian is

$$\hat{H}_{SO} \propto \mu_\ell \cdot \mu_s$$

but $\mu_\ell \approx \mu_s$ and $\mu_{proton} \approx \dfrac{1}{700} \cdot \mu_{electron}$ so the spin–orbit interaction is the order of one thousand times that of the hyperfine interaction. In Chapter 7 we calculate the magnitude of the spin–orbit interaction. For now, however, we concentrate on the hyperfine interaction as an exercise in angular momentum algebra. We would like to calculate the magnitude of the hyperfine interaction, that is, find out how much the energy of the ground state of the hydrogen atom is altered by this spin–spin interaction. To do this, we must calculate the eigenvalues of $\hat{H}_{HF}$. The wave functions for the spin–spin interaction will involve two spins and may therefore be regarded as 4-component spinors. The operator may thus be represented by a 4×4 matrix.

To determine the components of this matrix we (unwisely) choose to use the uncoupled basis set with the notation

$$|1\rangle = |\alpha_1\rangle|\alpha_2\rangle$$
$$|2\rangle = |\alpha_1\rangle|\beta_2\rangle$$
$$|3\rangle = |\beta_1\rangle|\alpha_2\rangle$$
$$|4\rangle = |\beta_1\rangle|\beta_2\rangle \tag{3.44}$$

Because of symmetry, we need not specify which number, 1 or 2, corresponds to the electron and which to the proton. The matrix form of the Hamiltonian is

$$\hat{H}_{HF} = \begin{pmatrix} \langle 1|\hat{H}_{HF}|1\rangle & \langle 1|\hat{H}_{HF}|2\rangle & \langle 1|\hat{H}_{HF}|3\rangle & \langle 1|\hat{H}_{HF}|4\rangle \\ \langle 2|\hat{H}_{HF}|1\rangle & \langle 2|\hat{H}_{HF}|2\rangle & \langle 2|\hat{H}_{HF}|3\rangle & \langle 2|\hat{H}_{HF}|4\rangle \\ \langle 3|\hat{H}_{HF}|1\rangle & \langle 3|\hat{H}_{HF}|2\rangle & \langle 3|\hat{H}_{HF}|3\rangle & \langle 3|\hat{H}_{HF}|4\rangle \\ \langle 4|\hat{H}_{HF}|1\rangle & \langle 4|\hat{H}_{HF}|2\rangle & \langle 4|\hat{H}_{HF}|3\rangle & \langle 4|\hat{H}_{HF}|4\rangle \end{pmatrix} \tag{3.45}$$

We must now compute the matrix elements. We begin with $\langle 1|\hat{H}_{HF}|1\rangle$.

$$\langle 1|\hat{H}_{HF}|1\rangle = K\langle\alpha_1|\langle\alpha_2|\hat{\mathbf{S}}_1 \cdot \hat{\mathbf{S}}_2|\alpha_1\rangle|\alpha_2\rangle \tag{3.46}$$

Again we exploit the properties of the raising and lowering operators and cast $\hat{\mathbf{S}}_1 \cdot \hat{\mathbf{S}}_2$ in terms of them. We have

$$\hat{\mathbf{S}}_1 \cdot \hat{\mathbf{S}}_2 = \hat{S}_{1x}\hat{S}_{2x} + \hat{S}_{1y}\hat{S}_{2y} + \hat{S}_{1z}\hat{S}_{2z}$$
$$= \left(\frac{1}{4}\right)[(\hat{S}_{1+} + \hat{S}_{1-})(\hat{S}_{2+} + \hat{S}_{2-}) - (\hat{S}_{1+} - \hat{S}_{1-})(\hat{S}_{2+} - \hat{S}_{2-})] + \hat{S}_{1z}\hat{S}_{2z}$$
$$= \left(\frac{1}{2}\right)(\hat{S}_{1+}\hat{S}_{2-} + \hat{S}_{1-}\hat{S}_{2+}) + \hat{S}_{1z}\hat{S}_{2z} \tag{3.47}$$

where we have used

$$\hat{S}_x = \left(\frac{1}{2}\right)(\hat{S}_+ + \hat{S}_-)$$

and

$$\hat{S}_y = \left(\frac{1}{2i}\right)(\hat{S}_+ - \hat{S}_-)$$

Then

$$\langle 1|\hat{H}_{HF}|1\rangle = K\langle\alpha_1|\langle\alpha_2|\hat{\mathbf{S}}_1 \cdot \hat{\mathbf{S}}_2|\alpha_1\rangle|\alpha_2\rangle$$
$$= K\langle\alpha_1|\langle\alpha_2|\left[\left(\frac{1}{2}\right)(\hat{S}_{1+}\hat{S}_{2-} + \hat{S}_{1-}\hat{S}_{2+}) + \hat{S}_{1z}\hat{S}_{2z}\right]|\alpha_1\rangle|\alpha_2\rangle$$
$$= K\langle\alpha_1|\langle\alpha_2|S_{1z}S_{2z}|\alpha_1\rangle|\alpha_2\rangle$$
$$= K\left(\frac{1}{2}\hbar\right)^2$$
$$= \frac{K\hbar^2}{4} \tag{3.48}$$

Note that the raising and lowering operators always change the basis state and therefore cannot contribute to any diagonal matrix elements. Thus, the diagonal elements will all have the same absolute value.

Identical computations show that all off-diagonal elements vanish except $\langle 3|\hat{H}_{HF}|2\rangle = \langle 3|\hat{H}_{HF}|2\rangle$. Evaluating this matrix element we have

$$\langle 3|\hat{H}_{HF}|2\rangle = K\langle\beta_1|\langle\alpha_2|\left\{\left(\frac{1}{2}\right)(\hat{S}_{1+}\hat{S}_{2-} + \hat{S}_{1-}\hat{S}_{2+}) + \hat{S}_{1z}\hat{S}_{2z}\right\}|\alpha_1\rangle|\beta_2\rangle \quad (3.49)$$

Operating to the right with $\hat{S}_{1+}\hat{S}_{2-}$ gives zero because $\hat{S}_{1+}|\alpha_1\rangle \equiv 0$. The last term is also zero because all kets are eigenvectors of $\hat{S}_{1z}$ and $\hat{S}_{2z}$. That is,

$$\langle\beta_1|\langle\alpha_2|\hat{S}_{1z}\hat{S}_{2z}|\alpha_1\rangle|\beta_2\rangle = \frac{\hbar^2}{2}\langle\beta_1 |.\alpha_1\rangle\langle\alpha_2 |.\beta_2\rangle$$

$$= 0$$

Then, only the $\hat{S}_{1-}\hat{S}_{2+}$ term survives and the matrix element $\langle 3|\hat{H}|2\rangle$ is given by

$$\langle 3|\hat{H}_{HF}|2\rangle = \frac{K}{2}\langle\beta_1|\langle\alpha_2|\hat{S}_{1-}\hat{S}_{2+}|\alpha_1\rangle|\beta_2\rangle \quad (3.50)$$

But, recalling that $\hat{S}_+|\beta\rangle = \hbar|\alpha\rangle$ and $\hat{S}_-|\alpha\rangle = \hbar|\beta\rangle$ we have

$$\langle 3|\hat{H}_{HF}|2\rangle = \frac{K}{2}\langle\beta_1|\langle\alpha_2|\hat{S}_{1-}\hat{S}_{2+}|\alpha_1\rangle|\beta_2\rangle$$

$$= \frac{K}{2}\langle\beta_1|\langle\alpha_2|\hbar^2|\beta_1\rangle|\alpha_2\rangle$$

$$= \frac{K}{2}\hbar^2 \quad (3.51)$$

The matrix of the Hamiltonian is then

$$\hat{H}_{HF} = \frac{K\hbar^2}{4}\begin{pmatrix} 1 & 0 & 0 & 0 \\ 0 & -1 & 2 & 0 \\ 0 & 2 & -1 & 0 \\ 0 & 0 & 0 & 1 \end{pmatrix} \quad (3.52)$$

on the basis set $|1\rangle$, $|2\rangle$, $|3\rangle$, $|4\rangle$. Notice that the matrix for $\hat{H}_{HF}$, Equation (3.52), is not diagonal so we must solve the eigenvalue equation using general methods of matrix algebra. Letting

$$|\psi\rangle = a_1|1\rangle + a_2|2\rangle + a_3|3\rangle + a_4|4\rangle \quad (3.53)$$

which, in matrix form is

$$|\psi\rangle = \begin{pmatrix} a_1 \\ a_2 \\ a_3 \\ a_4 \end{pmatrix}$$

the eigenvalue equation in matrix form is

$$\frac{K\hbar^2}{4}\begin{pmatrix} 1 & 0 & 0 & 0 \\ 0 & -1 & 2 & 0 \\ 0 & 2 & -1 & 0 \\ 0 & 0 & 0 & 1 \end{pmatrix}\begin{pmatrix} a_1 \\ a_2 \\ a_3 \\ a_4 \end{pmatrix} = E \begin{pmatrix} a_1 \\ a_2 \\ a_3 \\ a_4 \end{pmatrix}$$

This may be rewritten as

$$\begin{pmatrix} (K\hbar^2/4) - E & 0 & 0 & 0 \\ 0 & -(K\hbar^2/4) - E & \left(\dfrac{K\hbar^2}{2}\right) & 0 \\ 0 & \left(\dfrac{K\hbar^2}{2}\right) & -(K\hbar^2/4) - E & 0 \\ 0 & 0 & 0 & (K\hbar^2/4) - E \end{pmatrix}\begin{pmatrix} a_1 \\ a_2 \\ a_3 \\ a_4 \end{pmatrix} = 0$$

(3.54)

This matrix equation, Equation (3.54), constitutes a set of four homogeneous simultaneous equations in the a_i. Such a set of equations has a nontrivial solution, that is, all a_is vanish, only if the determinant of the coefficients vanishes. That is, if

$$\begin{vmatrix} (K\hbar^2/4) - E & 0 & 0 & 0 \\ 0 & -(K\hbar^2/4) - E & \left(\dfrac{K\hbar^2}{2}\right) & 0 \\ 0 & \left(\dfrac{K\hbar^2}{2}\right) & -(K\hbar^2/4) - E & 0 \\ 0 & 0 & 0 & (K\hbar^2/4) - E \end{vmatrix} = 0$$

which leads to

$$\left(\frac{K\hbar^2}{4} - E\right)^2\left\{\left(\frac{K\hbar^2}{4} + E\right)^2 - \left(\frac{K\hbar^2}{2}\right)^2\right\} = 0 \qquad (3.55)$$

which is referred to as the secular equation.

Letting $\kappa = (\hbar^2/2)K$ we have

$$\left(\frac{\kappa}{2} - E\right)^2\left(E^2 + 2\kappa E - \frac{3}{4}\kappa^2\right) = 0 \qquad (3.56)$$

the roots of which are

$$E = \frac{\kappa}{2}, \frac{\kappa}{2}, \frac{\kappa}{2}, -\frac{3\kappa}{2} \qquad (3.57)$$

Because $\kappa > 0$ the first three eigenvalues must be associated with the excited (degenerate) triplet state and the last eigenvalue is that of the ground state, a singlet.

Successively inserting these eigenvalues in the matrix eigenvalue equation we obtain the matrix representation of the eigenfunctions. For example, using $E = \kappa/2$

we have

$$\kappa \begin{pmatrix} (1/2) & 0 & 0 & 0 \\ 0 & -(1/2) & 1 & 0 \\ 0 & 1 & -(1/2) & 0 \\ 0 & 0 & 0 & (1/2) \end{pmatrix} \begin{pmatrix} a_1 \\ a_2 \\ a_3 \\ a_4 \end{pmatrix} = \frac{\kappa}{2} \begin{pmatrix} a_1 \\ a_2 \\ a_3 \\ a_4 \end{pmatrix} \tag{3.58}$$

Multiplying, we obtain four equations for the a_is.

$$\frac{1}{2}a_1 = \frac{1}{2}a_1$$

$$-\frac{1}{2}a_2 + a_3 = \frac{1}{2}a_2 \Rightarrow a_2 = a_3$$

$$a_2 - \frac{1}{2}a_3 = \frac{1}{2}a_3 \Rightarrow a_2 = a_3$$

$$\frac{1}{2}a_4 = \frac{1}{2}a_4 \tag{3.59}$$

Now, the eigenvectors corresponding to a degenerate eigenvalue (such as $\kappa/2$) are not necessarily orthogonal.[2] For convenience we would like them to be orthogonal. This can be accomplished by selecting the coefficients to be consistent with Equations (3.59) and then find the others. Subsequently we can use the Gram–Schmidt orthogonalization process. In this case, however, it is simple enough to form the eigenvectors by inspection. The spinors

$$\begin{pmatrix} 1 \\ 0 \\ 0 \\ 0 \end{pmatrix} \quad \text{and} \quad \begin{pmatrix} 0 \\ 0 \\ 0 \\ 1 \end{pmatrix} \tag{3.60}$$

fit the criteria of Equations (3.59) and are orthogonal to each other. The spinor

$$\frac{1}{\sqrt{2}} \begin{pmatrix} 0 \\ 1 \\ 1 \\ 0 \end{pmatrix} \tag{3.61}$$

also fits these criteria. Furthermore, Equation (3.61) is orthogonal to the spinors in Equation (3.60) and is an eigenvector of $\hat{H}_{HF}$ with eigenvalue $\kappa/2$. Thus, we have the three eigenvectors

$$\begin{pmatrix} 1 \\ 0 \\ 0 \\ 0 \end{pmatrix} ; \frac{1}{\sqrt{2}} \begin{pmatrix} 0 \\ 1 \\ 1 \\ 0 \end{pmatrix} ; \begin{pmatrix} 0 \\ 0 \\ 0 \\ 1 \end{pmatrix} \tag{3.62}$$

that correspond to the eigenvalue $\kappa/2$. To find the remaining eigenvector we solve the eigenvalue equation with $E = -3\kappa/2$. Designating this remaining eigenvector by

$$\begin{pmatrix} a_1 \\ a_2 \\ a_2 \\ a_4 \end{pmatrix}$$

we have (with $E = -3\kappa/2$)

$$\kappa \begin{pmatrix} (1/2) & 0 & 0 & 0 \\ 0 & -(1/2) & 1 & 0 \\ 0 & 1 & -(1/2) & 0 \\ 0 & 0 & 0 & (1/2) \end{pmatrix} \begin{pmatrix} a_1 \\ a_2 \\ a_3 \\ a_4 \end{pmatrix} = -\frac{3\kappa}{2} \begin{pmatrix} a_1 \\ a_2 \\ a_3 \\ a_4 \end{pmatrix}$$

which leads to

$$\frac{1}{2}a_1 = -\frac{3}{2}a_1$$

$$-\frac{1}{2}a_2 + a_3 = -\frac{3}{2}a_2 \Rightarrow a_2 = -a_3$$

$$a_2 - \frac{1}{2}a_3 = -\frac{3}{2}a_3 \Rightarrow a_2 = -a_3$$

$$\frac{1}{2}a_4 = -\frac{3}{2}a_4 \tag{3.63}$$

In contrast to the relations we found for $E = \kappa/2$ the signs of a_2 and a_3 must be different. Thus, the remaining eigenvector has the form

$$\frac{1}{\sqrt{2}} \begin{pmatrix} 0 \\ \pm 1 \\ \mp 1 \\ 0 \end{pmatrix}$$

We choose the upper signs and obtain for the eigenvector corresponding to the eigenvalue $-3\kappa/2$. Note that the signs are irrelevant because we have not even specified which subscript on the operators and uncoupled kets corresponds to the proton and which to the electron. It is clear that the four eigenvectors

$$\begin{pmatrix} 1 \\ 0 \\ 0 \\ 0 \end{pmatrix} ; \frac{1}{\sqrt{2}} \begin{pmatrix} 0 \\ 1 \\ 1 \\ 0 \end{pmatrix} ; \begin{pmatrix} 0 \\ 0 \\ 0 \\ 1 \end{pmatrix} ; \frac{1}{\sqrt{2}} \begin{pmatrix} 0 \\ 1 \\ -1 \\ 0 \end{pmatrix} \tag{3.64}$$

are mutually orthogonal.

The triplet state eigenvectors are the following three linear combinations of the uncoupled spin wave functions.

$$|\text{triplet}\rangle_1 = |\alpha_1\rangle|\alpha_2\rangle$$

$$|\text{triplet}\rangle_0 = \frac{1}{\sqrt{2}}|\alpha_1\rangle|\beta_2\rangle + \frac{1}{\sqrt{2}}|\beta_1\rangle|\alpha_2\rangle$$

$$|\text{triplet}\rangle_{-1} = |\beta_1\rangle|\beta_2\rangle \tag{3.65}$$

ENERGY LEVEL DIAGRAM
(not to scale)

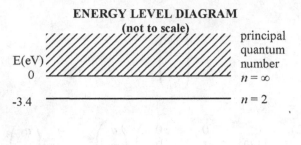

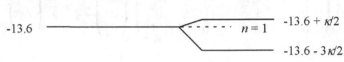

FIGURE 3.9. Energy levels of the hydrogen atom showing hyperfine structure of the ground state.

and the singlet is

$$|\text{singlet}\rangle_0 = \frac{1}{\sqrt{2}}|\alpha_1\rangle|\beta_2\rangle - \frac{1}{\sqrt{2}}|\beta_1\rangle|\alpha_2\rangle \tag{3.66}$$

where the subscripts on these kets have been chosen to be the z-component of the *total* spin. This is discussed shortly.

The hyperfine splitting is shown schematically in FIGURE 3.9.

Notice that the splitting is exaggerated in the diagram because it is so small that, compared with energies on the order of electron volts, it is less than the thickness of a line.

The form of the eigenvectors that we obtained suggests that we might not have made the wisest choice when we elected to use the uncoupled basis set. Although we started with uncoupled kets, we found that the coupled kets are the eigenkets of the hyperfine Hamiltonian.

Suppose we had begun with coupled kets $|SM\rangle$. The total angular momentum operator is

$$\hat{J}^2 = \hat{L}^2 + \hat{S}^2 + 2\hat{L}\cdot\hat{S} \tag{3.67}$$

which, in our problem, becomes

$$\hat{S}^2 = \hat{S}_1^2 + \hat{S}_2^2 + 2\hat{S}_1\cdot\hat{S}_2 \tag{3.68}$$

But, retaining the use of κ

$$\hat{H}_{HF} = \frac{2\kappa}{\hbar^2}\hat{S}_1\cdot\hat{S}_2 \tag{3.69}$$

so that we may write

$$\hat{H}_{FS} = \frac{\kappa}{\hbar^2}\left(\hat{S}^2 - \hat{S}_1^2 - \hat{S}_2^2\right) \tag{3.70}$$

Now, the coupled kets are, as usual, $|SM\rangle = |s_1, s_2; S, M\rangle$. In short, the coupled kets $|SM\rangle$ are eigenkets of the spin–spin Hamiltonian. If we use them to construct the matrix representation of $\hat{H}_{HF}$ it would already be diagonal. To see this, we operate on each of the $|SM\rangle$ kets with $\hat{H}_{HF}$ and examine each term independently. We obtain

$$\hat{S}^2|SM\rangle = S(S+1)\hbar^2|SM\rangle$$
$$\hat{S}_1^2|SM\rangle = S_1(S_1+1)\hbar^2|SM\rangle$$
$$\hat{S}_2^2|SM\rangle = S_2(S_2+1)\hbar^2|SM\rangle \tag{3.71}$$

Notice that nothing in the Hamiltonian produces any eigenvalue that depends on M. Therefore, all kets $|1M\rangle$ have the same eigenvalue and all three triplet states have the same energy which, of course, we already found using the uncoupled basis set. To find this energy we simply find the eigenvalue by operating on one of the kets with $\hat{H}_{HF}$.

$$\hat{H}_{HF}|1M\rangle = \frac{\kappa}{\hbar^2}\left\{1(1+1)\hbar^2 - \frac{1}{2}\left(\frac{1}{2}+1\right)\hbar^2 - \frac{1}{2}\left(\frac{1}{2}+1\right)\hbar^2\right\}|1M\rangle$$
$$= \frac{\kappa}{2}|1M\rangle \tag{3.72}$$

Therefore, $E_{\text{triplet}} = \kappa/2$ as we found using the unwieldy uncoupled ket as a basis.

Now, how about the singlet? Again we must solve the eigenvalue equation, this time using the ket $|SM\rangle = |00\rangle$.

$$\hat{H}_{HF}|00\rangle = \frac{\kappa}{\hbar^2}\left\{0(0+1)\hbar^2 - \frac{1}{2}\left(\frac{1}{2}+1\right)\hbar^2 - \frac{1}{2}\left(\frac{1}{2}+1\right)\hbar^2\right\}|00\rangle$$
$$= -\frac{3}{2}\kappa|00\rangle \tag{3.73}$$

Therefore, $E_{\text{singlet}} = -3\kappa/2$, again as we found before. Notice that we got the energies without having to solve the eigenvalue problem because, this time, we chose the more convenient kets, the coupled kets. They are more convenient because they happen to be eigenkets of the spin–spin Hamiltonian.

The coupled ket $|00\rangle$ clearly corresponds to the singlet state. Therefore,

$$|00\rangle = |\text{singlet}\rangle_0$$
$$= \frac{1}{\sqrt{2}}|\alpha_1\rangle|\beta_2\rangle - \frac{1}{\sqrt{2}}|\beta_1\rangle|\alpha_2\rangle \tag{3.74}$$

We may easily correlate the uncoupled triplet states with their coupled counterparts using results previously obtained in this chapter. From Equation (3.23) it is clear that

$$|11\rangle = |\text{triplet}\rangle_1 = |\alpha_1\rangle|\alpha_2\rangle$$
$$|10\rangle = |\text{triplet}\rangle_0 = \frac{1}{\sqrt{2}}|\alpha_1\rangle|\beta_2\rangle + \frac{1}{\sqrt{2}}|\beta_1\rangle|\alpha_2\rangle$$
$$|1-1\rangle = |\text{triplet}\rangle_{-1} = |\beta_1\rangle|\beta_2\rangle \tag{3.75}$$

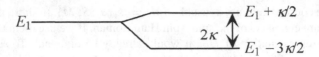

FIGURE 3.10. Hyperfine splitting of the ground state of the hydrogen atom.

The splitting of the ground state of hydrogen due to the hyperfine interaction is shown schematically in FIGURE 3.10.

Experimentally, it is found that the hyperfine splitting of the ground state 2κ is

$$2\kappa = \hbar\omega = 5.9 \times 10^{-6}\, eV$$

Thus, a transition from the upper to the lower hyperfine levels of the ground state of hydrogen emits a photon of frequency

$$f = \frac{\omega}{2\pi} = 1420.405751768 \pm 0.001\, \text{MHz}$$

which corresponds to a wavelength

$$\lambda = \frac{c}{f} = 21\, \text{cm} \tag{3.76}$$

This transition between hyperfine levels of the hydrogen atom ground state has an important astrophysical application. Detection of this 21 cm radiation using radio telescopes has been used to map galaxies. It also permits Doppler shift measurements of astrophysical objects to be made in the radio region of the electromagnetic spectrum. The frequency of this transition is the most accurately known physical quantity today!

Problems

3.1. An electron in an atom is in an uncoupled spin and orbital eigenstate given by $|\ell\, m_\ell; m_s\rangle = |1\ -1; \alpha\rangle$ where we have omitted $s = 1/2$ from the ket because all electrons have $s = 1/2$. What is the probability that a measurement of the total angular momentum will yield the value $j = 3/2$? $j = 1/2$? $j = 5/2$?

3.2. An electron in an atom is in an coupled eigenstate

$$|j\, m_j ; \ell s\rangle = \left|\frac{1}{2}\ -\frac{1}{2}; 1\ \frac{1}{2}\right\rangle$$

What is the probability that a measurement of the z-component of the orbital angular momentum will yield the value $m_\ell = 0$? $m_\ell = +1$? $m_\ell = -1$?

3.3. Show that $\hat{J}_1 \cdot \hat{J}_2 = \hat{J}_{1z}\hat{J}_{2z} + \left(\frac{1}{2}\right)(\hat{J}_{1+}\hat{J}_{2-} + \hat{J}_{1-}\hat{J}_{2+})$.

Clebsch–Gordan coefficients for $j_1 = \tfrac{3}{2}$, $j_2 = 1$:

m_1	m_2	$j=\tfrac{5}{2}$ $m=\tfrac{5}{2}$	$j=\tfrac{5}{2}$ $m=\tfrac{3}{2}$	$j=\tfrac{5}{2}$ $m=\tfrac{1}{2}$	$j=\tfrac{5}{2}$ $m=-\tfrac{1}{2}$	$j=\tfrac{5}{2}$ $m=-\tfrac{3}{2}$	$j=\tfrac{5}{2}$ $m=-\tfrac{5}{2}$	$j=\tfrac{3}{2}$ $m=\tfrac{3}{2}$	$j=\tfrac{3}{2}$ $m=\tfrac{1}{2}$	$j=\tfrac{3}{2}$ $m=-\tfrac{1}{2}$	$j=\tfrac{3}{2}$ $m=-\tfrac{3}{2}$	$j=\tfrac{1}{2}$ $m=\tfrac{1}{2}$	$j=\tfrac{1}{2}$ $m=-\tfrac{1}{2}$
$\tfrac{3}{2}$	1	1											
$\tfrac{3}{2}$	0		$\sqrt{2/5}$					$\sqrt{3/5}$					
$\tfrac{3}{2}$	-1			$\sqrt{1/10}$					$\sqrt{2/5}$			$\sqrt{1/2}$	
$\tfrac{1}{2}$	1		$\sqrt{3/5}$					$-\sqrt{2/5}$					
$\tfrac{1}{2}$	0			$\sqrt{3/5}$					$\sqrt{1/15}$			$-\sqrt{1/3}$	
$\tfrac{1}{2}$	-1				$\sqrt{3/10}$					$-\sqrt{8/15}$			$\sqrt{1/6}$
$-\tfrac{1}{2}$	1			$\sqrt{3/10}$					$-\sqrt{8/15}$			$\sqrt{1/6}$	
$-\tfrac{1}{2}$	0				$\sqrt{3/5}$					$\sqrt{1/15}$			$-\sqrt{1/3}$
$-\tfrac{1}{2}$	-1					$\sqrt{3/5}$					$\sqrt{2/5}$		
$-\tfrac{3}{2}$	1				$\sqrt{1/10}$					$\sqrt{2/5}$			$\sqrt{1/2}$
$-\tfrac{3}{2}$	0					$\sqrt{2/5}$					$-\sqrt{3/5}$		
$-\tfrac{3}{2}$	-1						1						

3.4. An eigenfunction of $\hat{J}^2$ and $\hat{J}_z$ (total angular momentum and its z-component) is given in terms of the uncoupled eigenfunctions $|j_1\, m_{j1};\, j_2\, m_{j2}\rangle$ as

$$|\psi\rangle = \sqrt{\frac{3}{5}}\left|\frac{3}{2}\; -\frac{1}{2};1\; -1\right\rangle + \sqrt{\frac{2}{5}}\left|\frac{3}{2}\; -\frac{3}{2};10\right\rangle$$

Using the operator identity of the previous problem, find the value of the total angular momentum and its z-component.

3.5. An eigenfunction of $\hat{J}^2$ and $\hat{J}_z$ (total angular momentum and its z-component) is given in terms of the uncoupled eigenfunctions $|j_1\, m_{j1};\, j_2\, m_{j2}\rangle$ as

$$|\psi\rangle = \sqrt{\frac{2}{5}}\left|\frac{3}{2}\,\frac{3}{2};1\; -1\right\rangle + \sqrt{\frac{1}{15}}\left|\frac{3}{2}\,\frac{1}{2};10\right\rangle - \sqrt{\frac{8}{15}}\left|\frac{3}{2}\; -\frac{1}{2};11\right\rangle$$

Using the table of Clebsch–Gordan coefficients on the previous page, find the value of the total angular momentum and its z-component.

References

1. G.W. Dunnington, *Carl Friedrich Gauss: Titan of Science* (Exposition Press, New York, 1955).
2. D.J. Griffiths, *Introduction to Quantum Mechanics* (Prentice-Hall, Upper Saddle River, NJ, 1995).

4
The Quantum Mechanical
Hydrogen Atom

4.1. The Radial Equation for a Central Potential

After writing the Schrödinger equation in spherical coordinates for an arbitrary central potential and separating variables it is found that the eigenfunctions may be written as

$$\psi(r, \theta, \phi) = Y_{\ell m}(\theta, \phi) R_{n\ell}(r) \tag{4.1}$$

where n is the principal quantum number. Because the Schrödinger equation does not contain spin, the only angular momentum to be considered here is orbital angular momentum. As is customary, we use ℓ as the angular momentum quantum number and m as the azimuthal or magnetic quantum number that defines the z-component of the angular momentum. The $Y_{\ell m}(\theta, \phi)$ are the spherical harmonics and $R_{n\ell}(r)$ is the radial part of the wave function. $Y_{\ell m}(\theta, \phi)$ is universal for any central potential and $R_{n\ell}(r)$ depends on the specific form of the central potential. Of course, the energy eigenvalues also depend on the specific form of the potential and, in general, depend upon both n and ℓ.

Solution of the angular part of the Schrödinger equation yields the magnitude of the total angular momentum, $\sqrt{\ell(\ell+1)}\hbar^2$, and the eigenvalues of the z-component of the angular momentum $m\hbar$. The radial equation that remains depends upon the potential energy function $V(r)$,

$$\left\{ -\frac{\hbar^2}{2\mu} \left[\left(\frac{d^2}{dr^2} + \frac{2}{r} \frac{d}{dr} \right) - \frac{\ell(\ell+1)}{r^2} \right] + V(r) \right\} R_{n\ell}(r) = E_{n\ell} R_{n\ell}(r) \tag{4.2}$$

where μ is the reduced mass of a presumed two-particle system. We are here interested in the hydrogen atom for which μ is effectively the electronic mass, which we designate by m_e to avoid confusion with the quantum number m. Inasmuch as the energy eigenvalues are to be determined solely by the radial equation, we see that, in general, these eigenvalues will depend on both n and ℓ so we write them as $E_{n\ell}$ to show this explicitly.

Although the universal angular solutions, the $Y_{\ell m}(\theta, \phi)$, depend on both the m and ℓ quantum numbers, the radial functions do not depend upon m.

Recalling that there are $(2\ell + 1)$ values of m for every ℓ, we see that there is a $(2\ell + 1)$-fold degeneracy in m for *any* central potential. Degeneracies are always associated with an intrinsic symmetry in the potential. In this case it is the spherical symmetry (sometimes referred to as "isotropy") that is inherent in a central potential. In other words, the energy levels do not depend on the orientation in space because the potential is the same for all points on a given sphere centered at the force center.

To solve the radial equation we make, as in classical mechanics, the convenient substitution

$$R_{n\ell}(r) = \frac{u_{n\ell}(r)}{r} \tag{4.3}$$

which leads to

$$\left[-\frac{\hbar^2}{2m_e} \frac{d^2}{dr^2} + \frac{\ell(\ell + 1)\hbar^2}{2m_e r^2} + V(r) \right] u_{n\ell}(r) = E_{n\ell} u_{n\ell}(r) \tag{4.4}$$

This equation has the same form as the one-dimensional Schrödinger equation with an effective potential

$$V_{eff}(r) = V(r) + \frac{\ell(\ell + 1)\hbar^2}{2m_e r^2} \tag{4.5}$$

Therefore, we may think of the radial equation as being the solution to a one-dimensional problem, provided that the potential for this one-dimensional problem is $V_{eff}(r)$. Clearly, there are many different values of $V_{eff}(r)$ for a given $V(r)$. Thus, $V_{eff}(r)$ accounts for the parameter that is absent in one-dimensional problems, angular momentum. The second term in Equation (4.5) is referred to as the centrifugal term. FIGURE 4.1 shows the effective potential of a Coulomb potential $V(r) = -e^2/(4\pi \varepsilon_0 r)$ for several different values of ℓ.

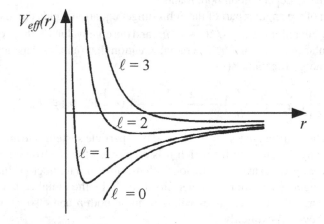

FIGURE 4.1. Effective potential for the Coulomb potential for several values of the angular momentum quantum number ℓ.

Examination of these effective potentials shows that the wave functions for $\ell > 0$ must vanish at $r = 0$ because the effective potential becomes infinite as r approaches zero. Classically, this means that if the angular momentum of a particle is nonzero, then the particle can never go through the origin. This is sensible because, classically, the angular momentum is given by $\boldsymbol{L} = \boldsymbol{r} \times \boldsymbol{p}$.

We wish to examine the behavior of $u_{n\ell}(r)$ near the origin for any central potential $V(r)$, but one for which the centrifugal term dominates near $r = 0$. Because the Coulomb potential $\sim 1/r$ it is included in this category. Near $r = 0$ Equation (4.4) becomes for $\ell > 0$ (when $u_{n\ell}(r)$ must vanish)

$$\frac{d^2 u_{n\ell}(r)}{dr^2} - \frac{\ell(\ell+1)}{r^2} u_{n\ell}(r) = 0 \tag{4.6}$$

The two solutions to Equation (4.6) are $r^{\ell+1}$ and $r^{-\ell}$, the latter of which must be discarded as $r \to 0$. Thus,

$$\lim_{r \to 0} u_{n\ell}(r) = r^{\ell+1} \tag{4.7}$$

Notice that as ℓ increases, the particle is less and less likely to be found in the vicinity of the origin. This is consistent with a classical notion of angular momentum.

4.2. Solution of the Radial Equation in Spherical Coordinates—The Energy Eigenvalues

For the hydrogen atom, the potential is the familiar Coulomb potential

$$V(r) = -\frac{e^2}{4\pi\varepsilon_0} \cdot \frac{1}{r} \tag{4.8}$$

where e is the magnitude of the electronic charge. For convenience, we scale r in terms of the Bohr radius $a_0 = 4\pi\varepsilon_0 \hbar^2/(m_e e^2)$ and use the dimensionless quantity ρ given by

$$r = \sqrt{\frac{-\hbar^2}{8m_e E}} \cdot \rho \tag{4.9}$$

We also make the substitution

$$\lambda = \frac{e^2}{4\pi\varepsilon_0 \hbar} \sqrt{-\frac{m_e}{2E}}$$

$$= \alpha \sqrt{-\frac{m_e c^2}{2E}} \tag{4.10}$$

where E is the total energy, a negative number, and α is the fine structure constant. Note that $-E$ is the ionization potential of hydrogen, that is, the minimum energy required to liberate the electron from the proton. We have temporarily dropped the n and ℓ subscripts on E because we do not as yet even know that E is quantized.

With these substitutions the radial equation for the hydrogen atom in terms of the $u_{n\ell}(r) = r R_{n\ell}(r)$ becomes

$$\left[\frac{d^2}{d\rho^2} - \frac{\ell(\ell+1)}{\rho^2} + \frac{\lambda}{\rho} - \frac{1}{4} \right] u_{n\ell}(\rho) = 0 \qquad (4.11)$$

Asymptotically, that is, as $\rho \to \infty$, this becomes

$$\left[\frac{d^2}{d\rho^2} - \frac{1}{4} \right] u_{n\ell}(\rho) = 0 \qquad (4.12)$$

the solutions of which are $e^{\pm\rho/2}$. As usual, we must discard the plus sign on physical grounds so that

$$\lim_{\rho \to \infty} u_{n\ell}(\rho) = \text{constant} \cdot e^{-\rho/2} \qquad (4.13)$$

Equation (4.13) and Equation (4.7) specify the nature of the hydrogen atom wave functions for small and large values of r.

To solve Equation (4.11) we try a solution of the form

$$u_{n\ell}(\rho) = F(\rho) \exp(-\rho/2) \qquad (4.14)$$

and, using the asymptotic form Equation (4.7) as a guide, we write $F(\rho)$ as a power series expansion

$$F(\rho) = \rho^{\ell+1} \left(a_0 + a_1 \rho^1 + a_2 \rho^2 + \cdots \right)$$

$$= \rho^{\ell+1} \sum_{j=0}^{\infty} a_j \rho^j \qquad (4.15)$$

Substituting this in Equation (4.11) we obtain

$$\sum_{j=0}^{\infty} \left\{ [j(j+1) + 2(\ell+1)(j+1)] a_{j+1} + (\lambda - \ell - 1 - j) a_j \right\} \rho^j = 0 \quad (4.16)$$

Because the ρ^j are linearly independent, their coefficients must separately vanish which leads to a recursion relation between successive coefficients in the expansion, Equation (4.15),

$$a_{j+1} = \frac{j + \ell + 1 - \lambda}{(j+1)(j+2\ell+2)} a_j \qquad (4.17)$$

Now, examine convergence of the series for which (4.17) is the recursion relation.

$$\lim_{j \to \infty} \left(\frac{a_{j+1}}{a_j} \right) = \frac{1}{j} \qquad (4.18)$$

This series diverges! The only way to obtain a physically acceptable solution is to force the series to terminate; that is, we drop the assumption that $F(\rho)$ is an infinite power series and force it to terminate. In order for the series to terminate, the numerator of the recursion relation must vanish for some value of ℓ equal to a positive integer. Recall that j is, by definition, an integer. This is a very important step. To this point there was no hint that the energy had to be "quantized". It is seen then that it is the necessity of convergence of the series which is based on the requirement that the wave function be bounded, that forces quantized levels upon the hydrogen atom. It is physics, not mathematics!

For the series to converge, λ must be an integer that is equal to the rest of the numerator. We designate this value of λ by n which will, of course, turn out to be the usual principal quantum number. We also replace j, the index, by n_r, the "radial" quantum number. We have then

$$\lambda = n$$
$$= n_r + \ell + 1 \tag{4.19}$$

so that, returning to the definition of λ, Equation (4.10), we have

$$\lambda = \alpha \sqrt{-\frac{m_e c^2}{2E}}$$
$$= n \tag{4.20}$$

Solving for E and noting that it depends upon the quantum number n we have

$$E_n = -(m_e c^2)\alpha^2 \cdot \frac{1}{2n^2} \tag{4.21}$$

which is precisely the Bohr energy.

We may also obtain a relationship between the principal quantum number and the angular momentum quantum number. Solving Equation (4.19) for n_r and noting that it is manifestly positive or zero leads to

$$n \geq \ell + 1 \tag{4.22}$$

It is often convenient to express the hydrogen eigenenergies in atomic units. Noting that the $c\alpha \equiv 1$ in atomic units, it is clear that

$$E_n = -\frac{1}{2n^2} \tag{4.23}$$

in atomic units. If the one-electron atom is not hydrogen, but is an atom with Z protons in the nucleus, for example, He^+, Li^{++}, and so on, then the energy is given by

$$E_n = -(m_e c^2)\alpha^2 \cdot \frac{Z^2}{2n^2} \tag{4.24}$$

For example, the ionization energy of the He^+ ion is $2^2 \times 13.6\,eV = 54.4\,eV$.

4.3. The Accidental Degeneracy of the Hydrogen Atom

Perhaps the most intriguing feature of the hydrogenic energy is that it depends only on the principal quantum number n, and not on the angular momentum quantum number ℓ, as is expected for a general central potential. As remarked previously, degeneracies are always associated with symmetries. We have discussed the spatial symmetry associated with a *central* potential and the consequent degeneracy in m, but to what symmetry does this "accidental degeneracy" correspond? The answer is that this symmetry is not apparent to ordinary humans, but, as discussed later, is associated with an additional constant of the classical motion that exists for the attractive $1/r$ potential. This constant of the motion does not exist for other central potentials although the attractive r^2 potential, the three-dimensional isotropic harmonic oscillator has yet another classical constant of the motion. It therefore possesses additional symmetry and, as a result, its own accidental degeneracy.

For a central potential we expect a $(2\ell + 1)$-fold degeneracy because there are $(2\ell + 1)$ values of m for each value of ℓ. That is, for any central potential there should be $(2\ell + 1)$ energies corresponding to given values of n and ℓ because the energy does not depend on m. To find the degree of the degeneracy for the hydrogen atom we sum the ℓ-states over all of the possible m-values. The degree of degeneracy is then given by

$$
\begin{aligned}
\text{Degree of degeneracy} &= \sum_{\text{all } \ell} m \\
&= \sum_{\ell=0}^{n-1} (2\ell + 1) \\
&= \sum_{\ell=0}^{n-1} 2\ell + \sum_{\ell=0}^{n-1} 1 \\
&= 2 \cdot (n - 1) \cdot \left(\frac{n}{2}\right) + n \\
&= n^2
\end{aligned}
\tag{4.25}
$$

where the term $(n - 1)(n/2)$ is the sum of the integers from zero to $(n - 1)$. The Gauss trick has been used in evaluating the sum in Equation (4.25). Spin adds a factor of 2; that is, the degeneracy is $2n^2$, but two proton spin states add another two. Therefore, the total degeneracy of the hydrogen atom is $4n^2$.

It is remarkable that, although the effective potential is different for different values of ℓ, the energy of the hydrogen atom remains independent of ℓ. It depends only on the principal quantum number n. FIGURE 4.2 illustrates this nicely. The hydrogenic energies for various values of n are superimposed on a plot of effective potentials versus r, measured in Bohr radii, for various values of ℓ. Notice that this diagram also makes clear the restriction on ℓ; that is, $0 \le \ell \le (n - 1)$. If a given level lies lower than the minimum of a particular effective potential well then the value of ℓ that is characteristic of that effective potential is forbidden. For example, there can be no $\ell = 2$ level having $n = 2$ because the $n = 2$ energy is below the minimum in $V_{eff}(r)$ for $\ell = 2$.

The cause of the accidental degeneracy is a subject of considerable interest which is discussed in detail in Chapter 7. For now it is sufficient to note that classical constants of the motion correspond to quantum mechanical operators that commute with the Hamiltonian. We show that, in addition to $\hat{L}^2$ and $\hat{L}_z$, there is a third operator that commutes with the Hamiltonian for the pure Coulomb

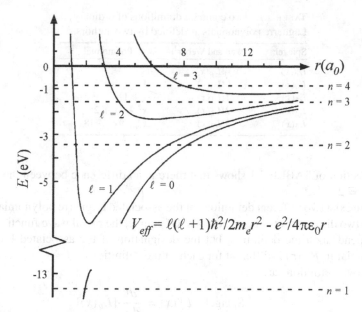

FIGURE 4.2. Energy levels of the hydrogen superimposed on different effective potentials illustrating the accidental degeneracy.

potential. This commuting operator does indeed correspond to another constant of the classical motion.

4.4. Solution of the Hydrogen Atom Radial Equation in Spherical Coordinates—The Energy Eigenfunctions

The finite power series

$$\sum_{j=0}^{\infty} a_j \rho^j$$

that is part of the solution to the radial part of the Schrödinger equation is, in fact, the associated Laguerre polynomials, designated $L_\alpha^\beta(x)$. These polynomials may be obtained by differentiation of the (ordinary) Laguerre polynomials $L_\gamma(x)$. Unfortunately, the definitions of neither the ordinary nor the associated Laguerre polynomials are universal. We attempt to clarify the situation.

Because the associated Laguerre polynomials depend upon the particular choice of ordinary Laguerre polynomial we first discuss the ordinary functions $L_p(x)$. The difference between the common definitions of the ordinary Laguerre polynomial definitions is a factor of $p!$ Two often used mathematical references[1,2] illustrate the differences in the definitions. TABLE 4.1 is a listing of the first few Laguerre polynomials and the two designations.

TABLE 4.1. Two common definitions of ordinary
Laguerre polynomials as defined by two authors.

Spiegel[2]	Arfken and Weber[1]	Polynomial
$L_p(x)$	$p!L_p(x)$	
$L_0(x)$	$0!L_0(x)$	1
$L_1(x)$	$1!L_1(x)$	$-x+1$
$L_2(x)$	$2!L_2(x)$	$x^2 - 4x + 2$
$L_3(x)$	$3!L_3(x)$	$-x^3 + 9x^2 - 18x + 6$

Inspection of TABLE 4.1 shows that there is no difference between the $L_p(x)$ until $p = 2$.

There exist two different definitions of the associated Laguerre polynomials that lead to two different sets of polynomials. Of course, the radial wave function does not depend upon the definition, but the designation of the associated Laguerre polynomial in $R_{n\ell}(r)$ is different for each of the definitions.

The two definitions are:

$$\text{Spiegel:}^2 \ L_q^p(x) = \frac{d^p}{d\rho^p}[L_q(x)]$$

$$\text{Arfken and Weber:}^1 \ L_{q-p}^p(x) = (-)^p \frac{d^p}{d\rho^p}[L_q(x)]$$

where, for this purpose, p and q are integers. One attraction of the Arfken and Weber definition is that it produces polynomials of degree $(q - p)$, that is, the lower index. The upper index is referred to as the order. TABLE 4.2 shows some of the associated Laguerre polynomials for both definitions.

Because of these different definitions of the associated Laguerre polynomials the radial wave function $R_{n\ell}(r)$ encountered in the literature takes two different forms. For example, Bethe and Salpeter[3] and Schiff[4] use Spiegel's[2] definition so that

$$R_{n\ell}(r) = M_{n\ell} \left(\frac{2r}{na_0} \right)^\ell \cdot \exp\left(-\frac{r}{na_0} \right) \cdot L_{n+\ell}^{2\ell+1} \left(\frac{2r}{na_0} \right) \tag{4.26}$$

TABLE 4.2. Two common definitions of associated
Laguerre polynomials as defined by two authors.

Spiegel[2]	Arfken and Weber[1]	Polynomial
$L_1^1(x)$	$-L_0^1(x)$	-1
$L_2^1(x)$	$-L_1^1(x)$	$2x - 4$
$L_2^2(x)$	$L_0^2(x)$	2
$L_3^1(x)$	$L_2^1(x)$	$-3x^2 + 18x - 18$
$L_3^2(x)$	$L_1^2(x)$	$-6x + 18$
$L_3^3(x)$	$-L_0^3(x)$	-6

whereas Griffiths[5] and Gasiorowicz[6] use Arfken and Weber's[1] definition which leads to

$$R_{n\ell}(r) = N_{n\ell} \left(\frac{2r}{na_0}\right)^{\ell} \cdot \exp\left(-\frac{r}{na_0}\right) \cdot L_{n-\ell-1}^{2\ell+1}\left(\frac{2r}{na_0}\right) \tag{4.27}$$

To normalize the radial wave function, that is, to compute the value of $N_{n\ell}$ and $M_{n\ell}$, it is necessary to use the orthogonality integral. These integrals, however, differ for the two definitions of the associated Laguerre polynomials. They are

$$\text{Speigel:}^2 \int_0^\infty x^p e^{-x} L_q^p(x) L_{q'}^p(x)\, dx = \frac{q!^3}{(q-p)!}\delta_{qq'}$$

$$\text{Arfken and Weber:}^1 \int_0^\infty x^p e^{-x} L_q^p(x) L_{q'}^p(x)\, dx = \frac{(q+p)!}{q!}\delta_{qq'}$$

Using these integrals it can be shown that

$$\text{Speigel:}^2 \int_0^\infty x^{2\ell} e^{-x} \left[L_{n+\ell}^{2\ell+1}(x)\right]^2 x^2 dx = \frac{2n\,(n+\ell)!^3}{(n-\ell-1)!}$$

$$\text{Arfken and Weber:}^1 \int_0^\infty x^{2\ell} e^{-x} \left[L_{n-\ell-1}^{2\ell+1}(x)\right]^2 x^2 dx = \frac{2n\,(n+\ell)!}{(n-\ell-1)!}$$

Notice that these integrals differ by a factor of $(n+\ell)!^2$ because of the definitions of the ordinary Laguerre polynomials.

Using these integrals the normalization factors in the $R_{n\ell}(r)$ can be calculated. We obtain

$$M_{n\ell} = -\left(\frac{2}{na_0}\right)^{3/2} \left[\frac{(n-\ell-1)!}{2n}\right]^{1/2} \left[\frac{1}{(n+\ell)!}\right]^{3/2} \tag{4.28}$$

and

$$N_{n\ell} = -\left(\frac{2}{na_0}\right)^{3/2} \left\{\frac{(n-\ell-1)!}{2n\,[(n+\ell)!]^3}\right\}^{1/2} \tag{4.29}$$

For convenience the first six radial wave functions are listed in TABLE 4.3.

TABLE 4.3. Normalized radial wave functions $R_{n\ell}(r)$ for $n = 1$–3.

$R_{10}(r) = 2a_0^{-3/2} \exp(-r/a_0)$

$R_{20}(r) = 2(2a_0)^{-3/2} (1 - r/2a_0)\exp(-r/2a_0)$

$R_{21}(r) = 3^{-1/2} (2a_0)^{-3/2} (r/a_0)\exp(-r/2a_0)$

$R_{30}(r) = 2(3a_0)^{-3/2} (1 - 2r/3a_0 + 2r^2/27a_0^2)\exp(-r/3a_0)$

$R_{31}(r) = (4\sqrt{2}/9)(3a_0)^{-3/2} (1 - r/6a_0)(r/a_0)\exp(-r/3a_0)$

$R_{32}(r) = (4/27)10^{-1/2} (3a_0)^{-3/2} (r/a_0)^2 \exp(-r/3a_0)$

4.5. The Nature of the Spherical Eigenfunctions

It is instructive to examine the energy eigenfunctions that arise from the separation of variables in spherical coordinates because, as will be discussed, the hydrogen atom problem can also be solved using separation of variables in parabolic coordinates. (In fact, it can also be solved in spheroidal coordinates,[7] but this solution is of limited use.) This separability of the Schrödinger equation in more than one coordinate system is indicative of the additional symmetry that leads to the accidental degeneracy, that is, symmetry beyond the spherical symmetry of any central potential. This additional symmetry manifests itself by permitting separability in parabolic coordinates. It should be clear that the spatial symmetry of a central potential is the symmetry that permits separation of variables in spherical coordinates for any central potential.

The complete energy eigenfunctions in spherical coordinates, sometimes referred to as spherical eigenfunctions or "orbital" eigenfunctions, are

$$\psi_{n\ell m}(r, \theta, \phi) = R_{n\ell}(r) Y_{\ell m}(\theta, \phi) \tag{4.30}$$

The probability density is, as usual, $|\psi|^2$. The angular part of ψ is the spherical harmonics (see Section 2.6),

$$Y_{\ell m}(\theta, \phi) = P_{\ell m}(\theta) \cdot \exp(im\phi) \tag{4.31}$$

where the $P_{\ell m}(\theta)$ are the associated Legendre functions. Upon taking the absolute square, the ϕ-dependence disappears. The probability density is therefore cylindrically symmetric about the z-axis. Now $|\psi|^2$ represents the probability per unit volume so that $|\psi|^2$ multiplied by the volume element is the probability of finding the electron somewhere in the volume element. Thus, the probability density (multiplied by the electronic charge) is the charge density of a given eigenstate. A convenient way to depict the charge density is by using a density plot in which the regions of the highest density of dots are the locations at which the electron would be more likely to be found. FIGURE 4.3 is such a plot for the $n = 4$, $\ell = 2$, $m = 0$ state.

The cylindrical symmetry discussed above is apparent. In contrast with the charge densities that will be obtained using parabolic coordinates it should be noted that the charge distribution for orbital eigenstates is symmetric about the xy-plane.

4.6. Separation of the Schrödinger Equation in Parabolic Coordinates

We follow the same procedure to effect the separation of variables in parabolic coordinates that was used to separate the Schrödinger equation in spherical coordinates. The treatment is standard and can be found in the books by Bethe and Salpeter,[3] Landau and Lifshitz,[8] and by Schiff.[4] The relationship between

$$n = 4 \; ; \; \ell = 2 \; ; \; m = 0$$

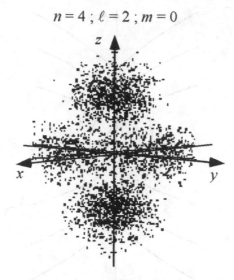

FIGURE 4.3. Density plot of the spherical wave function indicated. The maximimum on each axis is 50 a.u.

parabolic coordinates and spherical coordinates is given by the following transformation equations.

$$\xi = r - z = r(1 - \cos\theta)$$
$$\eta = r + z = r(1 + \cos\theta)$$
$$\phi = \phi \tag{4.32}$$

The angular coordinate ϕ is seen to be the same azimuthal angle as in spherical coordinates. This coordinate defines position with respect to the x-axis of a point in the xy-plane. The surfaces of constants ξ and η are paraboloids of revolution about the z-axis as shown in FIGURE 4.4.

The Coulomb potential in parabolic coordinates in SI units is

$$V(\xi, \eta) = -\left(\frac{e^2}{4\pi\varepsilon_0}\right)\left(\frac{2}{\xi + \eta}\right) \tag{4.33}$$

Using the Laplacian operator in parabolic coordinates,[1] the Schrödinger equation is

$$\left(-\frac{\hbar^2}{2m_e}\right)\left(\frac{4}{\xi + \eta}\right)\left[\frac{\partial}{\partial\xi}\left(\xi\frac{\partial\psi(\xi, \eta, \phi)}{\partial\xi}\right) + \frac{\partial}{\partial\eta}\left(\eta\frac{\partial\psi(\xi, \eta, \phi)}{\partial\eta}\right)\right]$$
$$+ \frac{1}{\xi\eta}\frac{\partial^2\psi(\xi, \eta, \phi)}{\partial\phi^2} - \left(\frac{e^2}{4\pi\varepsilon_0}\right)\left(\frac{2}{\xi + \eta}\right)\psi(\xi, \eta, \phi) = E\psi(\xi, \eta, \phi) \tag{4.34}$$

To separate this equation we let

$$\psi(\xi, \eta, \phi) = f(\xi)g(\eta)\Phi(\phi) \tag{4.35}$$

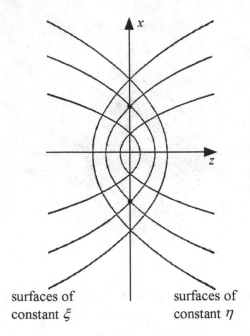

FIGURE 4.4. Parabolic coordinates and their relation to other coordinate systems.

After a considerable amount of algebra we obtain

$$F(\xi, \eta) = -\frac{1}{\Phi(\phi)} \frac{d^2 \Phi(\phi)}{d\phi^2}$$
$$= -m^2 \tag{4.36}$$

where the function $F(\xi, \eta)$ does not contain ϕ and m^2 is a separation constant that has been judiciously chosen to be the same azimuthal (magnetic) quantum number that represents the z-component of the angular momentum. This equation obviously leads to the same solution that was obtained in spherical coordinates for the Φ part of the wave function

$$\Phi_m(\phi) = \frac{1}{\sqrt{2\pi}} \exp(im\phi) \quad m = 0, \pm 1, \pm 2, \dots \tag{4.37}$$

The remaining equation, $F(\xi, \eta) = -m^2$, can also be separated. We obtain

$$\frac{1}{f(\xi)} \frac{d}{d\xi} \left(\xi \frac{df(\xi)}{d\xi} \right) - \frac{m^2}{4\xi} - \frac{m_e |E|}{2\hbar^2} \xi + \frac{m_e}{\hbar^2} \cdot \left(\frac{e^2}{4\pi \varepsilon_0} \right)$$
$$= -\left[\frac{1}{g(\eta)} \frac{d}{d\eta} \left(\eta \frac{dg(\eta)}{d\eta} \right) - \frac{m^2}{4\eta} - \frac{m_e |E|}{2\hbar^2} \eta \right] \tag{4.38}$$

Because the left side of Equation (4.38) is a function of ξ only and the right side a function of η only we set each side equal to a constant, ν. We have then

$$\frac{d}{d\xi}\left(\xi\frac{df(\xi)}{d\xi}\right) - \left(\frac{m^2}{4\xi} + \frac{m_e|E|}{2\hbar^2}\xi - \frac{m_e}{\hbar^2}\cdot\left(\frac{e^2}{4\pi\varepsilon_0}\right) + \nu\right)f(\xi) = 0 \quad (4.39)$$

and

$$\frac{d}{d\eta}\left(\eta\frac{dg(\eta)}{d\eta}\right) - \left(\frac{m^2}{4\eta} + \frac{m_e|E|}{2\hbar^2}\eta - \nu\right)g(\eta) = 0 \quad (4.40)$$

These equations differ only in that the equation for ξ contains an extra term, the term containing e^2. We thus need solve only one of these equations, the solution to the remaining one following from the first.

4.7. Solution of the Separated Equations in Parabolic Coordinates—The Energy Eigenvalues

We may cast these equations in a more convenient form. In Equation (4.39) let

$$\zeta = \varepsilon\xi$$

to obtain

$$\left[\frac{1}{\zeta}\frac{d}{d\zeta}\left(\zeta\frac{d}{d\zeta}\right) + \left(\frac{\lambda_1}{\zeta} - \frac{1}{4} - \frac{m^2}{4\zeta^2}\right)\right]f(\zeta) = 0 \quad (4.41)$$

where

$$\lambda_1 = \frac{1}{\varepsilon}\left[\frac{m_e}{\hbar^2}\left(\frac{e^2}{4\pi\varepsilon_0}\right) - \nu\right] \quad \text{and} \quad \varepsilon^2 = \frac{2m_e|E|}{\hbar^2} \quad (4.42)$$

Notice that this equation contains the ϕ quantum number m and that λ_1 contains the separation constant ν. The energy $|E|$ is contained in both λ_1 and ε. Notice also that, because we are interested in bound states, E will be negative thus necessitating the absolute value sign in the definition of ε^2.

Equation (4.40) can be similarly transformed. It has exactly the same form as Equation (4.41) with λ_1 replaced by λ_2 and

$$\zeta' = \varepsilon\eta \quad \text{and} \quad \lambda_2 = \nu/\varepsilon \quad (4.43)$$

We use the same technique to solve these equations as that employed to solve the radial equation in spherical coordinates. Asymptotically

$$f(\zeta) \to \exp\left(\pm\frac{1}{2}\zeta\right) \quad (4.44)$$

As usual, we must discard the plus sign. We try a solution of the form

$$F(\zeta) = \zeta^s(a_0 + a_1\zeta + a_2\zeta^2 + \cdots)$$

$$= \zeta^s L(\zeta)\exp\left(-\frac{1}{2}\zeta\right) \quad (4.45)$$

and find that $s = \pm(1/2)m$. Because m may be positive or negative, we must require that $s = -(1/2)|m|$ for correct asymptotic behavior. Also,

$$f(\zeta) = \zeta^{(1/2)|m|} \exp\left(-\frac{1}{2}\zeta\right) \cdot L(\zeta) \tag{4.46}$$

which leads to

$$\zeta L''(\zeta) + (|m| + 1 - \zeta) L'(\zeta) + \left[\lambda_1 - \frac{1}{2}(|m| + 1)\right] L(\zeta) = 0 \tag{4.47}$$

for the ξ equation.

As in the solution to the radial equation in spherical coordinates, we find that the series must be terminated to keep the wave function from blowing up. It follows then that the quantity $\lambda_1 - \frac{1}{2}(|m| + 1)$ must be an integer or zero. Denoting this integer by n_1 we have

$$n_1 = 0, 1, 2, \ldots$$

$$= \lambda_1 - \frac{1}{2}(|m| + 1) \tag{4.48}$$

Now, the η-equation is virtually identical to the ξ-equation. We may therefore obtain the solution from the solution to the ξ-equation by letting $\zeta \to \zeta'$, $n_1 \to n_2$, and $\lambda_1 \to \lambda_2$ in the equation that led to n_1. We have

$$n_2 = 0, 1, 2, \ldots$$

$$= \lambda_2 - \frac{1}{2}(|m| + 1) \tag{4.49}$$

The sum of λ_1 and λ_2 must also be an integer, an integer that will turn out to be the principal quantum number. We therefore judiciously denote the sum $\lambda_1 + \lambda_2$ by n.

$$n = \lambda_1 + \lambda_2$$

$$= n_1 + n_2 + |m| + 1 \tag{4.50}$$

From this equation, together with the equations that define λ_1, λ_2, and ε^2 we obtain the energy. We have

$$n = \lambda_1 + \lambda_2$$

$$= \frac{1}{\varepsilon}\left[\frac{m_e}{\hbar^2}\left(\frac{e^2}{4\pi\varepsilon_0}\right) - \nu\right] + \frac{\nu}{\varepsilon}$$

$$= \frac{m_e}{\hbar^2}\left(\frac{e^2}{4\pi\varepsilon_0}\right)\sqrt{\frac{\hbar^2}{2m_e(-E_n)}} \tag{4.51}$$

where we have replaced $\sqrt{|E|}$ by $\sqrt{-E_n}$ because we seek the quantized (negative) bound state energies. Solving Equation (4.51) for E_n we have

$$
\begin{aligned}
E_n &= -\left(\frac{1}{4\pi\varepsilon_0}\right)^2 \cdot \frac{m_e e^4}{2\hbar^2 n^2} \\
&= -\frac{1}{2}\left(m_e c^2\right)\frac{\alpha^2}{n^2}
\end{aligned}
\tag{4.52}
$$

which is, of course, the same as that obtained using spherical coordinates because the eigenenergies cannot depend upon the coordinate system. Nor can the degree of degeneracy depend upon the coordinate system. To check that they are the same we must count the parabolic eigenstates. For $|m| = 0$ there are n ways of choosing n_1 and n_2. For $|m| > 0$ there are two ways of choosing $m(= \pm|m|)$ and $n - |m|$ ways of choosing n_1 and n_2. The degeneracy is therefore given by

$$
\begin{aligned}
\text{degree of degeneracy} &= n + 2 \cdot \sum_{|m|=1}^{n-1} (n - |m|) \\
&= n + 2n \cdot \sum_{|m|=1}^{n-1} 1 - 2 \cdot \sum_{|m|=1}^{n-1} |m| \\
&= n + 2n \cdot (n - 1) - 2 \cdot \frac{n(n-1)}{2} \\
&= n^2
\end{aligned}
\tag{4.53}
$$

where Gauss' trick was used to evaluate the second sum.

4.8. Solution of the Separated Equations in Parabolic Coordinates—The Energy Eigenfunctions

To put the ξ- and η-equations in comparable forms we changed the variables to $\zeta = \varepsilon\xi$ and $\zeta' = \varepsilon\eta$ from which we obtained Equation (4.46) for ζ. This led us to Equation (4.47) which we must now solve to obtain the energy eigenfunctions in parabolic coordinates.

The solutions of Equation (4.47) for which the series is forced to terminate are the associated Laguerre polynomials $L_{n_1+|m|}^{|m|}(\zeta)$ with an associated solution for the η-equation. We are here using associated Laguerre polynomials as defined by Spiegel[2] and used by Bethe and Salpeter[3] and Schiff[4] (among many others). The (unnormalized) wave function $\psi(\xi, \eta, \phi)$ is then given by

$$
\begin{aligned}
\psi_{n_1 n_2 m}(\xi, \eta, \phi) = N \cdot \exp\left[-\frac{\varepsilon(\xi+\eta)}{2}\right] \cdot (\xi\eta)^{|m|/2} \\
\times L_{n_1+|m|}^{|m|}(\varepsilon\xi) \cdot L_{n_2+|m|}^{|m|}(\varepsilon\eta) \cdot \exp(im\phi)
\end{aligned}
\tag{4.54}
$$

$$n_1 = 3, n_2 = 0; m = 0; n = 4$$

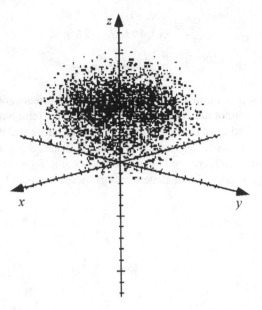

FIGURE 4.5. Density plot of the spherical wave function indicated. Full scale on each axis is 50 a.u.

Calculations using the parabolic eigenfunctions are frequently carried out using atomic units for which the normalization constant is

$$N = \frac{1}{\sqrt{\pi n}} \cdot \frac{\sqrt{n_1! n_2!} \varepsilon^{m+3/2}}{[(n_1 + m)!]^{3/2} [(n_2 + m)!]^{3/2}} \tag{4.55}$$

Upon taking the absolute square of ψ, the ϕ-dependence disappears just as it did in spherical coordinates so the probability density is cylindrically symmetric about the z-axis. As was done for the spherical wave functions, we may make a density plot of $|\psi|^2$. FIGURE 4.5 is a density plot for the quantum numbers $n_1 = 3, n_2 = 0$; $m = 0; n = 4$.

The contrast between the charge density for the parabolic eigenstate shown in FIGURE 4.5 and that for the spherical eigenstate shown in Figure 4.3 is striking. Although the charge densities for spherical eigenstates must be symmetric about the origin, no such symmetry exists for the Stark eigenstates. As shown in FIGURE 4.5, the charge density can be quite asymmetric with respect to the Cartesian axes thus making it possible for hydrogen atoms in parabolic eigenstates to have permanent electric dipole moments. This is not possible for spherical eigenstates due to the symmetry requirements on the square of the wave function. Clearly the separability of the Schrödinger equation in both spherical and parabolic coordinates makes this possible. As noted previously, this separability in both coordinate systems is a result of the extra symmetry of the Coulomb potential, the symmetry that leads to the accidental degeneracy.

Problems

4.1. The wave function for a hydrogen atom is given by

$$\psi(r, \theta, \phi) \propto r^2 e^{-r/3a_0} (3 \cos^2 \theta - 1)$$

(a) What are the quantum numbers n, ℓ, m for this atom? No calculations allowed.
(b) What is the most probable value of r for the electron in this state? What is significant about the result?

4.2. The wave function for a hydrogen atom is given by

$$\psi(r, \theta, \phi) = R_{54}(r) Y_{\ell m}(\theta, \phi) + R_{53}(r) Y_{30}(\theta, \phi) + R_{52}(r) Y_{21}(\theta, \phi)$$

For what values of ℓ and m $(= m_\ell)$ will ψ be an eigenfunction of the Hamiltonian? What is the energy in electron volts? In Rydbergs? In a.u.? In cm^{-1}?

4.3. An electron is in the ground state of tritium, that is, a hydrogen atom with two neutrons the symbol for which is ^3H. A nuclear reaction instantaneously changes the nucleus to ^3He which retains the single electron from the tritium atom. That is, the product is a singly ionized ^3He which is a one-electron atom. Calculate the probability that the atomic electron remains in the ground state of the ^3He ion.

4.4. An electron in the Coulomb field of a proton is described by the normalized wave function

$$|\psi(r)\rangle = \tfrac{1}{6}\{4 |\psi_{100}(r)\rangle + 3 |\psi_{211}(r)\rangle - |\psi_{210}(r)\rangle + \sqrt{10} |\psi_{21-1}(r)\rangle\}$$

where the $|\psi(r)\rangle$ are spherical (orbital) eigenfunctions of the hydrogen atom Hamiltonian.
Find:
(a) The expectation value of the energy in a.u.
(b) The expectation value of $\hat{L}^2$. Keep $\hbar$ in the answer.
(c) The expectation value of $\hat{L}_z$.

4.5. A hydrogen atom is in an eigenstate of L^2 and L_z. Show that, although neither L_x nor L_y are well defined, their sum $\{L_x^2 + L_y^2\}$ is well defined. Find the value of $\{L_x^2 + L_y^2\}$ in terms of ℓ and m_ℓ. Ignore spin.

4.6. Show that the integrals

$$z_{n\ell m}^{n(\ell-1)m} = \langle n(\ell-1)m|z|n\ell m\rangle$$

$$= \int_0^\infty r^2 dr\, R_{n\ell}(r) R_{n(\ell-1)}(r) r dr \int_0^\pi P_{\ell m}(\theta) P_{(\ell-1)m}(\theta) \cos\theta \cdot \sin\theta d\theta$$

and

$$z_{n\ell m}^{n(\ell+1)m} = \langle n(\ell+1)m|z|n\ell m\rangle$$

$$= \int_0^\infty r^2 R_{n\ell}(r) R_{n(\ell+1)}(r) r dr \int_0^\pi P_{\ell m}(\theta) P_{(\ell+1)m}(\theta) \cos\theta \cdot \sin\theta d\theta$$

are given in atomic units by

$$z_{n\ell m}^{n(\ell-1)m} = \sqrt{\frac{(\ell^2 - m^2)}{(2\ell+1)(2\ell-1)}} R_{n\ell}^{n(\ell-1)}$$

and

$$z_{n\ell m}^{n(\ell+1)m} = \sqrt{\frac{[(\ell+1)^2 - m^2]}{(2\ell+3)(2\ell+1)}} R_{n\ell}^{n(\ell+1)}$$

where

$$R_{n\ell}^{n(\ell-1)} = \frac{3}{2}n\sqrt{n^2 - \ell^2} \quad \text{and} \quad R_{n\ell}^{n(\ell+1)} = \frac{3}{2}n\sqrt{n^2 - (\ell+1)^2}$$

4.7. A collection of hydrogen atoms is in a parabolic eigenstate $|\psi\rangle$ given by

$$|\psi\rangle = \frac{1}{\sqrt{2}}|200\rangle + \frac{1}{\sqrt{2}}|210\rangle$$

where the kets on the right are the spherical eigenkets $|n\ell m\rangle$.
(a) What is the energy eigenvalue, in atomic units, corresponding to $|\psi\rangle$?
(b) What are the possible sets of parabolic quantum numbers that characterize $|\psi\rangle$? Virtually no computation is required. Merely list the combinations that are possible from inspection of the wave function given.
4.8. The normalized energy eigenfunctions for the hydrogen atom in parabolic coordinates are given by

$$\psi_{n_1 n_2 m}(\xi, \eta, \phi) = \frac{\exp(im\phi)}{\sqrt{\pi n}} \frac{\sqrt{n_1!}\sqrt{n_2!}\varepsilon^{|m|+3/2}}{(n_1 + m)^{3/2}(n_2 + m)^{3/2}}$$
$$\times \exp[-\varepsilon(\xi + \eta)](\xi\eta)^{|m|/2} L_{n_1+|m|}^{|m|}(\varepsilon\xi) L_{n_2+|m|}^{|m|}(\varepsilon\eta)$$

where $\varepsilon = \sqrt{-2E}$ and the Ls are associated Laguerre polynomials.
(a) Find the normalized hydrogen atom energy eigenfunctions in parabolic coordinates for $n = 2$, $n_1 = 1$, $n_2 = 0 = m$ in terms of the hydrogen atom energy eigenfunctions in spherical coordinates.
   (b)–(h) What are the probabilities of measuring the following?
(b) The energy of the ground state.
(c) The energy of the first excited state.
(d) Total angular momentum zero.
(e) Total angular momentum $\sqrt{2}\hbar$.
(f) z-Component of angular momentum zero.
(g) z-Component of angular momentum $\hbar$.
(h) Find the normalized ground state eigenfunction in parabolic coordinates in terms of the spherical eigenfunctions. What is noteworthy about this eigenfunction and why?
4.9. A hydrogen atom is in an eigenstate that is characterized by the parabolic quantum numbers

$n_1 = 1; n_2 = 0 = m$. The wave function in *spherical coordinates* in atomic units is:

$$u_{n_1 n_2 m} = u_{100} = \frac{1}{\sqrt{2\pi}} \left(\frac{1}{2}\right)^{3/2} \left[-1 + \tfrac{1}{2}r\left(1 + \cos\theta\right)\right] \cdot e^{-r/2}$$

That is, you have been spared the trouble of writing the wave function in parabolic coordinates and converting to spherical coordinates. Some selected spherical harmonics and radial hydrogen atom wave functions are:

$$Y_{00} = \frac{1}{\sqrt{4\pi}}; Y_{10} = \sqrt{\frac{3}{4\pi}} \cos\theta; Y_{11} = \sqrt{\frac{3}{8\pi}} \sin\theta \, e^{i\phi}$$

$$R_{10} = 2e^{-r}; R_{20} = \frac{1}{\sqrt{2}} \left(1 - \tfrac{1}{2}r\right)e^{-r/2}; R_{21} = \frac{1}{2\sqrt{6}} r e^{-r/2}$$

(a) If a measurement of the energy of the atom in this state is made what are the possible values of the energy that could be measured? Give your answer in atomic units. What probabilities are associated with each value?

(b) If a measurement of the *total* orbital angular momentum of the atom in this state is made what are the possible values that could be measured? What probabilities are associated with each possible angular momentum?

(c) If a measurement of the z-component of the angular momentum of the atom in this state is made what are the possible values that could be measured? What probabilities are associated with each possible z-component?

4.10. Show that the expectation value of the electric dipole moment of an atom having N electrons in a state of well-defined parity vanishes. Use atomic units. Suppose a hydrogen atom is in a spherical eigenstate. What is the dipole moment? How about a parabolic eigenstate?

References

1. G.B. Arfken and H.J. Weber, *Mathematical Methods for Physicists* (Harcourt, New York, 2001).
2. M.R. Spiegel, *Mathematical Handbook of Formulas and Tables* (McGraw-Hill, New York, 1998).
3. H.A. Bethe and E.E. Salpeter, *Quantum Mechanics of One- and Two-Electron Atoms* (Springer-Verlag, Berlin, 1957).
4. L.I. Schiff, *Quantum Mechanics* (McGraw-Hill, New York, 1968).
5. D.J. Griffiths, *Introduction to Quantum Mechanics* (Prentice-Hall, Upper Saddle River, NJ, 1995).
6. S. Gasiorowicz, *Quantum Physics* (Wiley, New York, 2003).
7. C.A. Coulson and A. Joseph, Proc. Phys. Soc. **90**, 887 (1967).
8. L.D. Landau and E.M. Lifshitz, *Quantum Mechanics (Non-relativistic Theory)* (Pergamon Press, Oxford, 1977).

5
The Classical Hydrogen Atom

5.1. Introduction

The accidental degeneracy of the hydrogen atom is the result of symmetries inherent in the Coulomb potential. These symmetries must manifest themselves in a classical formulation of the problem, so we examine the hydrogen atom from a classical point of view.

The Coulomb potential and the Newtonian gravitational potential are both proportional to $1/r$, so the classical description of the hydrogen atom is directly comparable with the Kepler problem. From any classical mechanics book[1] we find the equation of the orbit in the Kepler problem with potential

$$V(r) = -\frac{k}{r} \tag{5.1}$$

Because angular momentum is conserved for any central potential, the motion is confined to a plane and we use plane polar coordinates r and ϕ. The equation of the orbit of a particle under the influence of this potential is a conic section with eccentricity ε and is given by

$$\frac{\alpha}{r} = 1 + \varepsilon \cos \phi \tag{5.2}$$

The α that appears in this equation is not the fine-structure constant, but because this designation is universal, we retain this symbol. It is, in fact, referred to as the *latus rectum* and is given by

$$\alpha = \frac{\ell^2}{\mu k} \tag{5.3}$$

where ℓ is the angular momentum and μ the reduced mass of the system. Although ℓ is the same symbol used for the (dimensionless) angular momentum quantum number, it can, in this context, take on continuous values and, of course, it has units of angular momentum. The orbital eccentricity ε is given by

$$\varepsilon = \sqrt{1 + \frac{2E\ell^2}{\mu k^2}} \tag{5.4}$$

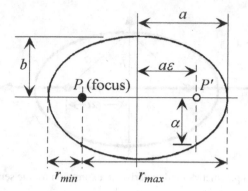

FIGURE 5.1. Notation and definitions of an elliptical orbit caused by Coulomb or gravitational potential.

where E is the total energy of the system. The nature of the trajectory of the particle of mass μ depends upon the value of ε in the following way.

$$\varepsilon = 1 \Rightarrow \text{orbit is a parabola}$$
$$\varepsilon > 1 \Rightarrow \text{orbit is a hyperbola}$$
$$\varepsilon < 1 \Rightarrow \text{orbit is an ellipse}$$

Clearly the only way the orbit can represent a bound state is if $E < 0$, thus making ε less than unity and the orbit an ellipse. In contrast to parabolas and hyperbolas an elliptical orbit is a closed orbit and is thus a bound state of the orbiting particle.

FIGURE 5.1 shows the quantities associated with the elliptical orbit caused by an attractive $1/r$ potential. The sun (proton) is located at the focus P and the earth (electron) executes the elliptical orbit; r_{min} and r_{max} are, respectively, the pericenter and apocenter of the orbit.

Using the notation and definitions in FIGURE 5.1 we may find an expression for a, the semi-major axis and b, the semi-minor axis. From the equation of the orbit, Equation (5.2), it is clear that r_{min} and r_{max} occur when $\theta = \pi$ and 0, respectively. Therefore,

$$r_{min} = \frac{\alpha}{1 + \varepsilon} \quad \text{and} \quad r_{max} = \frac{\alpha}{1 - \varepsilon} \tag{5.5}$$

Moreover,

$$a = \frac{r_{min} + r_{max}}{2} \tag{5.6}$$

so

$$a = \frac{\alpha}{1 - \varepsilon^2}$$
$$= \frac{\ell^2}{\mu k} \cdot \frac{1}{\left[1 - \left(1 + \frac{2|E|\ell^2}{\mu k^2} \right) \right]}$$
$$= \frac{k}{2|E|} \tag{5.7}$$

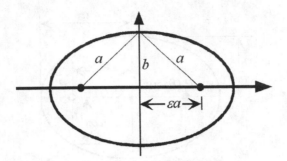

FIGURE 5.2. Orbital parameters used to derive an expression for the semi-minor axis of the elliptical orbit.

To find b, the semi-minor axis of the ellipse, we use the definition of an ellipse. That is, an ellipse is the locus of all points, each of which has the sum of its distances from the two points (the foci) equal to a constant. In view of this definition we specify the parameters shown in FIGURE 5.2.

From FIGURE 5.2 it is clear that

$$b^2 = a^2 - (\varepsilon a)^2$$
$$= a^2(1 - \varepsilon^2) \tag{5.8}$$

so that

$$b = \frac{\alpha}{(1 - \varepsilon^2)}\sqrt{1 - \varepsilon^2}$$
$$= \frac{\ell^2}{\mu k} \cdot \sqrt{\frac{\mu k^2}{2|E|\ell^2}}$$
$$= \frac{\ell}{\sqrt{2\mu|E|}} \tag{5.9}$$

Comparing the semi-major and semi-minor axes, Equations (5.7) and (5.9), we see that a is independent of the angular momentum, but b depends upon ℓ.

Solving Equation (5.7) for E we find that

$$E = -\frac{k}{2a} \tag{5.10}$$

where we have removed the absolute value signs and inserted the minus sign because the total energy must be negative for a bound orbit. Because a depends on the energy and not on the angular momentum, a given value of the energy uniquely determines the semi-major axis of the elliptical orbit. The energy is, however, independent of b, the semi-minor axis that does depend on the angular momentum as found in Equation (5.9). Evidently, for a given negative energy, there are an infinite number of elliptical orbits, each corresponding to a different

TABLE 5.1. Listing of classical parameters for an elliptical orbit in atomic units.

Quantity	Value
Energy	$-1/2n^2$
Semi-major axis	$a = n^2$
Semi-minor axis	$b = n\ell$
Orbital period	$\tau = 2\pi a^{3/2} = 2\pi n^3$
Orbital frequency	$\omega_n = 1/n^3$
Orbital eccentricity	$\varepsilon = \sqrt{1 - \ell^2/n^2}$
Pericenter (from focus)	$r_{\min} = n^2(1 - \varepsilon) = \ell^2/(1 + \varepsilon)$
Apocenter (from focus)	$r_{\max} = n^2(1 + \varepsilon) = \ell^2/(1 - \varepsilon)$

semi-minor axis (and thus angular momentum), but all having the same semi-major axis. The energy is thus independent of the angular momentum.

The independence of the energy on the angular momentum is reminiscent of the accidental degeneracy encountered in the quantum mechanical solution of the hydrogen atom. Of course, this is not an accident. In fact, this is a classical degeneracy. Obviously, the classical and quantum mechanical degeneracies are related.

To clarify this connection, recall that the Bohr radius for principle quantum number n is given by

$$r_n = n^2 a_0 \tag{5.11}$$

where a_0 is the first Bohr radius. For circular orbits the semi-major axis a is the radius so we may relate a to r_n. Substituting in the expression for the (classical) energy, we have

$$E = -\frac{k}{2a_0 n^2} \tag{5.12}$$

In atomic units $k = 1$ and $a_0 = 1$, so the classical energy is identical with the energy obtained by solving the Schrödinger equation. Classically, of course, n is continuously variable. For reference, TABLE 5.1 contains some parameters in atomic units for an electron in an elliptical orbit subject to a Coulomb potential. Of course, the "quantum numbers" in the listing are continuously variable.

5.2. The Classical Degeneracy

What is the origin of the degeneracy? Degeneracies are associated with symmetries. To investigate the origin of the degeneracy, we examine the nature of the orbits. Symmetries are always associated with constants of the associated motion. For example, let us inspect the geometric symmetry discussed above in more detail. If the potential is central, that is, $V(r) = V(r)$, then it is obvious that the problem has spherical symmetry. Classically this symmetry manifests itself as conservation of angular momentum. We can see this by inspection of the Lagrangian Λ. For a central potential the Lagrangian is

$$\Lambda = \left(\frac{1}{2}\right)\mu\left(\dot{r}^2 + r^2\dot{\phi}^2\right) - V(r) \qquad (5.13)$$

where r and ϕ are the usual spherical coordinates and the dots above them designate differentiation with respect to time. The angular momentum ℓ is p_ϕ, the momentum conjugate to the coordinate ϕ, and is given by

$$\ell = p_\phi$$
$$= \frac{\partial\Lambda}{\partial\dot{\phi}}$$
$$= \mu r^2\dot{\phi} \qquad (5.14)$$

Lagrange's equation is

$$\frac{\partial\Lambda}{\partial\phi} - \frac{d}{dt}\left(\frac{\partial\Lambda}{\partial\dot{\phi}}\right) = 0 \qquad (5.15)$$

Inspection of the Lagrangian shows that it is "cyclic" in ϕ; that is, it does not contain ϕ. Therefore $\dfrac{\partial\Lambda}{\partial\phi} = 0$ and

$$\frac{d}{dt}\left(\frac{\partial\Lambda}{\partial\dot{\phi}}\right) = 0 = \frac{dp_\phi}{dt} \quad \Rightarrow \quad p_\phi = \text{constant} \qquad (5.16)$$

Because $p_\phi = \dfrac{\partial\Lambda}{\partial\dot{\phi}}$

$$p_\phi = \mu r^2\dot{\phi}$$
$$= \ell \qquad (5.17)$$

and, indeed it is the angular momentum (in the classical sense) that is the conserved quantity.

Note that this is just Kepler's second law which, unlike Kepler's first and third laws, is valid for *any* central potential. The first and third laws are valid only for an attractive $1/r$ potential. We see then that

spherical symmetry $\Rightarrow$ conservation of ℓ $\Rightarrow$ motion in a plane

Because angular momentum is conserved classsically, the corresponding quantum mechanical operator for angular momentum must commute with the Hamiltonian. Thus, the spatial symmetry that causes the classical angular momentum to be conserved manifests itself quantum mechanically as $[\hat{H}, \hat{L}] = 0$. This spatial symmetry is the root of the degeneracy in energy with respect to the quantum number m. It is present for any central potential because the spherical coordinates θ and ϕ do not appear, only r.

For hydrogen, however, there is the additional degeneracy, the accidental degeneracy that causes the energy to be independent of the quantum number ℓ. It is natural to suspect that there is a symmetry other than that associated with the central potential that is responsible for this degeneracy. Such a symmetry would correspond classically to an additional constant of the motion. Quantum mechanically there would be an additional operator that commutes with the Hamiltonian, that is, commutes with $\hat{H}$ in addition to $\hat{L}^2$ and $\hat{L}_z$. Indeed, there is an additional constant of the motion. It is called the Lenz vector, a vector that, classically, points along the semi-major axis of the Keplerian ellipse that is the electronic orbit. The precise definition of this vector and, indeed, even its direction (toward or away from the apside), differ with different authors.

5.3. Another Constant of the Motion—The Lenz Vector

Following Goldstein,[1] the Lenz vector A is defined for a hydrogen atom as

$$A = \frac{1}{(e^2/4\pi\varepsilon_0)m_e} \cdot p \times L - \hat{r} \qquad (5.18)$$

where e is the electronic charge, m_e is the electronic mass, p is the linear momentum, L the angular momentum, and $\hat{r} = r/r$ is a unit vector that points from the force center (the focus at which the sun or proton lies) to the orbiting particle. The use of the "hat" to signify a unit vector as well as a quantum mechanical operator should cause no confusion.

The calculations are simplified if atomic units are used. In atomic units the Coulomb potential is

$$V(r) = -\frac{1}{r} \qquad (5.19)$$

and the Lenz vector is[2]

$$A = p \times L - \hat{r} \qquad (5.20)$$

Now, we wish to examine the time dependence of A so that we may determine the circumstances under which it is conserved. Taking the total time derivative of A, we have

$$\dot{A} = \dot{p} \times L - p \times \dot{L} - \frac{d}{dt}(\hat{r}) \qquad (5.21)$$

but, for a central potential $\dot{L} = 0$ and we have

$$\dot{A} = \dot{p} \times L - \frac{d}{dt}(\hat{r}) \qquad (5.22)$$

Now, the force is given by Newton's second law so

$$\dot{p} = -\frac{dV(r)}{dr}\hat{r}$$
$$= f(r)\frac{r}{r} \qquad (5.23)$$

where $f(r)$ is the the central force.

The first term in the definition of the Lenz vector may thus be written

$$\dot{p} \times L = \frac{f(r)}{r}[r \times (r \times p)] \tag{5.24}$$

But, $L = r \times p$ and in atomic units $p = \dot{r}$ so

$$\dot{p} \times L = \frac{f(r)}{r}[r \times (r \times \dot{r})]$$

$$= \frac{f(r)}{r}[r(r \cdot \dot{r}) - \dot{r}(r \cdot r)] \tag{5.25}$$

and

$$r \cdot \dot{r} = \frac{1}{2}\frac{d}{dt}(r \cdot r)$$

$$= \frac{1}{2}\frac{d}{dt}r^2$$

$$= \dot{r}r \tag{5.26}$$

Equation (5.25) becomes

$$\dot{p} \times L = \frac{f(r)}{r}[r(r \cdot \dot{r}) - \dot{r}(r \cdot r)]$$

$$= \frac{f(r)}{r}\left(rr\dot{r} - \dot{r}r^2\right)$$

$$= f(r)r\left(\frac{r}{r}\dot{r} - \dot{r}\right)$$

$$= f(r)r(\hat{r}\dot{r} - \dot{r}) \tag{5.27}$$

The last term in Equation (5.21) is $\dot{\hat{r}}$ which is

$$\frac{d}{dt}\hat{r} = \frac{d}{dt}\left(\frac{r}{r}\right)$$

$$= \frac{\dot{r}}{r} - \frac{1}{r^2}\dot{r}r$$

$$= -\frac{1}{r}(\dot{r}\hat{r} - \dot{r}) \tag{5.28}$$

so the term $(\dot{r}\hat{r} - \dot{r})$ in $\dot{p} \times L$ is simply $-r\dfrac{d}{dt}\hat{r}$ and Equation (5.21) becomes

$$\dot{A} = -f(r)r^2\frac{d}{dt}(\hat{r}) - \frac{d}{dt}(\hat{r})$$

$$= -\left[1 + r^2 f(r)\right]\frac{d}{dt}(\hat{r}) \tag{5.29}$$

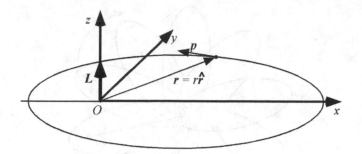

FIGURE 5.3. Elliptical orbit showing the linear and angular momenta vectors.

or, in terms of the potential energy

$$\dot{A} = -\left[1 - r^2\frac{dV(r)}{dr}\right]\frac{d}{dt}(\hat{r}) \tag{5.30}$$

This is the general form of $\dot{A}$ for any central potential. The uniqueness of the Coulomb potential lies in the fact that for it

$$\dot{A} \equiv 0 \tag{5.31}$$

This means that, for a Coulomb (or gravitational) potential, the vector A is a constant of the motion. It is fixed in magnitude and direction so the Keplerian orbit is fixed in space. Therefore, the elliptical orbit does not precess about the force center in the orbital plane. The orbiting particle retraces its path on successive orbits.

To find the direction in which the Lenz vector points, we note that the vector $p \times L$ lies in the xy-plane because it is perpendicular to p which is tangential to the orbit and perpendicular to L which, itself, is perpendicular to the plane of the orbit. Because $\hat{r}$ also lies in the xy-plane, A must lie in the same plane. We can determine where it lies in the xy-plane using the diagram in FIGURE 5.3.

For convenience, the x-axis is taken to be along the major axis and the y-axis along the minor axis. Therefore, the angular momentum is in the z-direction. We may compute the components of A from

$$A = p \times L - \hat{r}$$

$$= \begin{vmatrix} \hat{i} & \hat{j} & \hat{k} \\ p_x & p_y & p_z \\ 0 & 0 & L \end{vmatrix} - \frac{1}{r}(x\hat{i} + y\hat{j}) \tag{5.32}$$

where $\hat{i}, \hat{j}, \hat{k}$ are the unit vectors in the x, y, and z directions, respectively. We obtain

$$A_x = p_yL - \frac{x}{r} \quad \text{and} \quad A_y = -p_xL - \frac{y}{r} \tag{5.33}$$

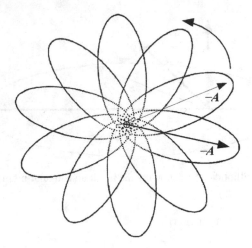

FIGURE 5.4. Precession of the nearly elliptical orbit that is caused by the small non-Coulombic term in the potential.

and, of course, $A_z = 0$. When the orbiting particle crosses the x-axis $y = 0$ and $p_x = 0$ so $A_y = 0$. Because $\dot{A} \equiv 0$ for all time, we must have $A \equiv A_x\hat{i}$. Therefore, A remains along the major axis and points toward the pericenter (see problem 5.4).

Suppose the potential energy is nearly a Coulomb potential, but contains a small non-Coulombic, but central, term which we write as

$$V(r) = -\frac{1}{r} + \Omega(r) \tag{5.34}$$

where $\Omega(r)$ is the non-Coulombic term. From Equation (5.30), the time derivative of the Lenz vector is

$$\dot{A} = -r^2\frac{d\Omega(r)}{dr} \cdot \frac{d}{dt}(\hat{r}) \tag{5.35}$$

so that, if $\Omega(r) \neq 0$, the Lenz vector conservation extant for the Kepler potential is destroyed.

If $\Omega(r)$ is small compared with the Coulomb term then the trajectory is a nearly Keplerian ellipse that revolves about the force center as shown in FIGURE 5.4. The Lenz vector A (which points along the major axis) then precesses about the force center.

By employing the properties of the Lenz vector, the equation of the orbit for a pure Coulomb potential may be easily derived. Moreover, we can show that the magnitude of the Lenz vector is the eccentricity of the orbit ε (in atomic units). Taking the dot product of A with r we have

$$A \cdot r = p \times L \cdot r - \hat{r} \cdot r \tag{5.36}$$

Now, for a cross product dotted into another vector we may cyclically permute the vectors so that

$$A \cdot r = (r \times p) \cdot L - r \tag{5.37}$$

Because $(r \times p) = L$, we have

$$Ar \cos \phi = L^2 - r \tag{5.38}$$

Recalling that $|L| = \ell$ we have

$$\frac{\ell^2}{r} = 1 + A \cos \phi \tag{5.39}$$

which is the equation of the Keplerian orbit provided $A = \varepsilon$, the eccentricity. We have thus derived the equation of the orbit in a very simple way using the Lenz vector. More importantly, however, we have shown that the magnitude of the Lenz vector is the eccentricity of the conic section that is the classical orbit.

From the standpoint of atomic physics, the significance of the Lenz vector lies in the fact that $\dot{A} = 0$; that is, A is a constant of the motion. Classically then, there are three constants of the motion:

1. Energy, E
2. Angular momentum, L
3. Lenz vector, A

Quantum mechanically, we expect that $\left[\hat{A}, \hat{H}\right] = 0$, but we know that we cannot simultaneously specify three components of a vector. When we solve the Schrödinger equation in spherical coordinates we use the operators corresponding to E (the Hamiltonian $\hat{H}$), $\hat{L}^2$, and $\hat{L}_z$ because $\hat{L}^2$ and only one component of $\hat{L}$ commutes with $\hat{H}$. We choose $\hat{L}_z$ as the commuting component. For the hydrogen atom, however, we are at liberty to choose a different set of commuting operators to solve the problem. If we choose $\hat{H}$, $\hat{L}_z$, and $\hat{A}_z$ it turns out that we are using parabolic coordinates.[3,4] Thus, the quantum numbers n and m are associated with $\hat{H}$ and $\hat{L}_z$, respectively. This leaves two quantum numbers n_1 and n_2 to correspond to the operator $\hat{A}_z$. Although it seems as if there is an "extra" quantum number, four rather than three obtained in spherical coordinates, the relationship between the quantum numbers

$$n = n_1 + n_2 + |m| + 1 \tag{5.40}$$

establishes a relationship between n_1 and n_2 if n and m are fixed.

Aside from its role as a classical vector pointing along the semi-major axis, the z-component of the Lenz vector has physical significance. It is proportional to the permanent electric dipole moment of an electron in a Keplerian orbit. As was seen in Chapter 4, hydrogen is unique among atoms in that it can have a permanent electric dipole moment, a property that is not immediately obvious from the symmetry of the charge distributions for the spherical coordinate eigenfunctions. On the other

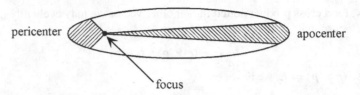

FIGURE 5.5. An elliptical orbit illustrating Kepler's second law of equal areas and how the orbiting particle must be moving more slowly at apocenter than at pericenter.

hand, the asymmetry of the charge distributions for parabolic eigenfunctions, as exemplified in Figure 4.5, makes it clear that hydrogen atoms can indeed have permanent electric dipole moments. The ground state is an exception because it is nondegenerate so that the spherical ground state eigenfunction and the parabolic ground state eigenfunction are identical.

The correspondence of the classical Lenz vector A with the quantum mechanical operator $\hat{A}$ makes the permanent electric dipole moment reasonable because A lies along the major axis. From Kepler's second law, the law of equal areas, the electron in the classical hydrogen atom is moving more slowly at apocenter than at pericenter as shown in FIGURE 5.5. Thus, averaged over a period, there is a net buildup of negative charge at the apocenter that produces a permanent electric dipole moment. By symmetry, it is along the major axis. Clearly it is in the direction of the pericenter.

The relation between the Lenz vector and the electric dipole moment can be deduced by noting that in atomic units the electric dipole moment for an electron in a Keplerian orbit is simply given by the average value of r over a single orbit. Thus, the dipole moment p is given by

$$\langle p \rangle = \langle r \cos \phi \rangle \left(\frac{A}{A} \right) \tag{5.41}$$

and its magnitude is the average over a period of $r \cos \phi$; that is,

$$\langle p \rangle = \frac{1}{\tau} \int_0^\tau r \cos \phi \, dt \tag{5.42}$$

Using the equation of the orbit and conservation of angular momentum it can be shown (see Problem 5.3) that $\langle p \rangle = (3/2) n^2 A$. We expect then that quantum mechanically the z-component of the electric dipole moment is given by

$$\hat{p}_z = \frac{3}{2} n^2 \hat{A}_z \tag{5.43}$$

These operators are useful later when we treat the quantum mechanical Stark effect, the effect of a constant electric field on a hydrogen atom.

Problems

5.1. Using SI units, find the period T_n of an electron in a hydrogen Bohr orbit of principal quantum number n. Show that this period is consistent with Kepler's third law. Show that it is the same as T_c, the period deduced from the correspondence principal. Hint: Eliminate the $\hbar$s so the constants match. Show that the result reduces to $T = 2\pi n^3$ in atomic units.

5.2. The (classical) Lenz vector (in a.u.) is

$$A = p \times L - \hat{r}$$

where $\hat{r}$ is the unit vector in the r direction. For a general central potential find a general expression for the time derivative of A, that is, $\dot{A}$, and show that A is a constant of the motion for a Coulomb potential.

5.3. (a) Derive the equation of a Keplerian ellipse in terms of the Lenz vector A and show that the eccentricity is $\varepsilon = |A|$.
(b) Show that

$$\varepsilon = \sqrt{1 - \frac{\ell^2}{n}}$$

Note that neither ℓ nor n are "quantum numbers" because, classically, they are continuously variable. ℓ is the classical angular momentum and n is a measure of the energy. Recall that, in atomic units $E = -1/(2n^2)$.
(c) By averaging over a period show that the electric dipole moment of a Keplerian hydrogen atom is

$$|\langle p \rangle| = \frac{3}{2}n^2\sqrt{1 - \ell^2/n^2}$$

(d) For what value of the ℓ is $|\langle p \rangle| = 0$? What is special about these orbits?
(e) Find the positions (or position) of the maxima (or maximum) in the radial probability density for hydrogen atoms having $\ell = n - 1$, the maximum angular momentum. What is special about these states? It may be helpful to recall that

$$R_{n\ell}(r) \propto \exp\left(-\frac{r}{n}\right) \cdot \left(\frac{2r}{n}\right) \cdot L_{n+\ell}^{2\ell+1}\left(\frac{2r}{n}\right)$$

in atomic units and that

$$L_p^q(\rho) = \frac{d^p}{d\rho^p}L_q(\rho)$$

5.4. This problem should show you why the conservation of the (classical) Lenz vector A implies closed orbits for the Kepler problem. Use A in a.u. so that $A = p \times L - \hat{r}$.
(a) Express A in terms of r and p alone; that is, eliminate L (no cross products).

(b) Show that

$$A = r_{max} \left(2E + \frac{1}{r_{max}} \right) = r_{min} \left(2E + \frac{1}{r_{min}} \right)$$

where r_{max} and r_{min} are the maximum and minimum values of r, that is, apocenter and pericenter.

(c) Show that A is parallel to r_{min} and antiparallel to r_{max}.

(d) From the answer to (c) it is clear that for a circular orbit $A = 0$. Prove this mathematically from the equations that you derived in (b).

References

1. H. Goldstein, *Classical Mechanics* (Addison-Wesley, Reading, MA, 1980).
2. T.P. Hezel, C.E. Burkhardt, M. Ciocca, et al., Am. J. Phys. **60**, 329 (1992).
3. V. Bargmann, Z. Phys. **99**, 576 (1936).
4. C.A. Coulson and A. Joseph, Proc. Phys. Soc. **90**, 887 (1967).

6
The Lenz Vector and the
Accidental Degeneracy

6.1. The Lenz Vector in Quantum Mechanics

To use the Lenz vector in quantum mechanics, an operator corresponding to this observable must be constructed. If it is constructed by simply taking the operator equivalent of the classical definition of the Lenz vector

$$A = p \times L - \hat{r} \tag{6.1}$$

it is found that such an operator is not Hermitian. Pauli[1] recognized that to convert A into $\hat{A}$, the classical definition must be modified to make the quantum mechanical operator Hermitian. Because it is to correspond to a classical constant of the motion, this operator must also commute with the Hamiltonian. Pauli deduced that the classical definition must be properly symmetrized. The correct form is

$$\hat{A} = \left(\frac{1}{2}\right)(\hat{p} \times \hat{L} - \hat{L} \times \hat{p}) - \hat{r} \tag{6.2}$$

where now

$$\hat{r} = \frac{x}{r}\hat{i} + \frac{y}{r}\hat{j} + \frac{z}{r}\hat{k}$$

may be regarded as both a unit vector and a quantum mechanical operator. Because they are operators, $\hat{p} \times \hat{L} \neq -\hat{L} \times \hat{p}$.

The proof that this operator is Hermitian need be done on only a single component of $\hat{A}$. The x-component is

$$\hat{A}_x = \frac{1}{2}\{(\hat{p}_y\hat{L}_z - \hat{p}_z\hat{L}_y) - (\hat{L}_y\hat{p}_z - \hat{L}_z\hat{p}_y)\} - \frac{x}{r} \tag{6.3}$$

and its Hermitian conjugate is

$$\hat{A}_x^\dagger = \frac{1}{2}\{(\hat{p}_y\hat{L}_z - \hat{p}_z\hat{L}_y) - (\hat{L}_y\hat{p}_z - \hat{L}_z\hat{p}_y)\}^\dagger - \left(\frac{x}{r}\right)^\dagger$$

$$= \frac{1}{2}\{(\hat{p}_y\hat{L}_z)^\dagger - (\hat{p}_z\hat{L}_y)^\dagger - (\hat{L}_y\hat{p}_z)^\dagger + (\hat{L}_z\hat{p}_y)^\dagger\} - \left(\frac{x}{r}\right)^\dagger$$

$$= \frac{1}{2}\{\hat{L}_z^\dagger\hat{p}_y^\dagger - \hat{L}_y^\dagger\hat{p}_z^\dagger - \hat{p}_z^\dagger\hat{L}_y^\dagger + \hat{p}_y^\dagger\hat{L}_z^\dagger\} - \left(\frac{x}{r}\right)^\dagger \tag{6.4}$$

TABLE 6.1. Some useful relations in
atomic units.

1. $[\hat{L}_i, \hat{p}_j] = i\hat{p}_k\varepsilon_{ijk}$

2. $[\hat{p}_i, \hat{L}_j] = i\hat{p}_k\varepsilon_{ijk}$

3. $[\hat{L}_i, \hat{r}_j] = i\hat{r}_k\varepsilon_{ijk}$

4. $[\hat{r}_i, \hat{L}_j] = i\hat{r}_k\varepsilon_{ijk}$

5. $\left[\hat{p}_i, \dfrac{1}{r}\right] = i\dfrac{r_i}{r^3}$

6. $\left[\dfrac{r_i}{r}, \hat{p}_i\right] = i\left(\dfrac{1}{r} - \dfrac{r_i^2}{r^3}\right)$

7. $[\hat{L}_i, \hat{L}_j] = i\hat{L}_k\varepsilon_{ijk}$

8. $[\hat{L}_i, \hat{A}_j] = i\hat{A}_k\varepsilon_{ijk}$

9. $[\hat{A}_i, \hat{A}_j] = -2i\hat{L}_k\hat{H}\varepsilon_{ijk}$

10. $\hat{A}^2 = 2(\hat{L}^2 + 1)\hat{H} + 1$

11. $\hat{A}_z = -\frac{1}{2}(\hat{L}_-\hat{A}_+ + \hat{A}_-\hat{L}_+) - \hat{A}_z\hat{L}_z$

12. $\hat{A}\cdot\hat{L} = 0 = \hat{L}\cdot\hat{A}$

But, all of the operators on the right-hand side are Hermitian operators so we have

$$\hat{A}_x^\dagger = \frac{1}{2}\{(\hat{p}_y\hat{L}_z - \hat{p}_z\hat{L}_y) - (\hat{L}_y\hat{p}_z - \hat{L}_z\hat{p}_y)\} - \frac{x}{r}$$

$$= \hat{A}_x \qquad (6.5)$$

Pauli's properly symmetrized Lenz vector operator does indeed commute with the Hamiltonian. Because of the importance of this relationship we show that $[\hat{A}, \hat{H}] = 0$ in some detail. It is sufficient to show that one component of $\hat{A}$ commutes with the Hamiltonian. In this calculation as well as others we make use of a number of commutator (and other) relations involving the quantum mechanical Lenz vector operator. Some of these relations are given as problems at the end of this chapter. TABLE 6.1 contains some relations that are useful in this chapter. Note that some of these are special cases of the more general commutator relations for vector operators discussed in Chapter 2.

Note that relation 8 in TABLE 6.1 assures us that $\hat{A}$ is a vector operator.

To prove that $[\hat{A}, H] = 0$ it is convenient to use a slightly different form of $\hat{A}$. It can be shown (see Problem 2) that

$$\hat{A} = \left(\frac{1}{2}\right)(\hat{p}\times\hat{L} - \hat{L}\times\hat{p}) - \hat{r}$$

$$= \hat{p}\times\hat{L} - i\hat{p} - \hat{r} \qquad (6.6)$$

We choose to work with the x-component for which

$$[\hat{A}_x, \hat{H}] = [(\hat{p}\times\hat{L})_x, H] - i[\hat{p}_x, H] - \left[\frac{x}{r}, H\right] \qquad (6.7)$$

We now evaluate each of the three terms individually. The first term is

$$
\begin{aligned}
[(\hat{\boldsymbol{p}} \times \hat{\boldsymbol{L}})_x, \hat{H}] &= [(\hat{p}_y \hat{L}_z - \hat{p}_z \hat{L}_y), \hat{H}] \\
&= [\hat{p}_y \hat{L}_z, \hat{H}] - [\hat{p}_z \hat{L}_y, \hat{H}] \\
&= [\hat{p}_y, \hat{H}]\hat{L}_z + \hat{p}_y[\hat{L}_z, \hat{H}] - [\hat{p}_z, \hat{H}]\hat{L}_y - \hat{p}_z[\hat{L}_y, \hat{H}] \\
&= [\hat{p}_y, \hat{H}]\hat{L}_z - [\hat{p}_z, \hat{H}]\hat{L}_y
\end{aligned}
\tag{6.8}
$$

where we have used the fact that the individual components of the angular momentum operator commute with the Hamiltonian for a central potential. For the hydrogen atom

$$
\hat{H} = \frac{1}{2}(\hat{p}_x^2 + \hat{p}_y^2 + \hat{p}_z^2) - \frac{1}{r}
\tag{6.9}
$$

so

$$
\begin{aligned}
[(\hat{\boldsymbol{p}} \times \hat{\boldsymbol{L}})_x, \hat{H}] &= [\hat{p}_y, \hat{H}]\hat{L}_z - [\hat{p}_z, \hat{H}]\hat{L}_y \\
&= \left[\hat{p}_y, \frac{-1}{r}\right]\hat{L}_z - \left[\hat{p}_z, \frac{-1}{r}\right]\hat{L}_y \\
&= -i\frac{y}{r^3}\hat{L}_z + i\frac{z}{r^3}\hat{L}_y \\
&= -i\frac{y}{r^3}(x\hat{p}_y - y\hat{p}_x) + i\frac{z}{r^3}(z\hat{p}_x - x\hat{p}_z) \\
&= -i\frac{xy}{r^3}\hat{p}_y + i\frac{y^2}{r^3}\hat{p}_x + i\frac{z^2}{r^3}\hat{p}_x - i\frac{xz}{r^3}\hat{p}_z
\end{aligned}
\tag{6.10}
$$

Adding and subtracting $i\dfrac{x^2}{r^3}\hat{p}_x$ we have

$$
\begin{aligned}
[(\hat{\boldsymbol{p}} \times \hat{\boldsymbol{L}})_x, \hat{H}] &= -i\frac{xy}{r^3}\hat{p}_y - i\frac{xz}{r^3}\hat{p}_z - i\frac{x^2}{r^3}\hat{p}_x + i\left(\frac{y^2}{r^3}\hat{p}_x + \frac{z^2}{r^3}\hat{p}_x + \frac{x^2}{r^3}\hat{p}_x\right) \\
&= -i\frac{xy}{r^3}\hat{p}_y - i\frac{xz}{r^3}\hat{p}_z - i\frac{x^2}{r^3}\hat{p}_x + \frac{i}{r}\hat{p}_x
\end{aligned}
\tag{6.11}
$$

The second term in the equation for $\lfloor\hat{A}_x, \hat{H}\rfloor$ is relatively simple. It is

$$
\begin{aligned}
i[\hat{p}_x, \hat{H}] &= i\left[\hat{p}_x, \frac{-1}{r}\right] \\
&= -i\left(i\frac{x}{r^3}\right) \\
&= \frac{x}{r^3}
\end{aligned}
\tag{6.12}
$$

The third term is somewhat more complicated. It is

$$
\begin{aligned}
\left[\frac{x}{r}, \hat{H}\right] &= \left[\frac{x}{r}, \left(\frac{1}{2}\right)(\hat{p}_x^2 + \hat{p}_y^2 + \hat{p}_z^2)\right] \\
&= \left(\frac{1}{2}\right)\left\{\left[\frac{x}{r}, \hat{p}_x^2\right] + \left[\frac{x}{r}, \hat{p}_y^2\right] + \left[\frac{x}{r}, \hat{p}_z^2\right]\right\}
\end{aligned}
\tag{6.13}
$$

Now we must evaluate $\left[\dfrac{x}{r}, \hat{p}_x^2\right]$ and $\left[\dfrac{x}{r}, \hat{p}_y^2\right]$; $\left[\dfrac{x}{r}, \hat{p}_z^2\right]$ can be inferred from $\left[\dfrac{x}{r}, \hat{p}_y^2\right]$. To evaluate these commutators we must operate on a function $f = f(x, y, z)$.

$$\left[\frac{x}{r}, \hat{p}_x^2\right] f = \frac{x}{r}\hat{p}_x^2 f - \hat{p}_x^2\left(\frac{x}{r} \cdot f\right)$$

$$= \frac{x}{r}\hat{p}_x^2 f - \hat{p}_x\left(-i\frac{\partial}{\partial x}\right)\left(\frac{x}{r} \cdot f\right)$$

$$= \frac{x}{r}\hat{p}_x^2 f + i\hat{p}_x\left[\frac{x}{r}\frac{\partial f}{\partial x} + \left(\frac{1}{r} - \frac{x^2}{r^3}\right)f\right]$$

$$= \frac{x}{r}\hat{p}_x^2 f + \frac{\partial}{\partial x}\left[\frac{x}{r}\frac{\partial f}{\partial x} + \left(\frac{1}{r} - \frac{x^2}{r^3}\right)f\right]$$

$$= 2 \cdot \left(\frac{1}{r} - \frac{x^2}{r^3}\right)\frac{\partial f}{\partial x} + \left(\frac{3x^3}{r^5} - \frac{3x}{r^3}\right)f$$

$$= 2 \cdot \left(\frac{1}{r} - \frac{x^2}{r^3}\right)(i\hat{p}_x)f + \left(\frac{3x^3}{r^5} - \frac{3x}{r^3}\right)f \qquad (6.14)$$

where use has been made of the relationship $\hat{p}_i \to -i\dfrac{\partial}{\partial r_i}$ so that $\hat{p}_i^2 \to -\dfrac{\partial^2}{\partial r_i^2}$.

Thus, the commutator $\left[\dfrac{x}{r}, \hat{p}_x^2\right]$ is

$$\left[\frac{x}{r}, \hat{p}_x^2\right] = 2 \cdot \left(\frac{1}{r} - \frac{x^2}{r^3}\right)(i\hat{p}_x) + \left(\frac{3x^3}{r^5} - \frac{3x}{r^3}\right) \qquad (6.15)$$

We also require $\left[\dfrac{x}{r}, \hat{p}_y^2\right]$ which is evaluated as follows.

$$\left[\frac{x}{r}, \hat{p}_y^2\right] f = \frac{x}{r}\hat{p}_y^2 f - \hat{p}_y^2\left(\frac{x}{r} \cdot f\right)$$

$$= \frac{x}{r}\hat{p}_y^2 f - x\hat{p}_y\left[\left(-i\frac{\partial}{\partial y}\right)\left(\frac{1}{r} \cdot f\right)\right]$$

$$= \frac{x}{r}\hat{p}_y^2 f + ix\hat{p}_y\left[\frac{1}{r}\frac{\partial f}{\partial y} + \left(\frac{y}{r^3}\right) \cdot f\right]$$

$$= \frac{x}{r}\hat{p}_y^2 f + ix\left(-i\frac{\partial}{\partial y}\right)\left[\frac{1}{r}\frac{\partial f}{\partial y} + \left(\frac{y}{r^3}\right) \cdot f\right]$$

$$= -2\frac{xy}{r^3} \cdot \frac{\partial f}{\partial y} + \left(3\frac{xy^2}{r^5} - \frac{x}{r^3}\right) \cdot f$$

$$= \left[-2\frac{xy}{r^3} \cdot (i\hat{p}_y) + \left(3\frac{xy^2}{r^5} - \frac{x}{r^3}\right)\right] \cdot f \qquad (6.16)$$

so that

$$\left[\frac{x}{r}, \hat{p}_y^2\right] = \left[-2\frac{xy}{r^3} \cdot (i\hat{p}_y) + \left(3\frac{xy^2}{r^5} - \frac{x}{r^3}\right)\right] \qquad (6.17)$$

with the companion expression $\left[\dfrac{x}{r}, \hat{p}_z^2\right]$ obtained by replacing y with z. Finally, the third term in the equation for $\lfloor \hat{A}_x, \hat{H} \rfloor$ becomes

$$\left[\frac{x}{r}, \hat{H}\right] = \left[\frac{x}{r}, \left(\frac{1}{2}\right)(\hat{p}_x^2 + \hat{p}_y^2 + \hat{p}_z^2)\right]$$

$$= \left(\frac{1}{2}\right)\left\{\left[\frac{x}{r}, \hat{p}_x^2\right] + \left[\frac{x}{r}, \hat{p}_y^2\right] + \left[\frac{x}{r}, \hat{p}_z^2\right]\right\}$$

$$= \left(\frac{1}{2}\right)\left[2 \cdot \left(\frac{1}{r} - \frac{x^2}{r^3}\right)(i\,\hat{p}_x) + \left(\frac{3x^3}{r^5} - \frac{3x}{r^3}\right)\right]$$

$$+ \left[-2\frac{xy}{r^3} \cdot (i\,\hat{p}_y) + \left(3\frac{xy^2}{r^5} - \frac{x}{r^3}\right)\right]$$

$$+ \left[-2\frac{xz}{r^3} \cdot (i\,\hat{p}_z) + \left(3\frac{xz^2}{r^5} - \frac{x}{r^3}\right)\right]$$

$$= 2i\left(\frac{1}{r}\hat{p}_x - \frac{x^2}{r^3}\hat{p}_x - \frac{xy}{r^3}\hat{p}_y - \frac{xz}{r^3}\hat{p}_z + i\frac{x}{r^3}\right) \tag{6.18}$$

Combining Equations (6.10), (6.12), and (6.18) we have

$$[\hat{A}_x, \hat{H}] = [(\hat{p} \times \hat{L})_x, H] - i[\hat{p}_x, H] - \left[\frac{x}{r}, H\right]$$

$$= -i\frac{xy}{r^3}\hat{p}_y - i\frac{x^2}{r^3}\hat{p}_x - i\frac{xz}{r^3}\hat{p}_z + \frac{i}{r}\hat{p}_x - \frac{x}{r^3}$$

$$- \left(\frac{1}{2}\right)2i\left(\frac{1}{r}\hat{p}_x - \frac{x^2}{r^3}\hat{p}_x - \frac{xy}{r^3}\hat{p}_y - \frac{xz}{r^3}\hat{p}_z + i\frac{x}{r^3}\right)$$

$$= 0 \tag{6.19}$$

This validates Pauli's form of the quantum mechanical operator corresponding to the Lenz vector. It is not only Hermitian, but it also commutes with the hydrogen atom Hamiltonian. It is shown below that using this operator together with angular momentum and the Hamiltonian, the hydrogenic energy $E_n = -1/2n^2$ can be obtained without recourse to a coordinate system.

6.2. Lenz Vector Ladder Operators; Conversion of a Spherical Eigenfunction into Another Spherical Eigenfunction

The fact that the orbital angular momentum raising and lowering operators operating on one of the $|n\ell m\rangle$ eigenfunctions for any central potential changes only the value of the m quantum number and neither ℓ nor n, is a manifestation of the degeneracy associated with the central nature of the potential. Recall that the angular parts of the spherical coordinate eigenfunctions for any central potential are the spherical harmonics, so this degeneracy is not restricted to the hydrogen atom. It exists for any central potential. We expect, therefore, that for the hydrogen atom with its additional degeneracy in ℓ there should be a similar operator that reflects this additional degeneracy. Furthermore, because the classical constancy of the Lenz vector (and the fact that $[\hat{A}, \hat{H}] = 0$) is the source of the "super" symmetry that leads to the additional degeneracy, we suspect that ladder operators involving the Lenz vector will change ℓ and possibly m, but leave n intact. We now develop the formalism that leads to these operators.

The commutator $[\hat{H}, \hat{A}] = 0$ so that the quantum mechanical operator $\hat{A}$, like its classical counterpart A, is a constant of the motion. But $[\hat{A}_i, \hat{A}_j] \neq 0$. $\hat{A}$ is therefore similar to, but not quite, an angular momentum. One component, usually $\hat{A}_z$, is chosen to be a mutually commuting operator with $\hat{H}$ and $\hat{L}_z$, not $\hat{L}^2$. In this case the "extra symmetry" permits the Schrödinger equation to be separated in parabolic coordinates as well as spherical coordinates.

We form the operators

$$\hat{A}_{\pm} = (\hat{A}_x \pm i\hat{A}_y) \tag{6.20}$$

Using the designation $|n\ell m\rangle$ for spherical hydrogen atom eigenfunctions we show that application of $\hat{A}_+$ raises *certain* spherical hydrogen atom eigenfunctions as follows.

$$\hat{A}_+|n\ell\ell\rangle = D_{\ell\ell}^+|n, (\ell+1), (\ell+1)\rangle \tag{6.21}$$

where $D_{\ell\ell}^{\pm}$ are constants analogous to the $C_{jm}^{\pm}$ of Chapter 2. We also show that

$$\hat{A}_z|n\ell\ell\rangle = \frac{-1}{\sqrt{2(\ell+1)}} D_{\ell\ell}^+|n, (\ell+1), \ell\rangle \tag{6.22}$$

Therefore, given $|n00\rangle$ we can generate the entire set of $|n\ell m\rangle$ by judicious application of $\hat{A}_+$, $\hat{A}_z$, and $\hat{L}_{\pm}$. Surprisingly, the operator $\hat{A}_-$, although lowering the value of m for $|n\ell(-\ell)\rangle$, raises the value of ℓ.

First we must show that $\{\hat{A}_+|n\ell m\rangle\}$ is an eigenfunction of $\hat{L}_z$ and $\hat{L}^2$ with eigenvalues $(m+1)$ and $(\ell+1)(\ell+2)$, respectively. Operating on $\{\hat{A}_+|n\ell m\rangle\}$ with $\hat{L}_z$ gives

$$\hat{L}_z\{\hat{A}_+|n\ell m\rangle\} = (\hat{L}_z\hat{A}_x + i\hat{L}_z\hat{A}_y)|n\ell m\rangle \tag{6.23}$$

But

$$[\hat{L}_z, \hat{A}_x] = i\hat{A}_y \Rightarrow \hat{L}_z\hat{A}_x = \hat{A}_x\hat{L}_z + i\hat{A}_y$$
$$[\hat{L}_z, \hat{A}_y] = -i\hat{A}_x \Rightarrow \hat{L}_z\hat{A}_y = \hat{A}_y\hat{L}_z - i\hat{A}_x \tag{6.24}$$

so

$$\begin{aligned}
\hat{L}_z\{\hat{A}_+|n\ell m\rangle\} &= (\hat{A}_x\hat{L}_z + i\hat{A}_y + i\hat{A}_y\hat{L}_z + \hat{A}_x)|n\ell m\rangle \\
&= \{\hat{A}_x(\hat{L}_z + 1) + i\hat{A}_y(\hat{L}_z + 1)\}|n\ell m\rangle \\
&= (m+1)\{\hat{A}_+|n\ell m\rangle\} \tag{6.25}
\end{aligned}$$

Therefore, $\{\hat{A}_+|n\ell m\rangle\}$ is an eigenfunction of $\hat{L}_z$ with eigenvalue $(m+1)$.

We now show that $\{\hat{A}_+|n\ell\ell\rangle\}$ is an eigenfunction of $\hat{L}^2$ with eigenvalue $(\ell+1)(\ell+2)$.

$$\hat{L}^2\{\hat{A}_+|n\ell m\rangle\} = \hat{L}^2(\hat{A}_x + i\hat{A}_y)|n\ell m\rangle \tag{6.26}$$

The commutators $\lfloor\hat{L}^2, \hat{A}_x\rfloor$ and $\lfloor\hat{L}^2, \hat{A}_y\rfloor$ must be evaluated. They can easily be shown to be

$$[\hat{L}^2, \hat{A}_x] = -i\hat{A}_z\hat{L}_y - i\hat{L}_y\hat{A}_z + i\hat{A}_y\hat{L}_z + i\hat{L}_z\hat{A}_y$$
$$[\hat{L}^2, \hat{A}_y] = i\hat{A}_z\hat{L}_x + i\hat{L}_x\hat{A}_z - i\hat{A}_x\hat{L}_z - i\hat{L}_z\hat{A}_x \tag{6.27}$$

so that

$$[\hat{L}^2, \hat{A}_+] = -\hat{A}_z\hat{L}_+ - \hat{L}_+\hat{A}_z + \hat{A}_+\hat{L}_z + \hat{L}_z\hat{A}_+ \tag{6.28}$$

Applying

$$[\hat{L}_z, \hat{A}_+] = \hat{A}_+ \tag{6.29}$$

and

$$[\hat{L}_+, \hat{A}_z] = -\hat{A}_+ \tag{6.30}$$

we find that

$$[\hat{L}^2, \hat{A}_+] = -2\hat{A}_z\hat{L}_+ + 2\hat{A}_+ + 2\hat{A}_+\hat{L}_z \tag{6.31}$$

so that

$$\begin{aligned}
\hat{L}^2\{\hat{A}_+|n\ell m\rangle\} &= (\hat{A}_+\hat{L}^2 - 2\hat{A}_z\hat{L}_+ + 2\hat{A}_+ + 2\hat{A}_+\hat{L}_z)|n\ell m\rangle \\
&= \ell(\ell+1)\{\hat{A}_+|n\ell m\rangle\} \\
&\quad - 2\hat{A}_z\sqrt{\ell(\ell+1) - m(m+1)}|n\ell\,(m+1)\rangle \\
&\quad + (2\hat{A}_+ + 2\hat{A}_+ m)|n\ell m\rangle
\end{aligned} \tag{6.32}$$

This does not appear to be helpful because the $|n\ell m\rangle$ are not eigenfunctions of $\hat{A}_z$. If, however, $m = \ell$ then the $\hat{A}_z$ term vanishes and we have

$$\begin{aligned}
\hat{L}^2\{\hat{A}_+|n\ell\ell\rangle\} &= (\ell(\ell+1) + 2 + 2\ell)\{\hat{A}_+|n\ell\ell\rangle\} \\
&= (\ell+1)(\ell+2)\{\hat{A}_+|n\ell\ell\rangle\}
\end{aligned} \tag{6.33}$$

Therefore, if $m = \ell$ then $\{\hat{A}_+|n\ell\ell\rangle\}$ is an eigenfunction of $\hat{L}^2$ with eigenvalue $(\ell+1)(\ell+2)$.

To obtain a complete expression for the action of $\hat{A}_+$ on a spherical eigenfunction $|n\ell\ell\rangle$ it is necessary to evaluate the constant $D_{\ell\ell}^+$. We begin by following a procedure similar to that used to evaluate the $C_{jm}^{\pm}$ in Chapter 2, the constants that result from the action of $\hat{L}_{\pm}$ on any spherical eigenfunction for a central potential. We examine the quantity

$$\langle n\ell\ell|\hat{A}_-\hat{A}_+|n\ell\ell\rangle = (D_{\ell\ell}^+)^* D_{\ell\ell}^+ \tag{6.34}$$

where the $(D_{\ell\ell}^+)^*$ results from operation to the left with $\hat{A}_-$ and the fact that the Hermitian conjugate of $\hat{A}_-$ is $\hat{A}_+$. As we did for the $C_{jm}^{\pm}$, for convenience we specify that $D_{\ell\ell}^+$ is real. Expanding the operator $\hat{A}_-\hat{A}_+$ we have

$$\hat{A}_-\hat{A}_+ = \hat{A}^2 - \hat{A}_z^2 + 2\hat{L}_z\hat{H}$$
$$= (2\hat{L}^2\hat{H} + 2\hat{H} + 1) - \hat{A}_z^2 + 2\hat{L}_z\hat{H} \tag{6.35}$$

where we have used relation number 10 of TABLE 6.1. The $|n\ell m\rangle$ are eigenfunctions of all operators on the right-hand side except $\hat{A}_z$. Using

$$\hat{L}^2|n\ell\ell\rangle = \ell(\ell+1)|n\ell\ell\rangle$$
$$\hat{L}_z|n\ell\ell\rangle = \ell|n\ell\ell\rangle$$
$$\hat{H}|n\ell\ell\rangle = \left(-\frac{1}{2n^2}\right)|n\ell\ell\rangle \tag{6.36}$$

we obtain

$$(D_{\ell\ell}^+)^2 = 2\ell(\ell+1)\left(\frac{-1}{2n^2}\right) + 2\left(\frac{-1}{2n^2}\right) + \frac{n^2}{n^2}$$
$$- 2\ell\left(\frac{-1}{2n^2}\right) - \langle n\ell\ell|\hat{A}_z^2|n\ell\ell\rangle$$
$$= \left\{\frac{n^2 - (\ell+1)^2}{n^2}\right\} - \langle n\ell\ell|\hat{A}_z^2|n\ell\ell\rangle \tag{6.37}$$

It is at this point that the derivation of $D_{\ell\ell}^+$ becomes slightly more difficult than the analogous derivation for the $C_{jm}^{\pm}$ because the $|n\ell m\rangle$ are not eigenfunctions of $\hat{A}_z$. Thus, evaluation of the matrix element $\langle n\ell\ell|\hat{A}_z^2|n\ell\ell\rangle$ is not straightforward.

To evaluate the matrix element $\langle n\ell\ell|\hat{A}_z^2|n\ell\ell\rangle$ we operate on $|n\ell\ell\rangle$ with $\hat{A}_z$ and use a relation number 11 from TABLE 6.1. We obtain

$$\hat{A}_z|n\ell\ell\rangle = -\left[\frac{1}{2}\hat{L}_-\hat{A}_+ - \frac{1}{2}\hat{A}_-\hat{L}_+ + \hat{A}_z\hat{L}_z\right]|n\ell\ell\rangle$$
$$= -\frac{1}{2}D_{\ell\ell}^+\sqrt{2(\ell+1)}|n(\ell+1)\ell\rangle - \ell\hat{A}_z|n\ell\ell\rangle, \tag{6.38}$$

Solving for $\hat{A}_z|n\ell\ell\rangle$ we have

$$\hat{A}_z|n\ell\ell\rangle = -\frac{1}{\sqrt{2(\ell+1)}}D_{\ell\ell}^+|n,(\ell+1),\ell\rangle \tag{6.39}$$

Clearly, $\hat{A}_z$ acting on $|n\ell\ell\rangle$ transforms it into another eigenfunction. In particular, it transforms it into $|n,(\ell+1),\ell\rangle$. In fact, $\hat{A}_z$ operating on $|n\ell\ell\rangle$ is itself a raising operator for ℓ, but not for m.

Because we now have an expression for $\hat{A}_z|n\ell\ell\rangle$ in terms of $D_{\ell\ell}^+$ we may evaluate the matrix element $\langle n\ell\ell|\hat{A}_z^2|n\ell\ell\rangle$ in terms of $(D_{\ell\ell}^+)^2$. It is

$$\langle n\ell\ell|\hat{A}_z^2|n\ell\ell\rangle = \frac{1}{2(\ell+1)}(D_{\ell\ell}^+)^2 \tag{6.40}$$

Inserting Equation (6.40) in Equation (6.37) and solving for $D_{\ell\ell}^+$ we have

$$D_{\ell\ell}^+ = -\frac{1}{n}\sqrt{\frac{2(\ell+1)}{(2\ell+3)}}[n^2 - (\ell+1)^2] \tag{6.41}$$

The final result for $\hat{A}_+$ is

$$\hat{A}_+|n\ell\ell\rangle = -\frac{1}{n}\sqrt{\frac{2(\ell+1)}{(2\ell+3)}}[n^2 - (\ell+1)^2]|n, (\ell+1), (\ell+1)\rangle \tag{6.42}$$

and for $\hat{A}_z$

$$\hat{A}_z|n\ell\ell\rangle = \frac{1}{n}\sqrt{\frac{[n^2 - (\ell+1)^2]}{(2\ell+3)}}|n(\ell+1)\ell\rangle \tag{6.43}$$

The action of $\hat{A}_-$ is similar to that of $\hat{A}_+$ with one variation; although application to $|n; \ell; -\ell\rangle$ lowers the $\hat{L}_z$ quantum number $-\ell$, to $-(\ell+1)$, it raises the angular momentum quantum number to $(\ell+1)$ as did $\hat{A}_+$. Following the same method employed for $\{\hat{A}_+|n\ell m\rangle\}$ it is found that $\{\hat{A}_-|n\ell m\rangle\}$ is an eigenfunction of $\hat{L}_z$ with eigenvalue $(m-1)$. We find, however, that $\hat{L}^2\{\hat{A}_-|n\ell m\rangle\}$ is

$$\hat{L}^2\{\hat{A}_-|n\ell m\rangle\} = \{\ell(\ell+1) - 2(m-1)\}\{\hat{A}_-|n\ell m\rangle\}$$
$$+ 2\hat{A}_z\sqrt{(\ell+m)(\ell-m+1)}|n, \ell, (m-1)\rangle \tag{6.44}$$

so that $\{\hat{A}_-|n\ell m\rangle\}$ is an eigenfunction of $\hat{L}^2$ with eigenvalue $(\ell+1)(\ell+2)$ only if $m = -\ell$ because $m = \ell+1$ is not permitted. Thus

$$\hat{A}_-|n, \ell, (-\ell)\rangle = D_{\ell-\ell}^-|n, (\ell+1), -(\ell+1)\rangle \tag{6.45}$$

Now, $D_{\ell-\ell}^-$ can be evaluated in the way that $D_{\ell\ell}^+$ was evaluated. We obtain

$$D_{\ell-\ell}^- = \frac{1}{n}\sqrt{\frac{2(\ell+1)}{(2\ell+3)}}[n^2 - (\ell+1)^2] \tag{6.46}$$

The sign of $D_{\ell-\ell}^-$ was chosen to be consistent with convention. Thus, using a combination of $\hat{A}_\pm$, $\hat{A}_z$, and $\hat{L}_\pm$, the complete set of spherical eigenfunctions for a given value of n can be generated provided $|n00\rangle$ is known. More important, however, is the fact that the $\hat{A}_\pm$ operating on certain hydrogen atom eigenfunctions transform them into others without changing the value of n. This means that the eigenfunctions are transformed into eigenfunctions having different values of ℓ without changing the energy, a result of the accidental degeneracy of the hydrogen atom.

6.3. Application of Lenz Vector Ladder Operators to a General Spherical Eigenfunction

We have exploited the properties of the Lenz vector operators to produce state-to-state conversions of spherical eigenfunctions. That is, we have obtained the conditions under which one eigenfunction is transformed into another eigenfunction as was done in Chapter 2 for generalized angular momentum ladder operators. These state-to-state conversions are, however, possible only for special initial eigenfunctions, in particular $|n\ \ell(\pm\ell)\rangle$. Application of the Lenz vector operators to a general spherical eigenfunction $|n\ell m\rangle$ produces a linear combination of spherical eigenfunctions. Using the results from Section 7.2 for the special case we can find the effects of application of $\hat{A}_{\pm}$ and $\hat{A}_z$ to a general spherical eigenfunction.

We begin by determining the effect of operating on $|n\ell m\rangle$ with $\hat{A}_z$. For this task it is convenient to recall that the spherical eigenfunctions of the hydrogen atom are

$$\psi_{n\ell m}(r,\theta,\phi) = Y_{\ell m}(\theta,\phi)R_{n\ell}(r) \tag{6.47}$$

Therefore, the inner product $\langle n'\ell'm'|\hat{A}_z|n\ell m\rangle$ may be written as

$$\langle n'\ell'm'|\hat{A}_z|n\ell m\rangle = \int_{\substack{\text{all} \\ \text{space}}} \psi^*_{n'\ell'm'}(r,\theta,\phi)|A|\cos\theta\,\psi_{n\ell m}(r,\theta,\phi)r^2 dr d\Omega$$

$$= \int_0^\infty R^*_{n'\ell'}(r)|A|R_{n\ell}(r)r^2 dr \int_\Omega Y^*_{\ell'm'}(\theta,\phi)\cos\theta\,Y_{\ell m}(\theta,\phi)d\Omega$$

$$= \langle n'\ell'\|\hat{A}\|n\ell\rangle \int_\Omega Y^*_{\ell'm'}(\theta,\phi)\cos\theta\,Y_{\ell m}(\theta,\phi)d\Omega \tag{6.48}$$

where we have defined

$$\langle n'\ell'\|\hat{A}\|n\ell\rangle = \int_0^\infty R^*_{n'\ell'}(r)|A|R_{n\ell}(r)r^2 dr \tag{6.49}$$

as the "radial matrix element" because it does not depend on m. It should be noted that this radial matrix element is similar to the "reduced matrix element" used in the formulation of the Wigner–Eckart theorem, but differs in numerical factors that depend on ℓ.

A well-known relation between spherical harmonics, Equation (2.116), is

$$\cos\theta\,Y_{\ell m}(\theta,\phi) = \sqrt{\frac{(\ell+m+1)(\ell-m+1)}{(2\ell+1)(2\ell+3)}}Y_{(\ell+1)m}(\theta,\phi)$$

$$+ \sqrt{\frac{(\ell+m)(\ell-m)}{(2\ell+1)(2\ell-1)}}Y_{(\ell-1)m}(\theta,\phi) \tag{6.50}$$

so, using the orthogonality relation for spherical harmonics, the angular integral in Equation (6.48) becomes

$$
\int_\Omega Y_{\ell'm'}^*(\theta, \phi) \cos\theta\, Y_{\ell m}(\theta, \phi) d\Omega = \sqrt{\frac{(\ell + m + 1)(\ell - m + 1)}{(2\ell + 1)(2\ell + 3)}} \delta_{m'm}\delta_{\ell'(\ell+1)}
$$

$$
+ \sqrt{\frac{(\ell + m)(\ell - m)}{(2\ell + 1)(2\ell - 1)}} \delta_{m'm}\delta_{\ell'(\ell-1)} \qquad (6.51)
$$

from which we deduce that the action of $\hat{A}_z$ on $|n\ell m\rangle$ is

$$
\hat{A}_z|n\ell m\rangle = \langle n\,(\ell + 1)\|\hat{A}\|n\,\ell\rangle \sqrt{\frac{(\ell + m + 1)(\ell - m + 1)}{(2\ell + 1)(2\ell + 3)}} |n,\,(\ell + 1),\,m\rangle
$$

$$
+ \langle n\,(\ell - 1)\|\hat{A}\|n\,\ell\rangle \sqrt{\frac{(\ell + m)(\ell - m)}{(2\ell + 1)(2\ell - 1)}} |n,\,(\ell - 1),\,m\rangle \qquad (6.52)
$$

Because the radial matrix element is independent of m we may obtain it by using the result for the $|n\ell\ell\rangle$. From Equation (6.52) the matrix element for $m = \ell$ is

$$
\langle n'\ell'm'|\hat{A}_z|n\ell\ell\rangle = \langle n'\ell'\|\hat{A}\|n\ell\rangle \sqrt{\frac{(2\ell + 1)(1)}{(2\ell + 1)(2\ell + 3)}} \delta_{m'\ell}\delta_{\ell',\ell+1}\delta_{n'n}
$$

$$
= \langle n'\ell'\|\hat{A}\|n\ell\rangle \sqrt{\frac{1}{(2\ell + 3)}} \delta_{m'\ell}\delta_{\ell',\ell+1}\delta_{n'n} \qquad (6.53)
$$

Therefore, this matrix element is nonzero only if $\ell' = \ell + 1$, $n' = n$, and $m' = \ell$. We have then

$$
\langle n\,(\ell + 1)\,\ell|\hat{A}_z|n\ell\ell\rangle = \langle n(\ell + 1)\|\hat{A}\|n\ell\rangle \sqrt{\frac{1}{(2\ell + 3)}} \qquad (6.54)
$$

From Equation (6.43) we know that

$$
\hat{A}_z|n\ell\ell\rangle = \frac{1}{n}\sqrt{\frac{[n^2 - (\ell + 1)^2]}{(2\ell + 3)}} |n,\,(\ell + 1),\,\ell\rangle \qquad (6.55)
$$

from which the inner product $\langle n\,(\ell + 1)\,\ell|\hat{A}_z|n\ell\ell\rangle$ is

$$
\langle n\,(\ell + 1)\,\ell|\hat{A}_z|n\ell\ell\rangle = \frac{1}{n}\sqrt{\frac{[n^2 - (\ell + 1)^2]}{(2\ell + 3)}} \qquad (6.56)
$$

Equating the two expressions for $\langle n\,(\ell + 1)\,\ell|\hat{A}_z|n\ell\ell\rangle$, Equations (6.54) and (6.56), we obtain the radial matrix element

$$
\langle n(\ell + 1)\|\hat{A}\|n\ell\rangle = \frac{1}{n}\sqrt{[n^2 - (\ell + 1)^2]}
$$

$$
= \sqrt{1 - (\ell + 1)^2/n^2} \qquad (6.57)
$$

The radial matrix element $\langle n(\ell - 1)\|\hat{A}\|n\ell \rangle$ may be obtained from Equation (6.57) by letting $\ell \to \ell - 1$

$$\langle n(\ell - 1)\|\hat{A}\|n\ell \rangle = \frac{1}{n}\sqrt{n^2 - \ell^2}$$

$$= \sqrt{1 - \ell^2/n^2} \tag{6.58}$$

Inserting Equations (6.57) and (6.58) into Equation (6.52) we arrive at the final result

$$\hat{A}_z|n\ell m\rangle = \frac{1}{n}\sqrt{\frac{(\ell - m + 1)(\ell + m + 1)\left[n^2 - (\ell + 1)^2\right]}{(2\ell + 1)(2\ell + 3)}}\,|n\,(\ell + 1)\,m\rangle$$

$$+ \frac{1}{n}\sqrt{\frac{(\ell - m)(\ell + m)(n^2 - \ell^2)}{(2\ell - 1)(2\ell + 1)}}\,|n\,(\ell - 1)\,m\rangle \tag{6.59}$$

Because it was found (in Chapter 5) that the magnitude of the classical Lenz vector is equal to the eccentricity of the Keplerian orbit it is interesting to compare the radial matrix elements of the magnitude of $|\hat{A}|$, Equations (6.57) and (6.58), with the expression for the eccentricity in atomic units $\varepsilon = \sqrt{1 - \ell^2/n^2}$. This is a further example of correlations that exist between classical and quantal quantities for the Kepler/Coulomb problem.

To determine the actions of $\hat{A}_\pm$ on $|n\ell m\rangle$ we use the fact that $\hat{A}$ is a vector operator and employ the last of the commutation relations in Table 2.2, $\left[\hat{A}_z, \hat{L}_\pm\right] = \pm\hat{A}_\pm$. Because we know the action of $\hat{A}_z$ on $|n\ell m\rangle$, Equation (6.59), and the action of the $\hat{L}_\pm$ on $|n\ell m\rangle$, Equations (2.61) and (2.62), it is simply a matter of applying these operators to $|n\ell m\rangle$ in the correct order (Problem 6.5). We obtain

$$\hat{A}_\pm|n\ell m\rangle = \mp\frac{1}{n}\sqrt{\frac{\left[n^2 - (\ell + 1)^2\right](\ell \pm m + 1)(\ell \pm m + 2)}{(2\ell + 1)(2\ell + 3)}}\,|n(\ell + 1)(m \pm 1)\rangle$$

$$\pm \frac{1}{n}\sqrt{\frac{\left[n^2 - \ell^2\right](\ell \mp m)(\ell \mp m - 1)}{(2\ell - 1)(2\ell + 1)}}\,|n(\ell - 1)(m \pm 1)\rangle \tag{6.60}$$

These general results for application of the $\hat{A}_\pm$ and $\hat{A}_z$ to spherical hydrogen atom eigenfunctions can also be obtained using angular momentum algebra and the Wigner–Eckart theorem.[2] In this book we choose to use the formalism developed here for that purpose and avoid the use of the Wigner–Eckart theorem.

6.4. A New Set of Angular Momentum Operators

Although the Lenz vector operator is not an angular momentum, it is possible to construct angular momentum operators by scaling it and forming linear combinations of this scaled Lenz vector with the orbital angular momentum. Using these operators, the energy eigenvalues for the hydrogen atom can be obtained without solving the Schrödinger equation. It is, of course, necessary to solve the Schrödinger equation to obtain the wave functions, but they depend upon the particular coordinate system chosen.

We begin by defining a new operator $\hat{A}'$

$$\sqrt{-2\hat{H}}\,\hat{A}' = \hat{A} \tag{6.61}$$

If this operator is to operate only on eigenfunctions of the hydrogen atom $\hat{A}'$ may be written

$$\hat{A}' = \sqrt{\frac{-1}{2E}}\,\hat{A} \tag{6.62}$$

where E is the energy eigenvalue (a negative number for bound states). We assume that E is, at this point, unknown. In fact, we do not even assume that it is quantized so we omit, temporarily, any subscripts (such as the quantum number n).

In terms of $\hat{A}'$ the commutators become (in a.u.)

$$[\hat{L}_i, \hat{A}'_j] = i\hat{A}'_k$$
$$[\hat{A}'_i, \hat{A}'_j] = i\hat{L}_k \tag{6.63}$$

Now define two new operators $\hat{I}$ and $\hat{K}$ as follows.

$$\hat{I} = \left(\frac{1}{2}\right)(\hat{L} + \hat{A}')$$

$$\hat{K} = \left(\frac{1}{2}\right)(\hat{L} - \hat{A}') \tag{6.64}$$

Using the commutation relations of Equation (6.63) we find that

$$[\hat{I}, \hat{K}] = 0$$
$$[\hat{I}_i, \hat{I}_j] = i\varepsilon_{ijk}\hat{I}_k$$
$$[\hat{K}_i, \hat{K}_j] = i\varepsilon_{ijk}\hat{K}_k$$
$$[\hat{I}, \hat{H}] = 0 = [\hat{K}, \hat{H}] \tag{6.65}$$

Amazingly, the components of $\hat{I}$ and $\hat{K}$ each obey the commutation rule that is the very definition of "angular momentum"! They therefore both qualify as angular momenta and we immediately know that the possible eigenvalues of the squares of $\hat{I}$ and $\hat{K}$ are given by the equations

$$\hat{I}^2|i\rangle = i(i+1)|i\rangle \quad \text{and} \quad \hat{K}^2|k\rangle = k(k+1)|k\rangle \tag{6.66}$$

where $|i\rangle$ and $|k\rangle$ are the eigenfunctions and i and k are quantum numbers that, in accordance with our previous finding, can take on only the values

$$i, k = 0, \frac{1}{2}, 1, \frac{3}{2}, 2, \frac{5}{2}, 3, \frac{7}{2}, \ldots \tag{6.67}$$

Moreover, because $\hat{I}$ and $\hat{K}$ commute with the Hamiltonian as well as each other it is possible to find a set of eigenfunctions that is common to these three operators. We may write these eigenfunctions symbolically as

$$|i, m_i; k, m_k\rangle \tag{6.68}$$

where m_i and m_k are the eigenvalues of $\hat{I}_z$ and $\hat{K}_z$, respectively. Because $\hat{I}$ and $\hat{K}$ are angular momenta we know also that the ladder operators $\hat{I}_\pm = \hat{I}_x \pm i\hat{I}_y$ and $\hat{K}_\pm = \hat{K}_x \pm i\hat{K}_y$ will have the same effect on the quantum numbers m_i and m_k as was derived in Chapter 2 for a general angular momentum $\hat{J}$ having quantum number m as the eigenvalue of $\hat{J}_z$.

6.5. Energy Eigenvalues

To obtain the energy eigenvalues we first square $\hat{I}$ and $\hat{K}$ to obtain

$$\hat{I}^2 = \left(\frac{1}{4}\right)\left(\hat{L}^2 + \hat{L} \cdot \hat{A}' + \hat{A}' \cdot \hat{L} + \hat{A}'^2\right) \tag{6.69}$$

and

$$\hat{K}^2 = \left(\frac{1}{4}\right)\left(\hat{L}^2 - \hat{L} \cdot \hat{A}' - \hat{A}' \cdot \hat{L} + \hat{A}'^2\right) \tag{6.70}$$

so that

$$\hat{I}^2 = \left(\frac{1}{4}\right)\left(\hat{L}^2 + \hat{A}'^2\right)$$

$$= \hat{K}^2 \tag{6.71}$$

because the $\hat{L} \cdot \hat{A}'$ and $\hat{A}' \cdot \hat{L}$ terms vanish (see relation 12 of TABLE 6.1). Therefore, the quantum numbers corresponding to $\hat{I}^2$ and $\hat{K}^2$ are the same; that is, $i = k$. Because the quantum numbers corresponding to the absolute value of an angular momentum can take on all positive integral and half-integral values, Equation (6.67), we have

$$|\hat{I}^2| + |\hat{J}^2| = i(i+1) + k(k+1)$$

$$= 2k(k+1); \quad k = 0, \frac{1}{2}, 1, \frac{3}{2}, 2, \ldots \tag{6.72}$$

From relation 10 of TABLE 6.1 the relationship among $\hat{A}^2$, $\hat{L}^2$, and $\hat{H}$ may be cast in terms of the scaled Lenz vector $\hat{A}'$ yielding

$$-2\hat{H}\hat{A}'^2 = 2\hat{H}(\hat{L}^2 + 1) + 1 \tag{6.73}$$

which may be rewritten as

$$\{1 + 2\hat{H}(\hat{A}'^2 + \hat{L}^2) + 2\hat{H}\} = 0 \tag{6.74}$$

But

$$(\hat{A}'^2 + \hat{L}^2) = 4\hat{K}^2$$
$$= 4\hat{I}^2 \tag{6.75}$$

so

$$\{1 + 2\hat{H}(4\hat{K}^2) + 2\hat{H}\} = 0 \tag{6.76}$$

Applying this operator to a function that is simultaneously an eigenfunction of $\hat{I}^2$, $\hat{K}^2$, and $\hat{H}$, which we designate by $|\Phi\rangle$, we obtain

$$\{1 + 2\hat{H}(4\hat{K}^2) + 2\hat{H}\}|\Phi\rangle = 0$$

Because $|\Phi\rangle$ is an eigenfunction of all operators in the bracket we have

$$\{1 + 2\hat{H}(4\hat{K}^2) + 2\hat{H}\}|\Phi\rangle = \{1 + 2E[4k(k+1)] + 2E\}|\Phi\rangle$$
$$= 0 \tag{6.77}$$

which implies that

$$\{1 + 2E[4k(k+1)] + 2E\} = 0 \tag{6.78}$$

Solving for E we have

$$E = -\left(\frac{1}{2}\right)\frac{1}{4k(k+1)+1}$$

$$= -\left(\frac{1}{2}\right)\frac{1}{4k^2 + 4k + 1}$$

$$= -\left(\frac{1}{2}\right)\frac{1}{(2k+1)^2} \tag{6.79}$$

Now, from Equation (6.67) we know that $(2k + 1)$ must be an integer because the eigenvalues of the square of an angular momentum (such as k) are 0, 1/2, 1, 3/2, . . . Therefore,

$$k = 0 \rightarrow 2k + 1 = 1$$
$$k = \frac{1}{2} \rightarrow 2k + 1 = 2$$
$$k = 1 \rightarrow 2k + 1 = 3$$
$$k = \frac{3}{2} \rightarrow 2k + 1 = 4$$
$$\text{etc.} \tag{6.80}$$

and we may let $2k + 1 = n$ where $n = 1, 2, 3, \ldots$ (which is, of course, the principal quantum number). Attaching the subscript n to the energy we find that the eigenvalue for the nth state is just the Bohr energy in a.u.

$$E_n = -\frac{1}{2n^2} \qquad n = 1, 2, 3, \ldots \tag{6.81}$$

Note that this derivation of the hydrogen atom energy eigenvalues was performed using only operator methods. We can deduce information about the quantum number ℓ by noting that, from the definition of $\hat{I}$ and $\hat{K}$,

$$\hat{L} = \hat{I} + \hat{K}$$

(which is reminiscent of $\hat{J} = \hat{J}_1 + \hat{J}_2$). The values of ℓ are thus restricted to

$$\ell = (i + k),\ (i + k - 1),\ (i + k - 2),\ \ldots |i - k|$$

But, $i = k$ so $(i + k) = 2k = n - 1$ and $|i - k|$ has minimum value of zero because of the absolute value. Furthermore, $\ell_{max} = (i + k) = n - 1$ as before. Thus, we have

$$\ell = (n - 1),\ (n - 2),\ (n - 3),\ \ldots,\ 0 \tag{6.82}$$

We see then that the operator solution for the energy eigenvalues gives the correct restrictions on ℓ, positive integral values in the range $0 < \ell < (n - 1)$.

The degree of degeneracy of a particular energy eigenvalue can also be computed. Again using the fact that $\hat{I}$ and $\hat{K}$ are angular momenta, $\hat{I}_z$ and $\hat{K}_z$ each have $(2k + 1)$ values (recall that $i = k$). Therefore, there are $(2k + 1) \cdot (2k + 1)$ states. But $(2k + 1) = n$ so, excluding spin, there are n^2 states that have the same energy eigenvalue.

6.6. Relations Between the Parabolic Quantum Numbers

It was remarked in Chapter 5 that separation of the Schrödinger equation in parabolic coordinates is equivalent to employing the mutually commuting operators[3,4] $\hat{H}$, $\hat{L}_z$, and $\hat{A}_z$. Although we did not explicitly use these operators, the technique employed to determine the energy eigenvalues in this chapter is, in fact, equivalent to separating the Schrödinger equation in parabolic coordinates. From their definitions, Equation (6.64), we may consider the commuting operators to be $\hat{I}_z$ and $\hat{K}_z$ and either $\hat{H}$ or $\hat{I}^2$. The relation among the Hamiltonian and the squares of $\hat{I}$ and $\hat{K}$, together with the fact that $\hat{I}^2 = \hat{K}^2$ means that either $\hat{I}^2$ or $\hat{H}$ will suffice. Although $\hat{I}$ and $\hat{K}$ are Hermitian and are "angular momenta", they do not correspond to "observables" in the sense that their eigenvalues are not recognizable physical quantities. Of course, they are constructions made up of the "true" angular momentum and the Lenz vector, both of which correspond to "legitimate" observables.

To show that this procedure is equivalent to separating the Schrödinger equation in parabolic coordinates we find the relationship between the two sets of quantum numbers $(i, m_i; k, m_k)$ and the usual parabolic quantum numbers (n, n_1, n_2, m). We show that the energy eigenkets $|i, m_i; k, m_k\rangle$ are disguised versions of the parabolic kets $|n_1 n_2 m\rangle$. We also show that these kets are also eigenkets of $\hat{L}_z$ and $\hat{A}_z$, thus establishing that they are indeed the parabolic eigenkets.

To obtain the relationship among the $(i, m_i; k, m_k)$ quantum numbers and the more commonly used (n, n_1, n_2, m) we use results from the derivation of the energy eigenvalues, $i = k = (n - 1)/2$. We write $\hat{L}_z$ and $\hat{A}'_z$ in terms of $\hat{I}_z$ and $\hat{K}_z$

$$\hat{L}_z = \hat{I}_z + \hat{K}_z$$

$$\hat{A}'_z = \hat{I}_z - \hat{K}_z \tag{6.83}$$

Therefore,

$$\hat{L}_z \, |i, m_i; k, m_k\rangle = \left(\hat{I}_z + \hat{K}_z\right) |i, m_i; k, m_k\rangle$$

$$= (m_i + m_k) \, |i, m_i; k, m_k\rangle \tag{6.84}$$

But $|n_1 \, n_2 \, m\rangle$ is an eigenfunction of $\hat{L}_z$.

$$\hat{L}_z \, |n_1 \, n_2 \, m\rangle = m \, |n_1 \, n_2 \, m\rangle \tag{6.85}$$

so we must have

$$m = m_i + m_k \tag{6.86}$$

which demonstrates that $|i, m_i; k, m_k\rangle$ and $|n_1 \, n_2 \, m\rangle$ are different notations for the eigenfunctions of $\hat{L}_z$. Because the $|n_1 \, n_2 \, m\rangle$ are known to be parabolic eigenfunctions, so too must the $|i, m_i; k, m_k\rangle$ be parabolic eigenfunctions.

Noting the symmetry between n_1 and n_2 and between m_i and m_k we write, using the relationship between the usual parabolic quantum numbers

$$n = 2n_1 + 2m_i + 1 \tag{6.87}$$

where we have considered only positive values of m. The eigenfunctions for negative values are merely their complex conjugates. From this equation we obtain

$$n_1 = \frac{(n - 1)}{2} - m_i$$

$$n_2 = \frac{(n - 1)}{2} - m_k \tag{6.88}$$

which also gives the proper limits on n_1 and n_2. Because $-i \le m_i \le i$ and $2i + 1 = n$

$$-\frac{(n - 1)}{2} \le \left\{ \frac{(n - 1)}{2} - n_1 \right\} \le \frac{(n - 1)}{2} \tag{6.89}$$

or

$$0 \le n_1 \le (n - 1) \tag{6.90}$$

Similarly,

$$0 \le n_2 \le (n - 1) \tag{6.91}$$

To find the eigenvalues of $\hat{A}'_z$ (and hence $\hat{A}_z$) we apply it to $|i, m_i; k, m_k\rangle$

$$\hat{A}'_z \, |i, m_i; k, m_k\rangle = \left(\hat{I}_z - \hat{K}_z\right) |i, m_i; k, m_k\rangle$$

$$= (m_i - m_k) \, |i, m_i; k, m_k\rangle$$

$$= (n_2 - n_1) \, |i, m_i; k, m_k\rangle \tag{6.92}$$

where m_i and m_k have been replaced in accord with Equation (6.88). Because $\hat{A}'_z = n\hat{A}_z$ it is clear that the parabolic eigenfunctions are indeed eigenfunctions of $\hat{A}_z$; moreover the eigenvalues of $\hat{A}_z$ are $(n_2 - n_1)/n$.

Knowledge of the relationship between the $|i, m_i; k, m_k\rangle$ and the $|n_1 n_2 m\rangle$ makes it possible to use ladder operators to convert one parabolic eigenstate into another. We are interested in conversion of those states that are designated by the commonly used quantum numbers (n, n_1, n_2, m).

First, apply $\hat{I}_-$ to $|i, m_i; k, m_k\rangle$

$$\hat{I}_- |i, m_i; k, m_k\rangle = \sqrt{i(i+1) - m_i(m_i - 1)} \, |i, (m_i - 1); k, m_k\rangle \quad (6.93)$$

From the relationships between the m_i and n_1 and between m_k and n_2, Equation (6.88), it is seen that lowering m_i (or m_k) actually raises n_1 (or n_2). The action of $\hat{I}_-$ on $|n_1 n_2 m\rangle$ in terms of n, n_1, and n_2 is then

$$\hat{I}_- |n_1 n_2 m\rangle = \sqrt{\frac{(n-1)}{2}\frac{(n+1)}{2} - \left[\frac{(n-1)}{2} - n_1\right]\left[\frac{(n-3)}{2} - n_1\right]}$$
$$\times |(n_1 + 1)n_2(m-1)\rangle$$
$$= \sqrt{(n_1 + 1)[n - (n_1 + 1)]} \, |(n_1 + 1)n_2(m-1)\rangle \quad (6.94)$$

By symmetry we have

$$\hat{K}_- |n_1 n_2 m\rangle = \sqrt{(n_2 + 1)[n - (n_2 + 1)]} |n_1 (n_2 + 1)(m-1)\rangle \quad (6.95)$$

If n_1 or n_2 has its maximum value $(n-1)$, then application of $\hat{I}_-$ or $\hat{K}_-$ yields zero as it must.

To lower n_1 or n_2 it is necessary to raise m_i or m_k. Applying $\hat{I}_+$ and $\hat{K}_+$ we find

$$\hat{I}_+ |n_1 n_2 m\rangle = \sqrt{n_1(n - n_1)} |(n_1 - 1)n_2(m+1)\rangle \quad (6.96)$$

and

$$\hat{K}_+ |n_1 n_2 m\rangle = \sqrt{n_2(n - n_2)} |n_1(n_2 - 1)(m+1)\rangle \quad (6.97)$$

It is seen that judicious application of these operators can produce the entire manifold of parabolic eigenstates for a given n from knowledge of only one of the eigenstates.

6.7. Relationship Between the Spherical and Parabolic Eigenfunctions

Inasmuch as the spherical eigenfunctions and the parabolic eigenfunctions are complete sets, a given eigenfunction may be written as a linear combination of the other set. Comparison of the spherical eigenfunctions $|n\ell m\rangle$ with the parabolic eigenfunctions written in the form $|i, m_i; k, m_k\rangle$ makes it clear that they are simply coupled (spherical) and uncoupled (parabolic) representations. Therefore, when expanding a given eigenfunction on the other set of eigenfunctions the expansion coefficients are just the Clebsch–Gordan coefficients.[5] It is possible to apply the methods of Chapter 2 to obtain these coefficients in certain cases. For example, if we start with the eigenfunction that is an eigenfunction in both coordinate systems we can generate expansions for the entire ladder of spherical states having the same values of n and ℓ. In particular, the kets

$$|n\ (n-1)(n-1)\rangle_{sph} = |n\,00\,(n-1)\rangle_{par} \tag{6.98}$$

represent eigenfunctions in both spherical and parabolic coordinates. In Equation (6.98) the quantum number m is explicit in the parabolic ket and subscripts have been added for clarity. That these kets are identical can be seen by noting that m is a good quantum number in both coordinate systems so the value of m must be the same in both kets. If $n_1 = 0 = n_2$ then $m = n - 1$. If $m = n - 1$ then there is only one possible value of ℓ; it too must be $n - 1$.

We apply $\hat{L}_-$ to the spherical eigenfunction and $\hat{L}_-$ in the form $\hat{L}_- = \hat{I}_- + \hat{K}_-$ to the parabolic eigenfunction and obtain

$$\hat{L}_-|n\,(n-1)(n-1)\rangle_{sph} = \sqrt{2(n-1)}|n\,(n-1)(n-2)\rangle_{sph} \tag{6.99}$$

and

$$\hat{L}_-|n\,00\,(n-1)\rangle_{par} = (\hat{I}_- + \hat{K}_-)|n\,00\,(n-1)\rangle_{par}$$
$$= \sqrt{(n-1)}|n\,1\,0\,(n-2)\rangle_{par} + \sqrt{(n-1)}|n\,0\,1\,(n-2)\rangle_{par} \tag{6.100}$$

Therefore

$$|n(n-1)(n-2)\rangle_{sph} = \frac{1}{\sqrt{2}}|n\,1\,0\,(n-2)\rangle_{par} + \frac{1}{\sqrt{2}}|n\,0\,1\,(n-2)\rangle_{par} \tag{6.101}$$

Successive application of $\hat{L}_- = \hat{I}_- + \hat{K}_-$ will generate the entire set of spherical eigenfunctions for $\ell = n - 1$. Other Clebsch–Gordan coefficients can be obtained, but with considerably more labor and are not within the scope of the work presented here.

6.8. Additional Symmetry Considerations

We have seen that the Coulomb potential possesses "super-symmetry". Another potential energy function possessing super symmetry is that of the isotropic harmonic oscillator oscillator, the potential energy for which is given by

$$V(r) = \frac{1}{2}kr^2$$
$$= \frac{1}{2}k\left(x^2 + y^2 + z^2\right) \tag{6.102}$$

This system is known to be highly degenerate and, in fact, has a higher degree of symmetry than even the Coulomb potential. Note that the Schrödinger equation can

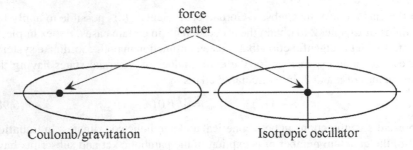

FIGURE 6.1. Elliptical orbits for Coulomb/gravitational potential and isotropic oscillator potential.

be separated in both spherical and Cartesian coordinates for this potential. It can be separated in spherical coordinates because it is a central potential and separation of variables in Cartesian coordinates is easily seen to lead to three one-dimensional harmonic oscillator equations.

Of course, angular momentum is conserved in the isotropic oscillator problem because it is a central potential. It is natural to ask if there is an additional conserved quantity, analogous to the Lenz vector for the Coulomb/gravitational potential. Indeed, there is such a constant of the motion, but, because the isotropic oscillator potential has a higher degree of symmetry than the Coulomb potential this constant is a tensor rather than a vector. We do not pursue this any further, but the additional symmetry of the isotropic oscillator can be seen by a simple graphical argument.[6]

For the classical Kepler problem the force center is at one focus of the elliptical orbit as illustrated in FIGURE 6.1. Because the center of attraction is at a focus, the major axis of the ellipse is an axis of symmetry, but not the minor axis. In contrast, elliptical orbits resulting from the isotropic oscillator potential have as the force center the geometric center of the ellipse. That is, the force center lies on the major axis midway between the foci as shown in FIGURE 6.1. Therefore, an elliptical orbit of a particle under the influence of the isotropic oscillator potential has two symmetry axes, the major and minor axes. We conclude that the isotropic oscillator is "more symmetric" than the Coulomb problem.

Problems

6.1. Find the probability of measuring the z-component of the Lenz vector to be $-1/2$ in atomic units for a hydrogen atom for which $n = 2$; $n_1 = 1$; $n_2 = 0 = m$ (parabolic coordinates).

6.2. Show that

$$\hat{A} = \left(\frac{1}{2}\right)(\hat{p} \times \hat{L} - \hat{L} \times \hat{p}) - \hat{r}$$

$$= \hat{p} \times \hat{L} - i\hat{p} - \hat{r}$$

6.3. (a) Show that the commutator $[\hat{A}_+, \hat{L}_+] = 0$.

(b) Assume that it is known that $\hat{A}_+ |n\ell\ell\rangle = D_{\ell\ell}^+ |n(\ell+1)m\rangle$. That is, it is known that $\hat{A}_+$ raises ℓ by unity, but we don't know what it does to $m (=\ell)$. Using the commutator $[\hat{A}_+, \hat{L}_+] = 0$, confirm that when operating on $|n\ell\ell\rangle$, $\hat{A}_+$ also raises $m (=\ell)$ to $(\ell+1)$.

6.4. Show that $\{\hat{A}_- |n\ell m\rangle\}$ is an eigenfunction of $\hat{L}_z$ with eigenvalue $(m-1)$ and of $\hat{L}^2$ with eigenvalue $(\ell+1)(\ell+2)$ provided $m = -\ell$.

6.5. Show that

$$
\hat{A}_\pm |n\ell m\rangle = \mp \frac{1}{n}\sqrt{\frac{\left[n^2 - (\ell+1)^2\right](\ell \pm m + 1)(\ell \pm m + 2)}{(2\ell+1)(2\ell+3)}}\, |n(\ell+1)(m\pm1)\rangle
$$
$$
\pm \frac{1}{n}\sqrt{\frac{\left[n^2 - \ell^2\right](\ell \mp m)(\ell \mp m - 1)}{(2\ell-1)(2\ell+1)}}\, |n(\ell-1)(m\pm1)\rangle
$$

using the known action of $\hat{A}_z$ on $|n\ell m\rangle$, Equation (6.59), and the commutation relation $\left[\hat{A}_z, \hat{L}_\pm\right] = \pm\hat{A}_\pm$ (see Table 2.2).

6.6. Find all parabolic eigenfunctions for $n=2$ as linear combinations of spherical eigenfunctions by applying $\hat{L}_- = \hat{I}_- + \hat{K}_-$ to $|211\rangle_{sp} = |2001\rangle_{par}$ where spherical eigenfunctions are designated $|n\ell m\rangle_{sp}$ and parabolic eigenfunctions $|nn_1n_2m\rangle_{par}$. Solve these simultaneous equations to obtain the parabolic eigenfunctions in terms of the spherical eigenfunctions.

Answer:

$$|211\rangle_{sp} = |2001\rangle_{par} \qquad\qquad |2001\rangle_{par} = |211\rangle_{sp}$$

$$|210\rangle_{sph} = \frac{1}{\sqrt{2}}|2100\rangle_{par} + \frac{1}{\sqrt{2}}|2010\rangle_{par} \quad |2100\rangle_{par} = \frac{1}{\sqrt{2}}|210\rangle_{sph} + \frac{1}{\sqrt{2}}|200\rangle_{sph}$$

$$|21-1\rangle_{sp} = |211-1\rangle_{par}$$

$$\qquad\qquad\qquad\qquad\qquad\qquad |2010\rangle_{par} = \frac{1}{\sqrt{2}}|210\rangle_{sph} - \frac{1}{\sqrt{2}}|200\rangle_{sph}$$

$$|200\rangle_{sph} = \frac{1}{\sqrt{2}}|2100\rangle_{par} - \frac{1}{\sqrt{2}}|2010\rangle_{par} \quad |211-1\rangle_{par} = |21-1\rangle_{sp}$$

References

1. W. Pauli, in *Sources of Quantum Mechanics*, edited by B.L.v.d. Waerden (1967), Vol. 36, p. 336.
2. L.C. Biedenharn and J.D. Louck, *Angular Momentum in Quantum Mechanics: Theory and Application* (Addison-Wesley, Reading, MA, 1981).
3. V. Bargmann, Z. Phys. **99**, 576 (1936).
4. C.A. Coulson and A. Joseph, Proc. Phys. Soc. **90**, 887 (1967).
5. D. Park, Z. Phys. **150**, 155 (196).
6. L.I. Schiff, *Quantum Mechanics* (McGraw-Hill, New York, 1968).

7
Breaking the Accidental Degeneracy

7.1. Introduction

To this point our discussion of the hydrogen atom has centered on the eigenstates and their eigenenergies, the Bohr energies. There are, however, corrections to these energies that are caused by effects not included in this Schrödinger equation. These corrections are conveniently characterized by their magnitudes in terms of the fine-structure constant α. We may write the total energy of the hydrogen atom as

$$E_{TOTAL} = E_n^{(0)} + E_{FS} + E_{Lamb} + E_{HF} \qquad (7.1)$$

where $E_n^{(0)}$ is the Bohr energy, $-\left(\frac{1}{2}\right) m_e c^2 \alpha^2 / n^2$. The remaining terms in Equation (7.1) are referred to as fine-structure, the Lamb shift, and the hyperfine structure, respectively. Hyperfine structure has already been discussed in relation to angular momentum in Chapter 2. In the present context we are, however, interested in the magnitudes of these effects. These corrections are given, roughly, by

$$E_{FS} \sim \alpha^2 E_n^{(0)}$$

$$E_{Lamb} \sim \alpha^3 E_n^{(0)}$$

$$E_{HF} \sim (3/1000) \alpha^2 E_n^{(0)} \qquad (7.2)$$

Because $\alpha^2 \sim 10^{-5}$ it is clear that even the fine-structure correction is a small, but easily observable, fraction of the Bohr energy.

The fine-structure correction can be obtained by solving the Dirac equation for the hydrogen atom. In fact, the Dirac equation can be solved *exactly* for the Coulomb potential. On the other hand, neither the Lamb shift nor the hyperfine corrections are inherent in the Dirac equation. The Lamb shift requires quantization of the electromagnetic field (QED) whereas the proton spin is absent from the Dirac Hamiltonian. We deal with these corrections later in this chapter.

One way to obtain the fine-structure corrections is to begin with the exact solution of the Dirac equation, expand it, and identify the interactions that constitute the terms proportional to α^2. Alternatively, we can expand the Dirac Hamiltonian, identify the first-order correction terms to the Schrödinger Hamiltonian and use perturbation theory. Neither method requires actual solution of the Dirac equation, merely acceptance that the Hamiltonian, and therefore the exact solution, is correct. Expansion of the Hamiltonian can, however, provide more insight into the physical origins of the corrections, so we elect to use the perturbation theory method.

Expanding the Dirac Hamiltonian for the hydrogen atom gives, in SI units[1]

$$\hat{H}_{rel} = m_e c^2 + \left[\frac{\hat{p}^2}{2m_e} - \left(\frac{e^2}{4\pi\varepsilon_0} \right) \cdot \frac{1}{r} \right]$$
$$- \frac{\hat{p}^4}{8m_e^3 c^2} + \left[\left(\frac{e^2}{4\pi\varepsilon_0} \right) \cdot \frac{1}{2m_e^2 c^2} \cdot \frac{1}{r^3} \left(\hat{L} \cdot \hat{S} \right) \right] + \left(\frac{e^2}{4\pi\varepsilon_0} \right) \frac{\hbar^2 \pi}{2m_e^2 c^2} \delta(r)$$

(7.3)

The first term in Equation (7.3) is, of course, the rest energy of the electron. The second and third comprise the nonrelativistic hydrogen atom (Schrödinger) Hamiltonian. The last three terms in this Hamiltonian make up the fine-structure correction and, as shown later, are proportional to α^2. Thus, the use of perturbation theory is justified.

The fine-structure terms are relativistic in nature. Each can be associated with a physical interaction and can be derived on physical grounds. For the last term, however, the one with the delta function, this association is equivocal.

7.2. Relativistic Correction for the Electronic Kinetic Energy

The most obvious relativistic correction is that due to the electronic motion. We already know that this correction will be small because the electron velocity in the lowest Bohr orbit is $\alpha c = c/137$. To find the correction due to the relativistic motion, we start with the expression for the relativistic kinetic energy

$$\hat{T}_{rel} = \sqrt{\hat{p}^2 c^2 + m_e^2 c^4} - m_e c^2$$

(7.4)

where m_e is the rest mass of the electron and c the speed of light. For convenience and to conform with most treatments, we continue to use SI units. It must be borne in mind that $\hat{T}_{rel}$ is an operator, written in terms of scalar quantities, m_e and c, and the operator $\hat{p}$. Note that we have ignored the motion of the nucleus about the center of mass of the electron-nucleus system.

Now, the rest energy $m_e c^2$ for the electron is ~ 0.5 MeV, whereas the most strongly bound state of the hydrogen atom, the ground state, is bound by only 13.6 eV! Therefore, the rest energy term must be dominant and we may expand the radical in powers of $\hat{p}^2 c^2 / m_e^2 c^4$.

$$\hat{T}_{rel} = m_e c^2 \left(1 + \frac{\hat{p}^2}{m_e c^2} \right)^{1/2} - m_e c^2$$
$$= m_e c^2 \left(1 + \frac{\hat{p}^2}{2m_e c^2} - \frac{\hat{p}^4}{8m_e^4 c^4} + \cdots \right) - m_e c^2$$
$$= \frac{\hat{p}^2}{2m_e} - \frac{\hat{p}^4}{8m_e^3 c^2} + \cdots$$

(7.5)

We see that the first two terms in $\hat{T}_{rel}$ are precisely the terms in Equation (7.3) that contain the momentum $\hat{p}$. The first, $\hat{p}^2/2m_e$, is the nonrelativistic kinetic energy. The second term represents the correction to the energy due to the relativistic motion of the electron. Letting $\hat{T}_0 = \hat{p}^2/2m_e$ we may rewrite $\hat{T}_{rel}$ as

$$\hat{T}_{rel} \approx \frac{\hat{p}^2}{2m_e}\left(1 - \frac{\hat{p}^2}{8m_e^2c^2}\right)$$

$$= \hat{T}_0\left(1 - \hat{T}_0\frac{1}{2m_ec^2}\right) \tag{7.6}$$

We see then that $\hat{T}_{rel}$ is equal to $\hat{T}_0$ plus a correction term. The magnitude of this correction term can be estimated by noting that, according to the virial theorem $\hat{T}_0$ is one-half the magnitude of the total energy. Thus,

$$\frac{T_0}{m_ec^2} \sim \frac{\left(\frac{1}{2}\alpha^2 m_ec^2\right)}{m_ec^2}$$

$$\sim \alpha^2 \tag{7.7}$$

The reason that α is called the fine-structure constant is apparent from Equation (7.7).

We have seen that the relativistic kinetic energy operator is given by the usual nonrelativistic kinetic energy operator plus the small correction term. Because this correction term, when operating on a suitable wave function, will return an energy that is only $\sim 10^{-4}$ of the Bohr energy, quantum mechanical perturbation theory may be used to evaluate this correction term. Thus, we designate the correction term in the Hamiltonian as $\hat{H}_T$ given by

$$\hat{H}_T = -\hat{T}_0^2\left(\frac{1}{2m_ec^2}\right)$$

$$= -\frac{\hat{p}^4}{8m_e^3c^2} \tag{7.8}$$

This is precisely the term in the expansion of the Dirac Hamiltonian presented earlier in this chapter. We defer the computation of the correction to the energy due to this term until all of the fine-structure operators have been derived.

7.3. Spin-Orbit Correction

The term "spin-orbit" refers to the interaction of the electron, envisioned as a bar magnet, with the magnetic field produced by the proton orbiting about the electron in the rest frame of the electron. The spin-orbit correction is therefore nothing more than the energy of a magnetic dipole immersed in a magnetic field. It should be noted, however, that the spin-orbit correction is a relativistic correction because electron spin is a relativistic characteristic. This is apparent from the fact that spin is *not* contained in the Schrödinger equation, but is inherent in the Dirac equation, an intrinsically relativistic equation.

The spin-orbit energy is

$$E_{SO} = -\mu_s \cdot B \tag{7.9}$$

where μ_s is the spin magnetic moment and B is the field due to the proton motion about the electron. We already know that μ_s is actually a quantum mechanical operator because it is proportional to the spin angular momentum and is given by

$$\hat{\mu}_S = -\frac{g_e \mu_B}{\hbar} \hat{S} \tag{7.10}$$

where μ_B is the Bohr magneton. We need only evaluate the magnetic induction B as "seen" by the electron.

We calculate the magnetic induction field at the location of the electron by assuming that the proton rotating about the electron with velocity v constitutes a circular current of radius r, as in the calculation of the orbital magnetic moment. This is, in fact, a bogus calculation because we are calculating the magnetic induction field in the rest frame of the electron. Because the electron is assumed in circular motion, it is not in an inertial frame of reference. We show that we get almost the correct answer. We will be off by a factor of $1/2$ from the term in the relativistic Hamiltonian. This correction is known as the Thomas precession.

The magnetic induction field at the center of a planar circular loop is easily calculated using the the law of Biot and Savard. The result is

$$B = \frac{\mu_0 i}{2r} \tag{7.11}$$

where $i = ev/(2\pi r)$. Therefore, the field due to the orbital motion $B_{orbital}$ is

$$\begin{aligned}
B_{orbital} &= \frac{\mu_0 e}{4\pi m_e r^3} \cdot (m_e v r) \\
&= \frac{\mu_0 e}{4\pi m_e r^3} \cdot L \\
&= \frac{1}{4\pi \varepsilon_0} \cdot \frac{e}{m_e c^2 r^3} \cdot L
\end{aligned} \tag{7.12}$$

where we have eliminated μ_0 using the relation $\mu_0 \varepsilon_0 = 1/c^2$. The correction term for spin-orbit coupling in the Hamiltonian is therefore

$$\hat{H}_{SO} = \frac{g_e}{4\pi \varepsilon_0} \cdot \frac{e^2}{2m_e^2 c^2 r^3} \left(\hat{S} \cdot \hat{L} \right) \tag{7.13}$$

As noted above, it is necessary to insert the Thomas correction factor, $1/2$, which cancels the g-factor. We obtain

$$\hat{H}_{SO} = \left(\frac{1}{4\pi \varepsilon_0} \right) \cdot \frac{e^2}{2m_e^2 c^2 r^3} \left(\hat{S} \cdot \hat{L} \right) \tag{7.14}$$

which is identical to the analogous term in the expansion of the Dirac Hamiltonian.

It is often convenient to simplify the notation in Equation (7.14) by writing it as

$$\hat{H}_{SO} = \xi(r) \left(\hat{S} \cdot \hat{L} \right) \tag{7.15}$$

where

$$\xi(r) = \left(\frac{1}{4\pi\varepsilon_0}\right) \cdot \frac{e^2}{2m_e^2 c^2 r^3} \tag{7.16}$$

7.4. The Darwin Term

The spin-orbit interaction is proportional to $\hat{S} \cdot \hat{L}$. Accordingly, there is no correction when the orbital angular momentum is zero. There is, however, an additional correction term that pertains *only* when the orbital angular momentum is zero. This effect, which arises naturally in the solution to the Dirac equation, has no classical analogue. It is caused by rapid oscillations of the electron that are referred to in the literature[2] as *zitterbewegung*, the translation of which is "shaking". The amplitude of these oscillations are of the order of the Compton wavelength,[2] $2\pi a_0 \alpha \approx 2 \times 10^{-12}$ m.

It should be emphasized that the Darwin correction to the Bohr energy is a natural consequence of the Dirac equation. Attempts to derive it on the basis of physical interactions are of dubious value. For this reason, we simply *use* the correction term in the relativistic Hamiltonian to compute its contribution to the fine-structure. From Equation (7.3), the Darwin Hamiltonian is

$$\hat{H}_D = \left(\frac{1}{4\pi\varepsilon_0}\right) \frac{e^2 \hbar^2 \pi}{2m_e^2 c^2} \delta(r) \tag{7.17}$$

7.5. Evaluation of the Terms That Contribute to the Fine-Structure of Hydrogen

From perturbation theory, the first-order correction to the energy is given by

$$E^{(1)} = \langle \hat{H}^{(1)} \rangle \tag{7.18}$$

where $\hat{H}^{(1)}$ is the perturbing Hamiltonian and the expectation value is taken using the unperturbed wave functions. In the present case $\hat{H}^{(1)} = \hat{H}_{FS}$ where $\hat{H}_{FS}$ is the Hamiltonian representing the total fine-structure correction to the Bohr energies

$$\hat{H}_{FS} = \hat{H}_T + \hat{H}_{SO} + \hat{H}_D \tag{7.19}$$

The unperturbed eigenfunctions are, of course, the nonrelativistic Schrödinger eigenfunctions for the hydrogen atom. We consider the constituent terms of $\hat{H}_{FS}$ separately.

The Relativistic Correction

From Equation (7.8) the first-order correction is given by

$$E_T^{(1)} = \langle \hat{H}_T \rangle$$

$$= -\frac{1}{8m_e^3 c^2} \langle \hat{p}^4 \rangle \tag{7.20}$$

Now, $\hat{H}_T$ contains only $\hat{p}^4 = \hat{p}^2 \hat{p}^2$, and $\hat{p}^2$ appears in the unperturbed Hamiltonian. That is, the unperturbed Hamiltonian $\hat{H}_0$ is

$$\hat{H}_0 = \frac{\hat{p}^2}{2m_e} - \frac{e^2}{4\pi\varepsilon_0 r} \tag{7.21}$$

Because $\hat{p}^2$ is present in $\hat{H}_T$, $\hat{H}_T$ is diagonal in the $|n\ell m\rangle$ basis set and

$$E_T^{(1)} = -\frac{1}{8m_e^3 c^2} \langle n\ell m| \hat{p}^2 \hat{p}^2 |n\ell m\rangle \tag{7.22}$$

From Equation (7.21) we have

$$\hat{p}^2 = 2m_e \left(\hat{H}_0 + \frac{1}{4\pi\varepsilon_0} \cdot \frac{e^2}{r} \right) \tag{7.23}$$

so that

$$
\begin{aligned}
E_T^{(1)} &= -\frac{1}{8m_e^3 c^2} \langle n\ell m| 4m_e^2 \left(\hat{H}_0 + \frac{1}{4\pi\varepsilon_0} \cdot \frac{e^2}{r} \right)^2 |n\ell m\rangle \\
&= -\frac{1}{2m_e c^2} \left[(E_n^{(0)})^2 + 2E_n^{(0)} \left(\frac{e^2}{4\pi\varepsilon_0} \right) \left\langle \frac{1}{r} \right\rangle_{n\ell m} + \left(\frac{e^2}{4\pi\varepsilon_0} \right)^2 \left\langle \frac{1}{r^2} \right\rangle_{n\ell m} \right]
\end{aligned}
\tag{7.24}
$$

where $E_n^{(0)}$ is the Bohr energy.

To complete the evaluation of $E_T^{(1)}$, the expectation values $\langle 1/r \rangle$ and $\langle 1/r^2 \rangle$ are required. These, and other, powers of r can be found the classic book by Bethe and Salpeter.[3] The two that are germane here are

$$\left\langle \frac{1}{r} \right\rangle_{n\ell m} = \frac{1}{a_0 n^2} \tag{7.25}$$

and

$$\left\langle \frac{1}{r^2} \right\rangle_{n\ell m} = \frac{1}{a_0^2} \cdot \frac{1}{n^3 (\ell + 1/2)} \tag{7.26}$$

each of which may be expressed in terms of $E_n^{(0)}$ giving

$$\left\langle \frac{1}{r} \right\rangle_{n\ell m} = -2 \left(\frac{4\pi\varepsilon_0}{e^2} \right) E_n^{(0)} \tag{7.27}$$

and

$$\left\langle \frac{1}{r^2} \right\rangle_{n\ell m} = 4 \cdot \frac{n}{(\ell + 1/2)} \left(\frac{4\pi\varepsilon_0}{e^2} \right)^2 (E_n^{(0)})^2 \tag{7.28}$$

We note parenthetically that the first expectation value, multiplied by $-e^2/(4\pi\varepsilon_0)$, is simply the average value of the potential energy, a quantity that can be evaluated using the virial theorem.

Inserting Equations (7.27) and (7.28) in Equation (7.24), we have

$$E_T^{(1)} = -\frac{\left(E_n^{(0)}\right)^2}{2m_e c^2}\left[-3 + \frac{4n}{(\ell + 1/2)}\right] \tag{7.29}$$

or, because $E_n^{(0)} = -\left(\frac{1}{2}\right)m_e c^2 \alpha^2/n^2$,

$$E_T^{(1)} = \left[-\frac{3}{4} + \frac{n}{(\ell + 1/2)}\right]\frac{\alpha^2}{n^2}E_n^{(0)} \tag{7.30}$$

Inasmuch as it is sometimes useful to have the Z-dependence of the fine-structure corrections for a one-electron atom having Z protons in the nucleus we note that for such an atom in Equation (7.30) $\alpha \to (Z\alpha)$ so that $E_T^{(1)} \propto Z^4$. This is reasonable because for high Z the electron will be moving faster than for low Z.

The Spin-Orbit Correction

We found that

$$\hat{H}_{SO} = \xi(r)\left(\hat{S} \cdot \hat{L}\right) \tag{7.31}$$

In contrast to the relativistic kinetic energy term, this relativistic perturbation contains the spin. When evaluating the perturbation due to $\hat{H}_T$ we were already working with a diagonal-perturbing Hamiltonian. To evaluate the spin-orbit contribution we must diagonalize $\hat{H}_{SO}$. But, we have already done this because

$$2\hat{S} \cdot \hat{L} = \hat{J}^2 - \hat{L}^2 - \hat{S}^2 \tag{7.32}$$

That is, the coupled states $|jm_j\ell s\rangle$ are eigenstates of $\hat{S} \cdot \hat{L}$. Therefore

$$\begin{aligned}
E_{SO}^{(1)} &= \langle \hat{H}_{SO}\rangle \\
&= \langle\xi(r)\rangle\langle jm_j\ell s|\left(\hat{J}^2 - \hat{L}^2 - \hat{S}^2\right)|jm_j\ell s\rangle \\
&= \left(\frac{1}{4\pi\varepsilon_0}\right)\cdot\frac{e^2}{4m_e^2 c^2}\left\langle\frac{1}{r^3}\right\rangle_{n\ell}\hbar^2\left[j(j+1) - \ell(\ell+1) - s(s+1)\right]
\end{aligned} \tag{7.33}$$

The expectation value of $1/r^3$ is given by[3]

$$\left\langle\frac{1}{r^3}\right\rangle_{n\ell} = \frac{1}{a_0^3}\cdot\frac{1}{n^3\ell(\ell+1/2)(\ell+1)} \tag{7.34}$$

which, because we would like to view the correction as a factor times the unperturbed hydrogen energy, we write in the form

$$\begin{aligned}
\left\langle\frac{1}{r^3}\right\rangle_{n\ell} &= \left(-\frac{4\pi\varepsilon_0}{e^2}\right)\cdot\left[\left(-\frac{e^2}{4\pi\varepsilon_0}\right)\cdot\frac{1}{2a_0 n^2}\right]\cdot\left(\frac{2}{a_0^2 n}\right)\cdot\frac{1}{\ell(\ell+1/2)(\ell+1)} \\
&= \left(-\frac{4\pi\varepsilon_0}{e^2}\right)E_n^{(0)}\left(\frac{2}{a_0^2 n}\right)\cdot\frac{1}{\ell(\ell+1/2)(\ell+1)} \\
&= \left(-\frac{4\pi\varepsilon_0}{e^2}\right)E_n^{(0)}\left(\frac{2}{n}\right)\left(\frac{\alpha^2 m_e^2 c^2}{\hbar^2}\right)\cdot\frac{1}{\ell(\ell+1/2)(\ell+1)}
\end{aligned} \tag{7.35}$$

The final result is

$$E_{SO}^{(1)} = -\frac{1}{2n}\alpha^2 E_n^{(0)} \frac{[j(j+1)-\ell(\ell+1)-3/4]}{\ell(\ell+1/2)(\ell+1)} \tag{7.36}$$

where, again, the magnitude of the correction is α^2 times the unperturbed hydrogen energy.

Because j can take on the values $\ell \pm 1/2$ we may write the correction in terms of ℓ.

$$E_{SO}^{(1)} = -\frac{1}{2}\alpha^2 E_n^{(0)} \frac{1}{n(\ell+1/2)(\ell+1)} \qquad \text{for } j = \ell+1/2$$

$$= \frac{1}{2}\alpha^2 E_n^{(0)} \frac{1}{n\ell(\ell+1/2)} \qquad \text{for } j = \ell-1/2 \tag{7.37}$$

Thus, as for the relativistic kinetic energy correction, the spin-orbit interaction produces a correction that is α^2 times the Bohr energy. As was the case for the relativistic kinetic energy correction, for a one-electron atom having Z protons in the nucleus $\alpha \to (Z\alpha)$ so the spin-orbit correction is proportional to Z^4.

The Darwin Term

The first-order correction to the energy due to the Darwin Hamiltonian, Equation (7.17), is given by

$$E_D^{(1)} = \langle n\ell m| \hat{H}_D |n\ell m\rangle$$

$$= \left(\frac{1}{4\pi\varepsilon_0}\right) \frac{e^2\hbar^2\pi}{2m_e^2c^2} \langle n\ell m|\delta(r)|n\ell m\rangle \tag{7.38}$$

Because of the δ-function, the expectation value of $E_D^{(1)}$ will be nonzero only for s-states because all radial wave functions vanish at $r = 0$ except those having $\ell = 0$. The integral $\langle n00|\delta(r)|n00\rangle$ is easily evaluated using the sifting property of the delta function.

$$\langle n00|\delta(r)|n00\rangle = \int\limits_{all\ space} Y_{00}^*(\theta,\phi)R_{n0}^*(r)\,\delta(r)\,R_{n0}(r)\,Y_{00}(\theta,\phi)$$

$$= \frac{1}{4\pi}|R_{n0}(0)|^2 \tag{7.39}$$

For $\ell = 0$ the radial part of the hydrogen atom wave functions, Equation (4.26), using Spiegel's definition of the associated Laguerre polynomials,[4] is

$$R_{n0}(r) = -\left(\frac{2}{na_0}\right)^{3/2} \left[\frac{(n-1)!}{2n}\right]^{1/2} \left[\frac{1}{n!}\right]^{3/2} \cdot L_n^1\left(\frac{2r}{na_0}\right) \tag{7.40}$$

which, for $r = 0$ is

$$R_{n0}(0) = -\left(\frac{2}{na_0}\right)^{3/2} \left[\frac{(n-1)!}{2n}\right]^{1/2} \left[\frac{1}{n!}\right]^{3/2} \cdot L_n^1(0) \tag{7.41}$$

The Laguerre polynomial is

$$L_n^1\left(\frac{2r}{na_0}\right) = \sum_{j=0}^{n-1} (-)^{j+1} \frac{(n!)^2}{(n-1-j)!\,(j+1)!\,j!} \cdot \left(\frac{2r}{na_0}\right)^j \quad (7.42)$$

But, the only nonzero term in the summation in Equation (7.42) is the $j = 0$ term because all other terms contain r^j. Therefore,

$$L_n^1(0) = \frac{(n!)^2}{(n-1)!\,1!\,0!}\rho^0$$

$$= n!n \quad (7.43)$$

and

$$\langle n00|\delta(r)|n00\rangle = \frac{1}{4\pi}\left(\frac{2}{na_0}\right)^3\left[\frac{(n-1)!}{2n}\right]\left[\frac{1}{n!}\right]^3 \cdot (n!)^2\,n^2$$

$$= \frac{1}{\pi a_0^3 n^3} \quad (7.44)$$

The energy correction to the $|n00\rangle$ state of the hydrogen atom due to the Darwin term is then

$$E_D^{(1)} = \left(\frac{1}{4\pi\varepsilon_0}\right)\frac{e^2\hbar^2\pi}{2m_e^2c^2}\langle n00|\delta(r)|n00\rangle$$

$$= \left(\frac{1}{4\pi\varepsilon_0}\right)\frac{e^2\hbar^2\pi}{2m_e^2c^2}\cdot\frac{1}{\pi a_0^3 n^3}$$

$$= -E_n^{(0)}\frac{\alpha^2}{n} \quad (7.45)$$

We see then that the Darwin term, as were the relativistic correction and the spin-orbit correction terms, is proportional to $\alpha^2 E_n^{(0)}$. For a one-electron atom having Z protons in the nucleus we again substitute $\alpha \to (Z\alpha)$.

Although the spin-orbit result is valid only for $\ell \neq 0$ we examine the limit of $E_{SO}^{(1)}(j = \ell + 1/2)$ as ℓ approaches zero. ($j = \ell - 1/2$ is unacceptable for $\ell = 0$ because j cannot be negative.) For $j = \ell + 1/2$ we obtain

$$\lim_{\ell\to 0} E_{SO}^{(1)} = -\frac{1}{2}\alpha^2 E_n^{(0)} \lim_{\ell\to 0}\left[\frac{1}{n(\ell+1/2)(\ell+1)}\right]$$

$$= -\alpha^2 E_n^{(0)}\cdot\frac{1}{n}$$

$$= E_D^{(1)} \quad (7.46)$$

Thus, the algebraic expression for $E_{SO}^{(1)}$, although not valid for $\ell = 0$, produces the correct answer for $E_D^{(1)}$ which is *only* valid for $\ell = 0$. This is not an accident as is discussed later in this chapter.

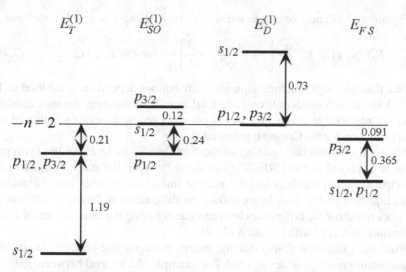

FIGURE 7.1. Individual contributions to the fine-structure splitting of the $n = 2$ state of the hydrogen atom. The energies are given in cm^{-1}. Adapted from B.H. Bransden and C.J. Joachain, *Physics of Atoms and Molecules: Second Edition* (Upper Saddle River, NJ: Prentice Hall, 2003), Fig. 5.2.

The magnitudes of the individual contributions to the fine-structure of the hydrogen atom are shown in FIGURE 7.1 for $n = 2$. Note that the spin-orbit energy for s-states is zero as are the Darwin energies for the p-states.

7.6. The Total Fine-Structure Correction

Because the Darwin correction to the Bohr energy can be included in the spin-orbit correction term we may write the total fine-structure correction as

$$E_{FS}^{(1)} = E_T^{(1)} + E_{SO}^{(1)} \tag{7.47}$$

where it is understood that if $\ell = 0$ then $E_{SO}^{(1)}$ represents the Darwin term. We have then

$$E_{FS}^{(1)}(n, \ell) = E_n^{(0)} \frac{\alpha^2}{n^2} \left[\frac{n}{(\ell + 1)} - \frac{3}{4} \right] \quad \text{for } j = \ell + 1/2$$

$$= E_n^{(0)} \frac{\alpha^2}{n^2} \left[\frac{n}{\ell} - \frac{3}{4} \right] \quad \text{for } j = \ell - 1/2 \tag{7.48}$$

Because the maximum value of ℓ is $(n - 1)$ it is clear that the terms in square brackets in Equations (7.48) can never be negative. Therefore, because $E_n^{(0)}$ is intrinsically negative, the fine-structure corrections will always lower the Bohr energy.

Equation (7.48) may be cast in terms of only j. Using $j = \ell \pm 1/2$ we have

$$E_{FS}^{(1)}(n, j) = E_n^{(0)}\frac{\alpha^2}{n^2}\left[\frac{n}{(j+1/2)} - \frac{3}{4}\right] \quad \text{for } j = \ell \pm 1/2 \qquad (7.49)$$

Notice that, although the three separate contributions depend on ℓ, the total shift, $E_{FS}^{(1)}$, does not. It depends only on j, the total angular momentum quantum number. This is a remarkable degeneracy that is present even in the exact solution of the Dirac equation for the Coulomb potential.

The total fine-structure splitting of the first three Bohr levels of the hydrogen atom is displayed in FIGURE 7.2. The shifts between the states designated by principal quantum numbers and the nearest fine-structure level were calculated from Equation (7.49). This figure shows the differences in levels of different j, but does not show the differences between states having the same values of j, but different ℓ in accord with Equation (7.49).

Note from Equation (7.49) that the energy between fine-structure states, the fine-structure interval, scales as $1/n^3$. For example, the interval between $j = 3/2$

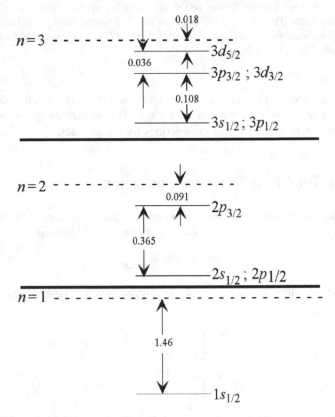

FIGURE 7.2. Total fine-structure splitting of the designated states for the first three Bohr levels of the hydrogen atom. The energies are given in cm^{-1}. Adapted from B.H. Bransden and C.J. Joachain, *Physics of Atoms and Molecules: Second Edition* (Upper Saddle River, NJ: Prentice Hall, 2003), Fig. 5.1.

and $j = 1/2$ states is

$$\Delta_n = \Delta E_{FS}(j = 1/2, 3/2)$$
$$= E_{FS}^{(1)}(n, j = 3/2) - E_{FS}^{(1)}(n, j = 1/2)$$
$$= \frac{1}{2} E_n^{(0)} \frac{\alpha^2}{n}$$
$$= \frac{1}{2} R_H \frac{\alpha^2}{n^3} \tag{7.50}$$

The fine-structure splitting therefore decreases dramatically with increasing principal quantum number. On the other hand, for atoms of nuclear charge Z the expressions for the hydrogen atom require the substitution $\alpha \to Z\alpha$ so the fine-structure interval increases with increasing nuclear charge.

7.7. The Lamb Shift

We have seen that, although fine-structure effects partially lift the degeneracy of the Bohr levels of hydrogen, some degeneracy remains. For example, the $2s_{1/2}$ and $2p_{1/2}$ levels are degenerate because the fine-structure correction depends only on the magnitude of j and not on how it is constructed. That is, $j = \ell \pm 1/2$ so that $\ell = 1$ and $m_s = -1/2$ gives $j = 1/2$. But, so too does $\ell = 0$ with $m_s = +1/2$ lead to $j = 1/2$. There is, however, a difference in the energies of these two states, a difference that is not predicted by even the Dirac equation. To obtain the energy difference between these two states, referred to as the Lamb shift, it is necessary to resort to quantum electrodymanics (QED). That is, it is necessary to quantize the electromagnetic field. We do not do this here, but full mathematical treatments are available in many textbooks.[2,3,5,6] The lifting of the j-degeneracy by QED was an important triumph for this theory.

Although we do not derive the expression for the Lamb shift we can attempt a qualitative explanation for it. It was remarked in Chapter 1 that a complete picture of a quantum mechanical system can only be obtained when both the particles and the fields to which they are subjected are quantized. When quantizing the electromagnetic field the basis set is that of a harmonic oscillator for which there is a zero-point energy. Thus, the absence of any field, the *vacuum state*, has nonzero energy. This means that even in the vacuum state charged particles are affected by an electromagnetic field. We show in Chapter 13 that this is the cause of spontaneous emission of a photon from an excited atom, even when no fields are present.

The interaction of the zero-point energy of the quantized field with the electron causes it to execute rapid oscillations. This shaking is a form of *zitterbewegung*, but its origin is different from that noted in the case of the Darwin effect for which the term *zitterbewegung* was coined by Schrödinger.[2] In this case, the amplitude of the shaking is roughly ten times smaller than the shaking that causes the Darwin effect. The electric field that binds the electron in an atom varies as $1/r^2$ so the electron experiences different potentials at the extremes of its oscillation, reminiscent of tidal forces that are proportional to $1/r^3$. Because of this nonuniformity of the electric field, the tidal force on this shaking electron is greatest when the electron is near the nucleus so that s-states (which have nonzero probability of being found at $r = 0$) suffer the greatest Lamb shift. Because the magnitude of the effect clearly depends upon the orbital angular momentum state, the j-degeneracy that remains after the Dirac theory is lifted.

Our primary interest is to compare the magnitude of the Lamb shift with the magnitudes of the fine-structure corrections. From quantum electrodynamics, the Lamb shift for states with $\ell = 0$ is given approximately by[7]

$$E_{Lamb} = \alpha^5 m_e c^2 \cdot \frac{1}{4n^3} \{k(n, \ell = 0)\}$$

$$= \alpha^3 \frac{1}{2n} E_n^{(0)} \{k(n, \ell = 0)\} \tag{7.51}$$

where the numerical factor depends only weakly on n and is

$$k(n, \ell = 0) = 12.7 \quad \text{for } n = 1$$

$$= 13.2 \quad \text{for } n \to \infty \tag{7.52}$$

For $\ell \neq 0$ the Lamb shift is

$$E_{Lamb} = \alpha^5 m_e c^2 \cdot \frac{1}{4n^3} \left\{ k(n, \ell) \pm \frac{1}{\pi(j + 1/2)(\ell + 1/2)} \right\} \quad j = \ell \pm 1/2$$

$$= \alpha^3 \frac{1}{2n} E_n^{(0)} \left\{ k(n, \ell) \pm \frac{1}{\pi(j + 1/2)(\ell + 1/2)} \right\} \quad j = \ell \pm 1/2 \tag{7.53}$$

The numerical factor in Equation (7.53), $k(n, \ell) \sim 0.05$, is two orders of magnitude smaller than that in Equation (7.51). Based on the discussion of the sensitivity of s-states to the *zitterbewegung* this difference is expected, so the Lamb shift of s-states is much greater than the Lamb shift for states for which $\ell \neq 0$.

The Lamb shift for the ground state is about eight times larger than that for the $n = 2$ states as predicted by Equation (7.51). The original experiment[8] that verified QED was, however, performed using the $2s_{1/2} - 2p_{1/2}$ interval because the precision available with microwaves was employed. There is no such nearby level of the $1s_{1/2}$ state so the measurement of the ground state Lamb shift was made much later[9,10] using two-photon laser spectroscopy on the $1s - 2s$ interval.

The splitting of the three $|n = 2; \ell\rangle$ states of hydrogen is shown in FIGURE 7.3. Notice that the magnitudes of the Lamb shift intervals are smaller than the fine-structure splittings by $\sim \alpha$.

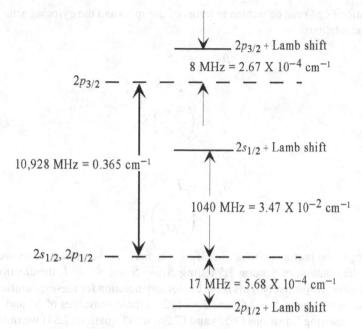

FIGURE 7.3. Schematic diagram of the Lamb shift of the $n = 2$ levels of the hydrogen atom (not to scale). Adapted from B.H. Bransden and C.J. Joachain, *Physics of Atoms and Molecules: Second Edition* (Upper Saddle River, NJ: Prentice Hall, 2003), Fig. 5.17.

7.8. Hyperfine Structure

In general, the hyperfine splitting of the energy is caused by the interaction of the intrinsic magnetic moment of the proton, another spin $1/2$ particle, with the magnetic fields created by both the orbital motion of the electron and the electron spin. We, however, consider only states having $\ell = 0$ so there is no magnetic field caused by the orbital motion of the electron. A special case, the ground state, was treated in Section 3.5. Our goal is to compare the magnitude of the hyperfine energy splitting, $\Delta E_{HF}(n)$, for $|n00\rangle$ states with the magnitude of the fine-structure corrections. In the notation of Section 3.5, $\Delta E_{HF}(n = 1) = 2\kappa$.

The Hamiltonian may be regarded as the energy of orientation of one magnetic dipole in the field of the other. The general expression for this interaction[11] for $\ell = 0$ reduces to

$$\hat{H}_{HF} = -\left(\frac{1}{4\pi\varepsilon_0}\right)\left(\frac{8\pi}{3c^2}\right)\hat{\mu}_S \cdot \hat{\mu}_p \delta(r) \tag{7.54}$$

where $\hat{\mu}_S$ and $\hat{\mu}_p$ are the magnetic dipole moments of the electron and proton, respectively. Equation (7.54) is referred to as the contact term because the delta function requires the electron and the proton to occupy the same point in space for a nonzero interaction.

Equation (7.54) can be written in terms of the spins and the gyromagnetic ratios using the relations

$$\hat{\mu}_S = -\gamma_e \hat{S}$$
$$= \frac{g_e \mu_B}{\hbar} \hat{S}$$
$$= \left(\frac{g_e e}{2 m_e} \right) \hat{S} \qquad (7.55)$$

and

$$\hat{\mu}_p = \gamma_p \hat{I}$$
$$= \left(\frac{g_p e}{2 M_p} \right) \hat{I} \qquad (7.56)$$

where g_p is the proton g-factor and M_p is the mass of the proton. Also, we have altered the notation of Section 3.5 letting $\hat{S}_1 \to \hat{S}$ and $\hat{S}_2 \to \hat{I}$, the electron and nuclear spins, respectively. This is the customary notation for these quantities. The quantum numbers are $S = 1/2$ and $I = 1/2$ so the eigenvalues of $\hat{S}^2$ and $\hat{I}^2$ are $(3/4)\hbar^2$. Inserting Equations (7.55) and (7.56) into Equation (7.54) we obtain

$$\hat{H}_{HF} = \left(\frac{1}{4\pi \varepsilon_0} \right) \left(\frac{8\pi}{3c^2} \right) \gamma_e \gamma_p \left(\hat{S} \cdot \hat{I} \right) \delta(r) \qquad (7.57)$$

To apply perturbation theory we require the expectation value of $\hat{H}_{HF}$ using the unperturbed states, the ordinary spherical eigenkets $|n00\rangle$. Having learned a valuable lesson in Section 3.5, we know immediately that we should use the coupled wave functions. Thus, again using customary notation we let $\hat{F} = \hat{I} + \hat{S}$ so that the first-order correction due to the hyperfine interaction for $|n00\rangle$ states is

$$E_{HF}^{(1)}(n) = \langle n : F\,M | \hat{H}_{HF} | n : F\,M \rangle$$
$$= \left(\frac{1}{4\pi \varepsilon_0} \right) \left(\frac{8\pi}{3c^2} \right) \gamma_e \gamma_p \langle F\,M | (\hat{S} \cdot \hat{I}) | F\,M \rangle \, |\psi_{n00}(0)|^2 \qquad (7.58)$$

where $|F\,M\rangle = |I, S; F, M\rangle$; F is the quantum number associated with the square of the angular momentum $\hat{F}^2$. From Equation (7.44)

$$|\psi_{n00}(0)|^2 = \frac{1}{n^3 \pi a_0^3} \qquad (7.59)$$

Now I and S are both $1/2$, so F can take on the values 0 or 1 (0 if the spins are antiparallel and 1 if parallel). Because

$$\hat{I} \cdot \hat{S} = \frac{1}{2} \left(\hat{F}^2 - \hat{S}^2 - \hat{I}^2 \right) \qquad (7.60)$$

the expectation value $\langle F\,M|\left(\hat{\mathbf{S}}\cdot\hat{\mathbf{I}}\right)|F\,M\rangle$ is

$$\langle F\,M|\left(\hat{\mathbf{S}}\cdot\hat{\mathbf{I}}\right)|F\,M\rangle = \frac{1}{2}\left[F\left(F+1\right)-\frac{3}{2}\right]\hbar^2$$

$$= \frac{1}{4}\hbar^2 \qquad F=1$$

$$= -\frac{3}{4}\hbar^2 \qquad F=0 \qquad (7.61)$$

The energies of the singlet ($F=0$) and the triplet ($F=1$) states are thus

$$E_{HF} = \Delta E_{HF}\left(n\right)\times\left[\frac{1}{4}\right] \qquad F=1$$

$$= \Delta E_{HF}\left(n\right)\times\left[-\frac{3}{4}\right] \qquad F=0 \qquad (7.62)$$

where

$$\Delta E_{HF}\left(n\right) = \left[\left(\frac{1}{n^3}\right)\left(\frac{1}{4\pi\varepsilon_0}\right)\left(\frac{8\pi}{3}\right)\left(\frac{\hbar^2}{c^2\pi a_0^3}\right)\left(\frac{g_e e}{2m_e}\right)\left(\frac{g_p e}{2M_p}\right)\right] \quad (7.63)$$

To facilitate comparison with the fine-structure corrections we cast this quantity in terms of the fine-structure constant and the Bohr energy. This is most easily done recalling that $\alpha a_0 = (\hbar/m_e c)$. Equation (7.63) becomes

$$\Delta E_{HF}\left(n\right) = \frac{4}{3}\cdot\frac{1}{n^3}\left[\frac{m_e}{m_p}\cdot g_p\right]\left[\alpha^2 g_e\left(\frac{1}{2}m_e c^2\alpha^2\right)\right] \quad (7.64)$$

Equation (7.64) has been displayed so as to emphasize the nature of the factors that it comprises. The term in ordinary parentheses is simply the Bohr energy of the ground state, 13.6 eV. The square bracket on the right is the order of the fine-structure energy; that is, α^2 times the Bohr energy. The square bracket on the left is the factor by which the fine-structure energy is reduced in the hyperfine interaction. The ratio of the masses is $1/1836$ whereas $g_p \approx 5.6$. Thus, the hyperfine interval is roughly $3/1000$ of the fine-structure interval. As noted in Section 3.5, for the ground state of hydrogen, this expression leads to a transition frequency of 1420 MHz, the famous 21 cm line. In view of the n-dependence in Equation (7.64) it is expected that the hyperfine splitting of the $2s_{1/2}$ level of the hydrogen atom should be $1/8$ that of the $1s_{1/2}$ ground state. Indeed, it is approximately 177 MHz.

From the foregoing discussion, we see that there exist progressively finer corrections to the Bohr energies starting with fine-structure, the Lamb shift, and hyperfine structure. These corrections are illustrated in FIGURE 7.4 which is a schematic diagram of these energies. The energy intervals are exaggerated and not to scale. The diagram is meant as a guide only.

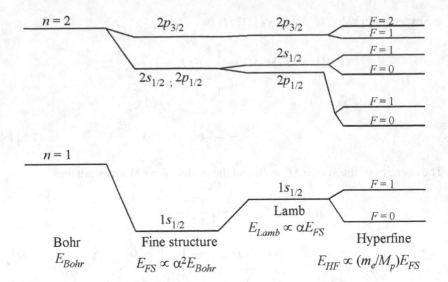

FIGURE 7.4. Schematic diagram showing the corrections to the Bohr energies of the hydrogen atom and the factors by which the magnitudes of the energies are altered. Adapted from B.H. Bransden and C.J. Joachain, *Physics of Atoms and Molecules: Second Edition* (Upper Saddle River, NJ: Prentice Hall, 2003), Fig. 5.18.

7.9. The Solution of the Dirac Equation

It was noted previously in this chapter that the Dirac equation can be solved exactly for the Coulomb potential. Quantized energies are obtained using boundary conditions in much the same way as the necessity of a bounded wave function led to quantized energies in the solution of the Schrödinger equation. Because the Dirac equation is inherently relativistic, it includes the effects of electron spin. It is therefore expected that the quantized hydrogen energies resulting from solution of the Dirac equation will contain the fine-structure effects. This is indeed the case. The energy obtained from the exact solution is

$$E(n, j) = m_e c^2 \cfrac{1}{\sqrt{1 + \cfrac{(Z\alpha)^2}{[n - (j + 1/2)] + \sqrt{(j + 1/2)^2 - (Z\alpha)^2}}}} \qquad (7.65)$$

which may be expanded as

$$E(n, j) = m_e c^2 - E_n^{(0)} - E_n^{(0)} \frac{(Z\alpha)^2}{n^2} \left[\frac{n}{(j + 1/2)} - \frac{3}{4} \right]$$

$$+ O\left[(Z\alpha)^4 \right] E_n^{(0)} \dots \qquad (7.66)$$

The third term in Equation (7.66) is the fine-structure correction obtained in Equation (7.49).

The energy eigenvalues of Equation (7.65) were obtained without expanding the Dirac Hamiltonian. It does not therefore permit identification of the individual terms as "relativistic kinetic energy", "spin orbit", or "Darwin". Indeed, the fine-structure corrections obtained in the exact solution may simply be regarded as relativistic in nature without any recognition of individual physical interactions. This is why the spin-orbit term reduces to the Darwin term in the limit $\ell \to 0$; they have the same source: relativity.

The corrections in Equations (7.49) and (7.65) depend only upon n and j. As discussed above, they do not include the Lamb shift or hyperfine structure. We might inquire about the additional terms in the expansion in Equation (7.66), those proportional to $(Z\alpha)^4 E_n^{(0)}$ and higher. There is, in fact, no need to compute these terms because they are smaller than both the Lamb shift and the hyperfine correction.

Problems

7.1. Show that no combination of quantum numbers can conspire to make the fine-structure correction to the Bohr energy vanish.

7.2. Suppose the proton is approximated as a spherical shell of radius $R \approx 10^{-4}$ nm. Calculate the first-order correction to the energy of the ground state, $E_0^{(1)}$, due to the finite size of the nucleus. Is it positive or negative? Use the fact that $R/a_0 \approx 10^{-5}$ to make the approximation $e^{-2R/a_0} \approx 1$. Estimate the value of $E_0^{(1)}$ in terms of the unperturbed ground state energy $E_0^{(0)}$ and compare with the fine-structure correction to the ground state.

7.3. A particle of rest mass m_0 is confined to an infinite one-dimensional potential well; that is,

$$V(x) = \infty \quad x < 0 \text{ and } x > L$$
$$= 0 \quad \text{otherwise}$$

Find the first-order correction to $E_n^{(1)}$, the energy of the nth level of this particle-in-a-box, due to the relativistic kinetic energy of the electron. State your answer in terms of the unperturbed energies and the rest energy of the particle. Under what circumstances will the validity of $E_n^{(1)}$ be questionable? Consider two cases, an electron in a 0.1 nm box (an atom) and a proton in a 10^{-4} nm box (a nucleus).

7.4. A particle of rest mass m_0 is confined to a one-dimensional harmonic oscillator potential $V(x) = \frac{1}{2}kx^2$. Find the first-order correction to $E_0^{(1)}$, the energy of the ground state due to the relativistic kinetic energy of the electron. State your answer in terms of the unperturbed energies and the rest energy of the particle. Suppose the vibrations of a diatomic molecule are approximated by the harmonic oscillator. Show that, because the separations between molecular vibrational levels are typically on the order of tenths of eV, the correction due to the relativistic motion of the electron is small. There are (at least) three ways to work this problem.

7.5. Starting with Equation (7.15), the Hamiltonian for the spin-orbit correction to the energy of the hydrogen atom

$$\hat{H}_{SO} = \xi(r) \cdot \left(\hat{S} \cdot \hat{L}\right)$$

find the energy difference between to the $j = \ell \pm \dfrac{1}{2}$ levels in terms of $\langle \xi(r) \rangle$.

7.6. An electron of mass m_e is bound by an isotropic harmonic oscillator potential

$$V(r) = \frac{1}{2}k\left(x^2 + y^2 + z^2\right) = \frac{1}{2}kr^2 = \frac{1}{2}m_e\omega^2 r^2$$

Find the corrections to the ground and first two excited state energies due to the spin-orbit correction

$$V_{SO} = \frac{\hbar^2}{2m_e^2 c^2} \cdot \frac{1}{r} \cdot \frac{dV(r)}{dr}\left(\hat{L} \cdot \hat{S}\right)$$

References

1. B.H. Bransden and C.J. Joachain, *Physics of Atoms and Molecules* (John Wiley, New York, 1983).
2. J.D. Bjorken and S.D. Drell, *Relativistic Quantum Mechanics* (McGraw-Hill, New York, 1964).
3. H.A. Bethe and E.E. Salpeter, *Quantum Mechanics of One- and Two-Electron Atoms* (Springer-Verlag, Berlin, 1957).
4. M.R. Spiegel, *Mathematical Handbook of Formulas and Tables* (McGraw-Hill, New York, 1998).
5. L.I. Schiff, *Quantum Mechanics* (McGraw-Hill, New York, 1968).
6. J.J. Sakurai, *Advanced Quantum Mechanics* (Pearson, Singapore, 1967).
7. D.J. Griffiths, *Introduction to Elementary Particles* (Wiley, New York, 1987).
8. W.E. Lamb and R.C. Retherford, Phys. Rev. **72**, 241 (1947).
9. D.J. Berkeland, E.A. Hinds, and M.G. Boshier, Phys. Rev. Lett. **75**, 2470 (1995).
10. M. Weitz, A. Huber, F. Schmidt-Kaler, et al., Phys. Rev. Lett. **72**, 328 (1994).
11. J.D. Jackson, *Classical Electrodynamics* (Wiley, New York, 1999).

8
The Hydrogen Atom in External Fields

8.1. Introduction

To this point we have dealt with only central potentials. If, however, a constant external electric or magnetic field is applied, the spherical symmetry of the potential is broken because the field establishes a direction in space. We expect that degeneracies that exist by virtue of the spatial symmetry will be lifted, at least partially, and that some of the (field-free) degenerate levels will be split by application of the field. Moreover, we expect the energy levels to depend on the magnetic quantum numbers m, a dependence that is necessarily absent for spherical symmetry. Although the spherical symmetry is broken, there still exists cylindrical symmetry, so the spatial degeneracy cannot be entirely broken.

We begin by noting that the total Hamiltonian for a hydrogen atom in the presence of an external field may be written as

$$\hat{H} = \hat{H}_{Coul} + \hat{H}_{FS} + \hat{H}_{field} \tag{8.1}$$

where

$$\hat{H}_{Coul} = \frac{\hat{p}^2}{2m_e} - \left(\frac{1}{4\pi\varepsilon_0}\right) \cdot \frac{e^2}{r} \tag{8.2}$$

is the field-free hydrogen atom Hamiltonian and

$$\hat{H}_{FS} = -\frac{\hat{p}^4}{8m_e^3 c^2} + \frac{1}{2} \cdot \left[\left(\frac{1}{4\pi\varepsilon_0}\right) \cdot \frac{e^2}{m_e^2 c^2} \cdot \frac{1}{r^3}\left(\hat{L} \cdot \hat{S}\right)\right]$$
$$+ \left(\frac{1}{4\pi\varepsilon_0}\right) \frac{e^2 \cdot \hbar^2 \pi}{2m_e^2 c^2} \delta(r) \tag{8.3}$$

is the Hamiltonian that describes the fine-structure corrections discussed in Chapter 7. The effects of the externally applied fields are contained in $\hat{H}_{field}$.

8.2. The Zeeman Effect—The Hydrogen Atom in a Constant Magnetic Field

When an external magnetic field is applied, it is necessary to consider only the spin-orbit portion of $\hat{H}_{FS}$ because this is the term that is sensitive to the atom's internal magnetic field. The effect of the external magnetic field depends on its magnitude relative to the internal field.

We designate the term representing the effects of the magnetic field by $\hat{H}_{field} = \hat{H}_B$. The energy associated with the interaction of the atom having total magnetic moment $\boldsymbol{\mu} = \boldsymbol{\mu}_\ell + \boldsymbol{\mu}_S$ with the external field is given by

$$
\begin{aligned}
\hat{H}_B &= -\hat{\boldsymbol{\mu}} \cdot \boldsymbol{B} \\
&= - \left(\hat{\boldsymbol{\mu}}_\ell + \hat{\boldsymbol{\mu}}_S \right) \cdot \boldsymbol{B} \\
&= \left(\frac{g_\ell \mu_B}{\hbar} \hat{L} + \frac{g_s \mu_B}{\hbar} \hat{S} \right) \cdot \boldsymbol{B} \\
&= \frac{\mu_B}{\hbar} \left(\hat{L} + 2\hat{S} \right) \cdot \boldsymbol{B}
\end{aligned}
\tag{8.4}
$$

where the g-factors, $g_\ell = 1$ and $g_s = 2$, have been used. Also, the spin of the proton has been ignored. This is justified because, as we saw in Chapter 7, the magnetic moment of the proton is roughly 2000 times smaller than either the spin magnetic moment or the electronic orbital magnetic moment. For convenience, we choose the direction of the external field to be the z-direction, $\boldsymbol{B} = B\hat{k}$, so that $\hat{H}_B$ becomes

$$
\hat{H}_B = \frac{\mu_B B}{\hbar} \left(\hat{L}_z + 2\hat{S}_z \right)
\tag{8.5}
$$

and the total Hamiltonian is

$$
\hat{H} = \hat{H}_{Coul} + \hat{H}_{FS} + \frac{\mu_B B}{\hbar} (\hat{L}_z + 2\hat{S}_z)
\tag{8.6}
$$

To decide which of the terms in the Hamiltonian is to be taken as the perturbation and which is to be considered as part of the unperturbed Hamiltonian we must examine the relative magnitudes of the internal and external magnetic fields. The internal field due to the orbital motion of the proton about the electron $B_{orbital}$ is given by Equation (7.12)

$$
B_{orbital} = \frac{1}{4\pi \varepsilon_0} \cdot \frac{e}{m_e c^2 r^3} \cdot L
\tag{8.7}
$$

which may be written in terms of the Bohr magneton as

$$
\begin{aligned}
B_{int} &= \frac{1}{4\pi \varepsilon_0} \cdot \frac{2}{c^2 r^3} \cdot \left(\frac{e\hbar}{2m_e} \right) \cdot \ell \\
&= \frac{1}{4\pi \varepsilon_0} \cdot \frac{2}{c^2 r^3} \cdot \mu_B \cdot \ell
\end{aligned}
\tag{8.8}
$$

where we have replaced the angular momentum with $\ell\hbar$. If we take $n = 2$ ($r \approx 2^2 a_0$) and $\ell = 1$ we find that $B_{int} \approx 1$ Tesla. Therefore, for $B \ll 1$ T we may consider $\hat{H}_B$ as the perturbation to the unperturbed Hamiltonian

$$\hat{H}_0 = \hat{H}_{Coul} + \hat{H}_{FS} \qquad (8.9)$$

On the other hand, for strong fields the unperturbed Hamiltonian should be considered to be

$$\hat{H}_0 = \hat{H}_{Coul} + \hat{H}_B \qquad (8.10)$$

with $\hat{H}_{FS}$ as the perturbation. In either case we write the total energy of the hydrogen atom, to first order, as

$$E_{TOTAL}^{(1)}(n, j, \ell, m_j, m_\ell, m_s) = E_n^{(0)} + E_Z^{(1)}(n, j, \ell, m_j, m_\ell, m_s) \qquad (8.11)$$

where $E_Z^{(1)}$ represents the total effect of the external B-field including the effect on the spin-orbit correction. As usual, $E_n^{(0)}$ is the Bohr energy. Because there is a magnetic moment associated with both the orbital motion of the electron and its spin, in this chapter we distinguish between their magnetic quantum numbers using the designations m_ℓ and m_s, respectively.

Not all quantum numbers $(n, j, \ell, m_j, m_\ell, m_s)$ in Equation (8.11) are present in each case that we treat, but we state them to make clear that they *may* be present in various combinations. Because the Bohr energy will always be part of the total energy we obtain expressions for the Zeeman energy $E_Z^{(1)}$ only, noting that it will contain the effects of the external B-field on the fine-structure as well as the direct effect of the field on the electron. Of course, only the spin-orbit portion of the fine-structure will be affected, but the relativistic correction, because it is the same order of magnitude as the spin-orbit correction, must be included.

We first examine the strong field case followed by the treatment for weak fields. We then solve the problem exactly. The exact solution is sometimes referred to as the intermediate field case for which $\hat{H}_B \sim \hat{H}_{FS}$. It would be aesthetically pleasing to work out the details for the intermediate field case first and then take limits to obtain the strong and weak field solutions. Unfortunately, as we show, the exact solution is sufficiently complicated so that it is only practical to work it out for a specific value of n. The strong and weak field cases are treated for any n. We use SI units for the treatment of the Zeeman effect because the energies are conveniently represented in terms of the Bohr magneton multiplied by the magnitude of the magnetic induction field B.

Strong Field Approximation

Although the spherical symmetry has been broken, there is cylindrical symmetry about the z-axis. In Equation (8.10), $\hat{H}_{coul}$ and $\hat{H}_B$ commute so, even though the spherical symmetry has been broken, ℓ is a good quantum number as are m_ℓ and m_s. The unperturbed energies are therefore

$$E_n^{(0)}(B) = E_n^{(0)} + \mu_B B(m_\ell + 2m_s) \qquad (8.12)$$

Equation (8.12) makes it clear that the original hydrogen atom energy levels are split and the magnitudes of the splittings depend on the magnetic quantum numbers

m_ℓ and m_s. This is, of course, the reason that they are called "magnetic" quantum numbers. The field-free degeneracy of $2(2\ell + 1)$ (the "extra" factor of 2 accounts for the two possible spin states) has been reduced by the breaking of the spherical symmetry. The only degeneracy that remains is that for states which have the same values of $(m_\ell + 2m_s)$. Because $2m_s = \pm 1$, states for which $[m_\ell; m_s = 1/2]$ and $[(m_\ell + 2); m_s = -1/2]$ will be degenerate where, of course, m_ℓ and $m_\ell + 2$ must be among the allowed values.

To evaluate $E_{FS}^{(1)}$, the contribution of fine-structure to the strong field energy, we must compute the expectation values of $\hat{H}_{FS}$ using the unperturbed wave functions, which in this case are the uncoupled set $|n\,\ell\,m_\ell\,m_s\rangle$. There is no change in the relativistic correction, but the spin-orbit term is different because our previous calculation was performed using the coupled set $|j\,m_j\,\ell s\rangle$. We begin by noting that $\hat{L} \cdot \hat{S}$ may be written as (see Problem 8.1)

$$\hat{L} \cdot \hat{S} = \hat{L}_z \hat{S}_z + \frac{1}{2}(\hat{L}_+ \hat{S}_- + \hat{L}_- \hat{S}_+) \tag{8.13}$$

Because both magnetic quantum numbers are good quantum numbers, even in the presence of the field, the raising and lowering operators cause that portion of the expectation value to vanish because of orthogonality. We are left with

$$\langle \hat{L} \cdot \hat{S} \rangle = m_\ell m_s \hbar^2 \tag{8.14}$$

From Equation (7.35) we obtain the coefficient of $\langle \hat{L} \cdot \hat{S} \rangle$ and find that the correction to the energy due to the spin-orbit interaction is

$$E_{SO}^{(1)} = -\frac{1}{n}\alpha^2 E_n^{(0)} \frac{m_\ell m_s}{\ell(\ell + 1/2)(\ell + 1)} \tag{8.15}$$

From Equation (7.29) the relativistic term is

$$E_T^{(1)} = \left[-\frac{3}{4} + \frac{n}{(\ell + 1/2)}\right] \frac{\alpha^2}{n^2} E_n^{(0)} \tag{8.16}$$

Adding Equations (8.15) and (8.16) we obtain the fine-structure correction in the case of a strong externally applied magnetic field

$$E_{FS}^{(1)} = -E_n^{(0)} \left(\frac{\alpha^2}{n}\right) \left\{\frac{3}{4n} - \left[\frac{\ell(\ell + 1) - m_\ell m_s}{\ell(\ell + 1/2)(\ell + 1)}\right]\right\} \tag{8.17}$$

Note that the term involving the quantum number ℓ is indeterminate for $\ell = 0$. It can be shown that in this case this term is unity, but we do not pursue this further.

The Zeeman energy $E_Z^{(1)}$ in the strong field limit is the sum of $E_{FS}^{(1)}$ in Equation (8.17) and the unperturbed energy $E_n^{(0)}(B)$ of Equation (8.12). Adding these contributions to the energy gives

$$E_Z^{(1)}(n, \ell, m_\ell, m_s) = -E_n^{(0)} \left(\frac{\alpha^2}{n}\right) \left\{\frac{3}{4n} - \left[\frac{\ell(\ell + 1) - m_\ell m_s}{\ell(\ell + 1/2)(\ell + 1)}\right]\right\}$$
$$+ \mu_B B (m_\ell + 2m_s) \tag{8.18}$$

Weak Field Approximation

In the weak field case it is assumed that $\hat{H}_B \ll \hat{H}_{FS}$ so that $\hat{H}_B$ is considered to be a perturbation to the unperturbed Hamiltonian $\hat{H}_0 = \hat{H}_{Coul} + \hat{H}_{FS}$. Because the eigenfunctions of this unperturbed Hamiltonian are the coupled states, we find the first-order shift in energy caused by application of the weak field by evaluating the expectation value of $\hat{H}_B$ on these coupled states.

Because the coupled eigenfunctions are eigenfunctions of neither $\hat{L}_z$ nor $\hat{S}_z$ we must express them in terms of the uncoupled eigenfunctions in order to evaluate $E_B^{(1)}$, the expectation value of $\hat{H}_B$. We can, however, simplify our task by noting that we can write $\hat{H}_B$ in terms of $\hat{J}_z$ and $\hat{S}_z$. We do this because the coupled eigenfunctions are indeed eigenfunctions of $\hat{J}_z$. We thus obtain

$$\hat{H}_B = \frac{\mu_B B}{\hbar}(\hat{J}_z + \hat{S}_z) \tag{8.19}$$

which leads to

$$E_B^{(1)} = \mu_B B \left\langle \left(j = \ell \pm \frac{1}{2}\right) m_j \left| \left(\hat{J}_z + \hat{S}_z\right) \right| \left(j = \ell \pm \frac{1}{2}\right) m_j \right\rangle \tag{8.20}$$

The expectation value of $\hat{J}_z$ is simply $m_j \hbar$. To evaluate $\langle \hat{S}_z \rangle$ we express the coupled wave functions in terms of the uncoupled using TABLE 8.1 which contains the Clebsch–Gordan coefficients for $j_2 = s = 1/2$. Regrettably, the symbol α is used for both the fine-structure constant and the spin-up spin state, but this should cause no confusion inasmuch as the latter occurs only as a bra or a ket.

Dropping the ℓ and the s from the designation of the coupled ket so the notation is $|j\, m_j\rangle = |(\ell \pm 1/2)\, m_j\rangle$ we have

$$\left|\left(\ell + \frac{1}{2}\right) m_j\right\rangle = \sqrt{\frac{\ell + m_j + \frac{1}{2}}{2\ell + 1}}\, |\ell\, m_\ell\, \alpha\rangle + \sqrt{\frac{\ell - m_j + \frac{1}{2}}{2\ell + 1}}\, |\ell\, m_\ell\, \beta\rangle \tag{8.21}$$

and

$$\left|\left(\ell - \frac{1}{2}\right) m_j\right\rangle = -\sqrt{\frac{\ell - m_j + \frac{1}{2}}{2\ell + 1}}\, |\ell\, m_\ell\, \alpha\rangle + \sqrt{\frac{\ell + m_j + \frac{1}{2}}{2\ell + 1}}\, |\ell\, m_\ell\, \beta\rangle \tag{8.22}$$

TABLE 8.1. Clebsch–Gordan coefficients for $j_1 = \ell$; $j_2 = s = 1/2$.

$$\left\langle j_1, \frac{1}{2}; m_1, m_2 \middle| j m_j \right\rangle$$

j	$m_2 = +1/2 = m_s = |\alpha\rangle$	$m_2 = -1/2 = m_s = |\beta\rangle$
$\ell + 1/2$	$\sqrt{(\ell + m_j + 1/2)/(2\ell + 1)}$	$\sqrt{(\ell - m_j + 1/2)/(2\ell + 1)}$
$\ell - 1/2$	$-\sqrt{(\ell - m_j + 1/2)/(2\ell + 1)}$	$\sqrt{(\ell + m_j + 1/2)/(2\ell + 1)}$

Note that the first basis ket on the right-hand side of Equation (8.22) contains α. Therefore, m_ℓ in that ket must be $m_\ell = m_j - 1/2$ so that $m_\ell + m_s = m_j$. Similarly, the m_ℓ in the second ket must be $(m_j + 1/2)$. We have

$$\left| \left(\ell \pm \frac{1}{2} \right) m_j \right\rangle = \sqrt{\frac{\ell \pm m_j + \frac{1}{2}}{2\ell + 1}} \left| \ell(m_{j-1/2}) \alpha \right\rangle$$

$$+ \sqrt{\frac{\ell \mp m_j + \frac{1}{2}}{2\ell + 1}} \left| \ell(m_{j+1/2}) \beta \right\rangle \qquad (8.23)$$

The expectation value of $\hat{S}_z$ is then

$$\langle \hat{S}_z \rangle = \frac{\hbar}{2} \left(\frac{\ell \pm m_j + \frac{1}{2}}{2\ell + 1} - \frac{\ell \mp m_j + \frac{1}{2}}{2\ell + 1} \right)$$

$$= \pm \frac{m_j \hbar}{2\ell + 1} \qquad (8.24)$$

where the plus sign refers to $j = \ell + 1/2$ and the minus sign to $j = \ell - 1/2$. From Equation (8.20), the first-order correction to the energy, the Zeeman splitting, is

$$E_B^{(1)} = \mu_B B m_j \left(1 + \frac{1}{2\ell + 1} \right) \qquad j = \ell + 1/2$$

$$= \mu_B B m_j \left(1 - \frac{1}{2\ell + 1} \right) \qquad j = \ell - 1/2 \qquad (8.25)$$

The quantity in brackets is called the Landé g-factor. It is usually written in terms of j and ℓ for which it is given by

$$g(j, \ell) = 1 + \frac{j(j+1) - \ell(\ell+1) + 3/4}{2j(j+1)} \qquad (8.26)$$

The Zeeman energy for the weak field case is the sum of the fine-structure energy given in Equation (7.48) and $E_B^{(1)}$ from Equation (8.25). We have

$$E_Z^{(1)}(n, \ell, m_j)$$

$$= E_n^{(0)} \frac{\alpha^2}{n^2} \left[\frac{n}{(\ell+1)} - \frac{3}{4} \right] + \mu_B B m_j \left(1 + \frac{1}{2\ell + 1} \right) \quad \text{for } j = \ell + 1/2$$

$$= E_n^{(0)} \frac{\alpha^2}{n^2} \left[\frac{n}{\ell} - \frac{3}{4} \right] + \mu_B B m_j \left(1 - \frac{1}{2\ell + 1} \right) \quad \text{for } j = \ell - 1/2$$

$$\qquad (8.27)$$

which may be written in terms of j as

$$E_Z^{(1)}(n, j, m_j)$$

$$= E_n^{(0)} \frac{\alpha^2}{n^2} \left[\frac{n}{(j + 1/2)} - \frac{3}{4} \right] + \mu_B B m_j \left(\frac{2j + 1}{2j} \right) \quad \text{for } j = \ell + 1/2$$

$$= E_n^{(0)} \frac{\alpha^2}{n^2} \left[\frac{n}{(j + 1/2)} - \frac{3}{4} \right] + \mu_B B m_j \left(\frac{2j + 1}{2j + 2} \right) \quad \text{for } j = \ell - 1/2$$

$$(8.28)$$

Notice that the first terms on the right-hand sides of Equations (8.27) and (8.28) are just the total fine-structure correction derived in Chapter 7. As such, they do not contain the field B. This is because this is the "weak field" result; that is, to first order, the fine-structure is unaffected by the presence of the (weak) external field. This is in contrast to the strong field case in which the fine-structure term is modified by the external field [see Equation (8.17)].

An interesting special case occurs when the energy of the "top of the ladder" state is investigated (the bottom of the ladder works too). In this case the coupled and uncoupled wave functions are the same and it is possible to find the expectation value of $(\hat{H}_{FS} + \hat{H}_B)$ immediately. At the top of the ladder the coupled and uncoupled wave functions are

$$|j; m_j = j; \ell; s\rangle_{\text{coupled}} = |\ell; s; m_\ell = \ell; m_s = 1/2\rangle_{\text{uncoupled}} \quad (8.29)$$

In terms of j, the quantum numbers have the values

$$m_j = j; \ m_\ell = \ell = j - 1/2; \ m_s = +1/2 \quad (8.30)$$

so that

$$m_\ell + 2m_s = j + 1/2 \quad (8.31)$$

Using Equations (7.49), (8.12), and (8.31) we have for the top of the ladder state

$$E_Z^{(1)} = E_n^{(0)} \frac{\alpha^2}{n^2} \left[\frac{n}{(j + 1/2)} - \frac{3}{4} \right] + \mu_B B (j + 1/2) \quad (8.32)$$

Note that for this state the strong and weak field expressions, Equations (8.18) and (8.28) should reduce to Equation (8.32). This is left as an exercise at the end of this chapter (see Problem 8.4).

As an example, we consider the effect of a weak field on the $n = 1$ and 2 states of hydrogen. Fine-structure corrections cause the $1s_{1/2}$ ground state energy to lower by 1.46 cm^{-1}, but it does not split. Because, for a given value of B, $E_B^{(1)}$ depends only on m_j it is clear that the $1s_{1/2}$ level splits into two states corresponding to $m_j = \pm 1/2$. The first excited state, $n = 2$, shifts *and* splits. Ignoring the Lamb shift, which is much smaller, the $n = 2$ level splits into three states, $2s_{1/2}$, $2p_{1/2}$,

TABLE 8.2. Landé g-factors for the fine-structure
states of the first two Bohr states of hydrogen.

State	ℓ	$g(j, \ell)$
$1s_{1/2}$	0	2
$2s_{1/2}$	0	2
$2p_{1/2}$	1	2/3
$2p_{3/2}$	1	4/3

and $2p_{3/2}$ (see Section 7.6). The $2s_{1/2}$ and $2p_{1/2}$ states are degenerate and of lower energy than the $2p_{3/2}$. As occurred for $1s_{1/2}$, $2s_{1/2}$, and $2p_{1/2}$ each split into two states corresponding to $m_j = \pm 1/2$ under the influence of the external B field. The $2p_{3/2}$ state splits into four levels corresponding to $m_j = \pm 1/2, \pm 3/2$. The magnitudes of the splittings differ though, depending on the Landé g-factor. The splitting will be a multiple of what we may consider to be the basic energy unit for a given value of the field, $E^{(1)}(B) = \mu_B B$, which, for simplicity, we denote as E_B. TABLE 8.2 is a listing of the $g(j, \ell)$ for each fine-structure state.

FIGURE 8.1 contains a diagram that shows the splitting for a fixed value of B. Note that the splittings of the $2s_{1/2}$ and $2p_{1/2}$ levels, which are degenerate in the absence of the field, are different. The vertical arrows in the diagram represent the $n = 2 \to 1$ radiative transitions that would be observed (in emission) in the laboratory. These transitions obey the selection rule $\Delta m_j = 0, \pm 1$. There are no transitisons connecting the $2s_{1/2}$ and $1s_{1/2}$ states because the selection rule $\Delta \ell = \pm 1$ forbids transitions between s-states. There could, however, be transitions between the $2s$ and $2p$ levels because there is no selection rule on n. Such transitions

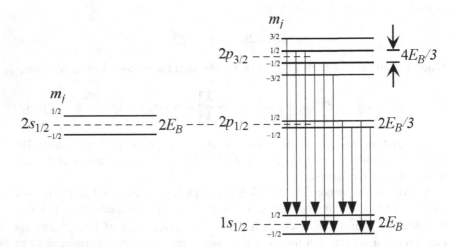

FIGURE 8.1. Weak field Zeeman splitting of the first two Bohr levels of hydrogen.

are omitted from the diagram. The origin of these "selection rules" is discussed in Chapter 13.

Intermediate Field—Exact Treatment

The correspondence between the weak and strong field energies in the special cases of the top and bottom of the ladder states indicates that these energies are not only the same, but that they are exact. Indeed, this is the case as is easily seen by noting that the states at the top and bottom of the ladder are eigenfunctions of the entire Hamiltonian, Equation (8.6). In the language of matrix mechanics this means that there are no off-diagonal elements involving these wave functions as basis vectors that "contaminate" the diagonal elements. To illustrate, consider the hypothetical Hamiltonian represented by the 3×3 matrix

$$\hat{H} = \begin{pmatrix} a & 0 & 0 \\ 0 & b_1 & b_2 \\ 0 & b_3 & b_4 \end{pmatrix} \tag{8.33}$$

where a and the bs are nonzero. The secular equation that results from the solution of the Schrödinger equation has the form

$$\begin{vmatrix} (a - E) & 0 & 0 \\ 0 & (b_1 - E) & b_2 \\ 0 & b_3 & (b_4 - E) \end{vmatrix} = 0 \tag{8.34}$$

from which it is clear that the matrix element a is indeed one of the eigenvalues. This is precisely the case that we have in the Zeeman effect for the top and bottom of the ladder states.

Solving for the energy shift in the presence of an external field for the intermediate case really amounts to solving the problem exactly. This would already have been done in the previous section if it were not for the degeneracy of the problem. It is this degeneracy that causes there to be nonzero off-diagonal matrix elements. Because the Bohr energy does not depend upon any quantum numbers other than n we take the Hamiltonian to be

$$\hat{H}_Z = \hat{H}_{FS} + \hat{H}_B \tag{8.35}$$

To keep the computation tractable we calculate the energies of the $n = 2$ states of hydrogen. We wish to determine the cases for which there will be degeneracy. If we examine the energy in the weak field case, Equation (8.28), it is not immediately obvious which states will be degenerate. If, however, we examine the strong field case, Equation (8.18), we see that at very high field the second term, $\mu_B B (m_\ell + 2m_s)$, dominates. For $n = 2$ there are eight states (two spin states for each spatial state), but some are degenerate. For insight into the degeneracy we first examine the strong field case. Usually, we imagine a perturbation due to an external field as breaking, at least partially, a degeneracy. If, on the other hand, we imagine the field to be so strong that only the second term in Equation (8.18)

TABLE 8.3. Possible quantum numbers for $n = 2$ used to reveal the degree of degeneracy in the strong field limit. Energies of a few states are included in the last column.

	m_ℓ	m_s	m_j	j	$(m_\ell + 2m_s)$	$E_Z^{(1)}(2, j, m_j)$
$\ell = 0$	0	1/2	1/2	1/2	1	$-5 \cdot \left(\dfrac{R_H \alpha^2}{64} \right) + \mu_B B$
	0	−1/2	−1/2	1/2	−1	$-5 \cdot \left(\dfrac{R_H \alpha^2}{64} \right) - \mu_B B$
$\ell = 1$	1	1/2	3/2	3/2	2	$-\left(\dfrac{R_H \alpha^2}{64} \right) + 2\mu_B B$
	0	1/2	1/2		1	
	−1	1/2	−1/2		0	
	1	−1/2	1/2		0	
	0	−1/2	−1/2		−1	
	−1	−1/2	−3/2	3/2	−2	$-\left(\dfrac{R_H \alpha^2}{64} \right) - 2\mu_B B$

determines the energy then we may, in this case, imagine the perturbation to occur when the magnetic field is *decreased*. This is, of course, equivalent to regarding the strong field states as the unperturbed states and the perturbation to be the fine-structure so the strong field degeneracy is removed by turning on the fine-structure.

We may find the degree of degeneracy for the strong field case by examining the possible combinations of the quantum numbers m_ℓ and m_s, that is, the energies for very strong fields. TABLE 8.3 is a listing of the possible quantum numbers and the combinations $(m_\ell + 2m_s)$ used to determine the degree of degeneracy. There are only five different values of $(m_\ell + 2m_s)$. Therefore, at very high fields there will be only five distinct energies so the matrix representing $\hat{H}_Z$, see Equation (8.35), will not be diagonal in either the coupled or uncoupled basis set.

Also included in TABLE 8.3 are the values of m_j and j (where possible). We notice that there are only four states for which it is possible to assign all of the uncoupled quantum numbers and all of the coupled quantum numbers. These are clearly the states at the tops and bottoms of the two ladders, the $j = 3/2$ and the $j = 1/2$ ladders. The energies of each of these states will change linearly with the applied B-field [see Equation (8.18)]. These energies are included in the last column of the table.

There are four other states for which the expansion of the coupled ket on the uncoupled basis set will contain more than one uncoupled ket (in this relatively simple case each contains two uncoupled kets). These are the four uncoupled states for which a definite value of j cannot be assigned. To solve the problem we must now actually compute the matrix elements. If we use the coupled kets as the basis set for the Hamiltonian then, when the proper quantum numbers

TABLE 8.4. Clebsch–Gordan coefficients for $\ell = 1$ and $s = 1/2$.

$$\left\langle \ell, \frac{1}{2}; m_\ell, m_s \middle| j m_j \right\rangle$$

j	$m_s = +1/2 \Leftrightarrow |\alpha\rangle$	$m_s = -1/2 \Leftrightarrow |\beta\rangle$
3/2	$\sqrt{(3/2 + m_j)/3}$	$\sqrt{(3/2 - m_j)/3}$
1/2	$-\sqrt{(3/2 - m_j)/3}$	$\sqrt{(3/2 + m_j)/3}$

are inserted, Equation (8.28) gives us the four energies listed in TABLE 8.3. These energies are the diagonal elements of the 4×4 diagonal submatrix of the 8×8 matrix that represents $\hat{H}_z$. The remaining 4×4 submatrix is not diagonal. To find the elements of this submatrix we use Clebsch–Gordan coefficients. This is easily done by modifying Table 3.2 for the present purpose. TABLE 8.4 is a listing of the Clebsch–Gordan coefficients for $j_1 = \ell = 1$ and $j_2 = s = 1/2$.

From TABLE 8.4 we can immediately write the required coupled kets in terms of the uncoupled kets. These are tabulated in TABLE 8.5 where the quantum number designations in the kets are displayed symbolically above the equations; C_α and C_β are constants.

The relevant submatrix is

$$
\begin{pmatrix}
\left\langle \frac{3}{2}\frac{1}{2} \middle| \hat{H}_z \middle| \frac{3}{2}\frac{1}{2} \right\rangle & \left\langle \frac{3}{2}\frac{1}{2} \middle| \hat{H}_z \middle| \frac{1}{2}\frac{1}{2} \right\rangle & 0 & 0 \\
\left\langle \frac{1}{2}\frac{1}{2} \middle| \hat{H}_z \middle| \frac{3}{2}\frac{1}{2} \right\rangle & \left\langle \frac{1}{2}\frac{1}{2} \middle| \hat{H}_z \middle| \frac{1}{2}\frac{1}{2} \right\rangle & 0 & 0 \\
0 & 0 & \left\langle \frac{3}{2}\left(-\frac{1}{2}\right) \middle| \hat{H}_z \middle| \frac{3}{2}\left(-\frac{1}{2}\right) \right\rangle & \left\langle \frac{3}{2}\left(-\frac{1}{2}\right) \middle| \hat{H}_z \middle| \frac{1}{2}\left(-\frac{1}{2}\right) \right\rangle \\
0 & 0 & \left\langle \frac{1}{2}\left(-\frac{1}{2}\right) \middle| \hat{H}_z \middle| \frac{3}{2}\left(-\frac{1}{2}\right) \right\rangle & \left\langle \frac{1}{2}\left(-\frac{1}{2}\right) \middle| \hat{H}_z \middle| \frac{1}{2}\left(-\frac{1}{2}\right) \right\rangle
\end{pmatrix}
$$

(8.36)

TABLE 8.5. Expansion of the coupled kets in terms of the uncoupled kets.

$$|j m_j\rangle = C_\alpha |m_\ell \alpha\rangle + C_\beta |m_\ell \beta\rangle$$

$$\left| \frac{3}{2}\frac{1}{2} \right\rangle = \sqrt{2/3} \, |0\alpha\rangle + \sqrt{1/3} \, |1\beta\rangle$$

$$\left| \frac{3}{2}\left(-\frac{1}{2}\right) \right\rangle = \sqrt{1/3} \, |-1\alpha\rangle + \sqrt{2/3} \, |0\beta\rangle$$

$$\left| \frac{1}{2}\frac{1}{2} \right\rangle = -\sqrt{1/3} \, |0\alpha\rangle + \sqrt{2/3} \, |1\beta\rangle$$

$$\left| \frac{1}{2}\left(-\frac{1}{2}\right) \right\rangle = -\sqrt{2/3} \, |-1\alpha\rangle + \sqrt{1/3} \, |0\beta\rangle$$

Inserting the matrix elements, the secular equation is found to be

$$
\begin{vmatrix}
-\gamma + \frac{2}{3}\mu_B B - E & -\frac{\sqrt{2}}{3}\mu_B B & 0 & 0 \\
-\frac{\sqrt{2}}{3}\mu_B B & -5\gamma + \frac{1}{3}\mu_B B - E & 0 & 0 \\
0 & 0 & -\gamma - \frac{2}{3}\mu_B B - E & -\frac{\sqrt{2}}{3}\mu_B B \\
0 & 0 & -\frac{\sqrt{2}}{3}\mu_B B & -5\gamma - \frac{1}{3}\mu_B B - E
\end{vmatrix} = 0
$$

(8.37)

where we have adopted the notation of Griffiths[1] in which

$$
\gamma = R_H \left(\frac{\alpha}{2^3}\right)^2
$$

$$
= \frac{1}{4}\Delta_2
$$

(8.38)

where Δ_2 is the separation between the fine-structure levels for $n = 2$ as given in Equation (7.50). From Equation (7.49), the $2p_{3/2}$ state is lower than the Bohr energy by γ, and the $2p_{1/2}$ state is lower by 5γ (see also Figure 7.2). This provides a convenient measure of "weak" and "strong" as they pertain to the $n = 2$ level of hydrogen. We wish to find the magnitude of B that makes the energy $\mu_B B$ comparable with the fine-structure splitting for $\Delta_2 = 4\gamma$. From Figure 7.2, $\gamma = 0.091\,\mathrm{cm}^{-1}$. Then

$$
B = \frac{4\gamma}{\mu_B}
$$

$$
= \frac{4\left(0.091\,\mathrm{cm}^{-1}\right)}{\left(0.46\,\mathrm{cm}^{-1}/\mathrm{T}\right)}
$$

$$
\approx 1\,\mathrm{T}
$$

(8.39)

Because the Earth's magnetic field is $\sim 10^{-4}\,\mathrm{T}$ we see that experimental studies of the Zeeman effect for low-lying excited states will be relatively unaffected by the Earth's field. If, however, higher lying states are being studied then effects of the Earth's field are important because of the $1/n^3$ dependence of the fine-structure intervals. Nonetheless, for the case of $n = 2$ we may regard a "weak" field to be $\sim 0.1\,\mathrm{T}$ and a "strong" field to be of magnitude $\sim 10\,\mathrm{T}$.

Returning to Equation (8.37), it is clear that it leads to a quartic equation, but, conveniently, this equation is already factored into the product of two quadratic

TABLE 8.6. Energies for each of the Zeeman states in coupled ket notation for $n = 2$.

$|jm_j\rangle$	ℓ	Energy (exact)	Energy ($B \to \infty$)
$\left|\frac{1}{2}\frac{1}{2}\right\rangle$	$\ell = 0$	$-5\gamma + \mu_B B$	$\mu_B B$
$\left|\frac{1}{2} -\frac{1}{2}\right\rangle$	$\ell = 0$	$-5\gamma - \mu_B B$	$-\mu_B B$
$\left|\frac{3}{2}\frac{3}{2}\right\rangle$	$\ell = 1$	$-\gamma + 2\mu_B B$	$2\mu_B B$
$\left|\frac{3}{2} -\frac{3}{2}\right\rangle$	$\ell = 1$	$-\gamma - 2\mu_B B$	$-2\mu_B B$
$\left|\frac{3}{2}\frac{1}{2}\right\rangle$	$\ell = 1$	$-3\gamma + \frac{1}{2}\mu_B B + \sqrt{4\gamma^2 + \frac{2}{3}\gamma\mu_B B + \frac{1}{4}(\mu_B B)^2}$	$\mu_B B$
$\left|\frac{1}{2}\frac{1}{2}\right\rangle$	$\ell = 1$	$-3\gamma + \frac{1}{2}\mu_B B - \sqrt{4\gamma^2 + \frac{2}{3}\gamma\mu_B B + \frac{1}{4}(\mu_B B)^2}$	0
$\left|\frac{3}{2} -\frac{1}{2}\right\rangle$	$\ell = 1$	$-3\gamma - \frac{1}{2}\mu_B B + \sqrt{4\gamma^2 - \frac{2}{3}\gamma\mu_B B + \frac{1}{4}(\mu_B B)^2}$	0
$\left|\frac{1}{2} -\frac{1}{2}\right\rangle$	$\ell = 1$	$-3\gamma - \frac{1}{2}\mu_B B - \sqrt{4\gamma^2 - \frac{2}{3}\gamma\mu_B B + \frac{1}{4}(\mu_B B)^2}$	$-\mu_B B$

equations. The quadratic equation from the first 2×2 submatrix is

$$E^2 + [6\gamma - \mu_B B]E + \left(5\gamma^2 - \tfrac{11}{3}\gamma\mu_B B\right) = 0 \qquad (8.40)$$

At this point it is worth noting which solutions will be obtained by solving Equation (8.40). Examination of the ordering of the coupled basis kets in the matrix of (8.36) makes it clear that the solutions to Equation (8.40) will yield the energies for $\left(j = 3/2\,;\, m_j = +1/2\right)$ and $\left(j = 1/2\,;\, m_j = +1/2\right)$, both of which were formed with $\ell = 1$ (we have already solved the case for $\ell = 0$). Moreover, examination of the second 2×2 submatrix reveals that the energies derived from it can be obtained from the solutions to Equation (8.40) by making the substitution $B \to -B$. All of the energies for the $n = 2$ state of hydrogen are listed in TABLE 8.6 together with the strong field limits of these energies.

Note that there are exactly five asymptotic energies, consistent with the number of different combinations of $(m_\ell + 2m_s)$ in TABLE 8.3. Of course, the last four entries in the middle column reduce to the weak field limit for $\mu_B B \ll \gamma$.

It is instructive to examine the graphs of eight Zeeman energy levels for $n = 2$ that are listed in TABLE 8.6. FIGURE 8.2 shows a graph of these eight energies for "weak" fields. The dashed lines represent the two states for which $\ell = 0$ and are two of the top of the ladder states. The other two are the $m_j = \pm 3/2$. The energies of these four states are linear in field strength B as shown in TABLE 8.6.

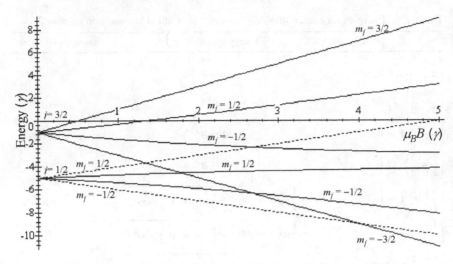

FIGURE 8.2. Exact Zeeman energies for $n = 2$ in the weak magnetic field regime.

As the field is increased so that $\mu_B B$ is comparable with the fine structure splitting, as may be seen in FIGURE 8.3, the eight distinct weak field energies tend toward the five strong field energies, but are nonetheless distinct.

If the field is increased to values at which $\mu_B B$ dominates the fine-structure splitting, the five distinct energies discussed in the strong field approximation are approached as is clearly shown in FIGURE 8.4.

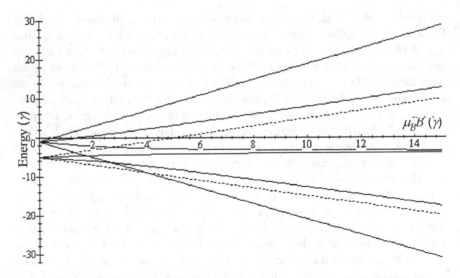

FIGURE 8.3. Exact Zeeman energies for $n = 2$ in the intermediate magnetic field regime.

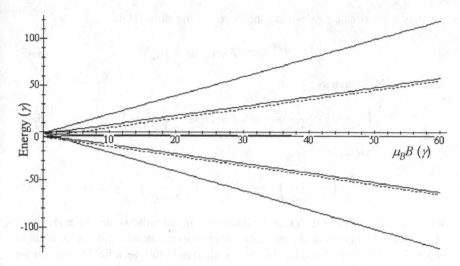

FIGURE 8.4. Exact Zeeman energies for $n = 2$ in the strong magnetic field regime.

8.3. Weak Electric Field—The Quantum Mechanical Stark Effect

The interaction of atoms with an external electric field is referred to as the Stark effect. When an atom is immersed in an electric field the Hamiltonian is the interaction energy of an electric dipole moment in the field. Thus, the Stark Hamiltonian $\hat{H}_S$ for an electric field in the z-direction $\boldsymbol{F} = F\hat{\boldsymbol{k}}$ is analogous to Equation (8.4) for a magnetic field. Because the spin is a magnetic property, it may be ignored so the Stark Hamiltonian is

$$\begin{aligned} \hat{H}_S &= -\hat{\boldsymbol{p}} \cdot \boldsymbol{F} \\ &= e\boldsymbol{r} \cdot \boldsymbol{F} \\ &= eFz \end{aligned} \tag{8.41}$$

where $\hat{\boldsymbol{p}} = e\boldsymbol{r}$ is the electric dipole moment of the atom. Recall that we are not using the "hat" for coordinate operators.

At this point it is necessary to discuss the meaning of the term "weak electric field". We wish to compare a typical Stark energy with the separations between adjacent n-states and with the fine-structure intervals Δ_n [see Equation (7.50)]. It is customary to define "weak electric field" to be a field such that the Stark energy is much less than the separation between adjacent n-states, but much larger than the fine-structure interval. We may thus write

$$\Delta_n \ll eFz \ll \Delta E_n^{(0)} \tag{8.42}$$

We may approximate z by $\sim n^2 a_0$, the "size" of the atom. Thus,

$$\frac{1}{2} \left| E_n^{(0)} \right| \frac{\alpha^2}{n} \ll e F n^2 a_0 \ll \frac{2}{n} \left| E_n^{(0)} \right| \tag{8.43}$$

which may be written as

$$\frac{1}{2} \left(\frac{e^2}{4\pi\varepsilon_0} \right) \cdot \frac{1}{2n^2 a_0} \frac{\alpha^2}{n} \ll e F n^2 a_0 \ll \frac{1}{n} \left(\frac{e^2}{4\pi\varepsilon_0} \right) \cdot \frac{1}{n^2 a_0} \tag{8.44}$$

Isolating F we have

$$\frac{\alpha^2}{4} \left(\frac{e}{4\pi\varepsilon_0 a_0^2} \right) \cdot \frac{1}{n^5} \ll F \ll \left(\frac{e}{4\pi\varepsilon_0 a_0^2} \right) \cdot \frac{1}{n^5} \tag{8.45}$$

We have written (8.45) such that the quantities in parenthesis are the atomic unit of electric field. In commonly used laboratory units the atomic unit of electric field is 5.142×10^9 V/cm. Thus, for the electric field to fit the weak field definition we must have

$$\frac{6.9 \times 10^4}{n^5} \text{ V/cm} \ll F \ll \frac{5.142 \times 10^9}{n^5} \text{ V/cm} \tag{8.46}$$

We note that from the point of view of the Bohr levels this criterion does indeed yield a weak field, but from the point of view of the fine-structure interval it is actually a strong field.

In accord with the weak field assumption in which $\hat{H}_{FS}$ is assumed small the unperturbed Hamiltonian is the Coulomb Hamiltonian; Equation (8.2) and the unperturbed eigenfunctions may be taken to be either the spherical hydrogen atom wave functions $|n\ell m\rangle$ or the parabolic eigenfunctions $|n\, n_1\, n_2\, m\rangle$. Because electron spin is being ignored, we forgo the use of subscripts such as m_ℓ and designate the magnetic quantum number by m in this discussion of the Stark effect.

The first-order change in the energy caused by the electric field is

$$E_S^{(1)} = \langle \hat{H}_S \rangle$$
$$= e F \langle z \rangle \tag{8.47}$$

The ground state is nondegenerate so the spherical and parabolic eigenfunctions for the ground state of hydrogen are the same. We elect to use spherical eigenfunction notation for the ground state so the first-order correction to the energy is

$$E_S^{(1)}(n = 1) = \langle \hat{H}_S \rangle$$
$$= e F \langle 100 | z | 100 \rangle \tag{8.48}$$

This integral vanishes because the nondegenerate ground state has definite parity. That is, the product of the square of the wave function (even) and z (odd) is odd so the integral vanishes. Because the first-order correction vanishes we must use second-order perturbation theory to obtain the magnitude of the effect of the field. From perturbation theory the second-order correction to the energy due to a perturbation $\hat{H}^{(1)}$ is

$$E_n^{(2)} = \sum_{\substack{k=1 \\ k \neq n}}^{\infty} \frac{|\langle n|\hat{H}^{(1)}|k\rangle|^2}{E_n^{(0)} - E_k^{(0)}} \tag{8.49}$$

where the bras and kets are the eigenkets of the unperturbed Hamiltonian.

Applying Equation (8.49) to the ground state hydrogen atom in a constant electric field we obtain

$$\begin{aligned}
E_1^{(2)} = \sum_{n \neq 1}^{\infty} \sum_{\ell=0}^{n-1} \sum_{m=0}^{\ell} e^2 F^2 \frac{|\langle 100|z|n\ell m\rangle|^2}{E_1^{(0)} - E_n^{(0)}} \\
+ \sum_{k} e^2 F^2 \frac{|\langle 100|z|k\rangle|^2}{E_1^{(0)} - E_k^{(0)}}
\end{aligned} \tag{8.50}$$

where the first sum in Equation (8.50) represents the sum over the bound states and the second the sum over continuum states as characterized by their momentum k. The reason the second sum is necessary is that the summation must be over a *complete* set of eigenfunctions. It is clear that $E_1^{(2)}$ will lower the ground state energy because both denominators are manifestly negative.

We may obtain an upper bound to $E_1^{(2)}$ by replacing all of the $E_n^{(0)}$ in the denominators of Equation (8.50) by $E_2^{(0)}$. This leads to

$$E_1^{(2)} < \frac{e^2 F^2}{E_1^{(0)} - E_2^{(0)}} \sum_{q, n \neq 1} |\langle 100|z|q\rangle|^2 \tag{8.51}$$

where we have used q to designate all quantum numbers, bound state, and continuum. We may further simplify the summation in Equation (8.51) using the previously established integral relation $\langle 100|z|100\rangle = 0$.

$$\begin{aligned}
\sum_{q, n \neq 1} |\langle 100|z|q\rangle|^2 &= \sum_{q} |\langle 100|z|q\rangle|^2 - |\langle 100|z|100\rangle|^2 \\
&= \sum_{q} \langle 100|z|q\rangle\langle q|z|100\rangle \\
&= \langle 100|z^2|100\rangle \\
&= \frac{1}{3}\langle 100|r^2|100\rangle \\
&= a_0^2
\end{aligned} \tag{8.52}$$

Also used in obtaining the result of Equation (8.52) was the spherical symmetry of the ground state, $\langle z^2 \rangle = \frac{1}{3}\langle r^2 \rangle$, the completeness relation[2]

$$\sum_{q} |q\rangle\langle q| = 1 \tag{8.53}$$

and the expectation value of r^2 in the ground state of hydrogen[3], $3a_0^2$.

Putting this in (8.51) and replacing the energies by their values we have

$$E_1^{(2)} < \frac{e^2 F^2}{\left(\frac{3}{4} E_1^{(0)}\right)} 3a_0^2$$

$$< \frac{4e^2 F^2}{3}\left[-\frac{(4\pi\varepsilon_0)\,2a_0}{e^2}\right] a_0^2$$

$$< -(4\pi\varepsilon_0)\left(\frac{8}{3}\right) F^2 a_0^3 \tag{8.54}$$

The sum in Equation (8.50) can, in fact, be evaluated exactly.[4] The answer is

$$E_1^{(2)} = -\frac{9}{4}\,(4\pi\varepsilon_0)\,a_0^3 F^2 \tag{8.55}$$

which authenticates that the result in (8.54) is indeed an upper bound.

Notice that $E_1^{(2)} \propto F^2$. In general, if there is no *permanent* electric dipole moment, a moment may be *induced* by the field. The relationship is linear so that the induced dipole moment p_{in}

$$p_{in} = \alpha_d F \tag{8.56}$$

where p_{in} represents an electric dipole moment and α_d is the dipole polarizability. The dipole polarizability is often represented simply by α; we use the subscript d to clearly differentiate it from the fine-structure constant. (The number of symbols available is limited and α seems to be a favorite among physicists.) According to Equation (8.56) α_d is the rate of change of the dipole moment with respect to the applied electric field. That is,

$$\alpha_d = \frac{dp_{in}}{dF} \tag{8.57}$$

In general, the dipole moment reflects the change in the energy with respect to the field F so

$$p = -\frac{dE}{dF} \tag{8.58}$$

and

$$\alpha_d = -\frac{d^2 E}{dF^2} \tag{8.59}$$

Therefore, from Equation (8.55) we obtain the exact value of the polarizability of the ground state of hydrogen

$$\alpha_d = \frac{9}{2}\,(4\pi\varepsilon_0)\,a_0^3 \tag{8.60}$$

Now, how about excited states? As we have seen previously, as a result of the accidental degeneracy, excited states of hydrogen can have permanent dipole moments, the z-components of which are given by $(3/2)\,n\,(n_1 - n_2)$. This is easily seen mathematically when it is realized that *any* linear combination of eigenfunctions having the same eigenvalue, that is, the same n, is an eigenfunction with energy $E_n^{(0)}$. Such a linear combination can be as skewed with respect to the xy-plane as we please and thus have a permanent electric dipole moment. Inspection of Equation (8.41) shows that the linear Stark effect will be observed for excited states of hydrogen.

The calculation of the Stark energies for excited states can be performed using spherical eigenfunctions and degenerate perturbation theory in a manner analogous to the method employed to solve for the Zeeman energies at intermediate magnetic fields. This calculation is carried out for the $n = 2$ states as an example of degenerate perturbation in textbooks on quantum mechanics.[5] The result is that, as expected, the accidental degeneracy is partially lifted by the electric field. States with the same value of the quantum number m are mixed and the energies corresponding to these mixed spherical states are shifted by an amount $\pm 3ea_0 F$ (note the correct units of energy, dipole moment times electric field). The wave function corresponding to the perturbed states are

$$|\psi_\pm\rangle = \frac{1}{\sqrt{2}}\left(|210\rangle \mp |200\rangle\right)$$
$$|\psi_1\rangle = |211\rangle$$
$$|\psi_2\rangle = |21-1\rangle \tag{8.61}$$

where the last two wave functions correspond to unshifted energies. The results of the degenerate perturbation theory calculation are summarized in FIGURE 8.5 where the energies are in atomic units.

It was remarked previously that for the hydrogen atom the Schrödinger equation is separable in parabolic coordinates even if there is an applied electric field. For this reason the parabolic eigenfunctions are sometimes referred to as Stark eigenfunctions. Indeed, the linear combinations of spherical eigenfunctions deduced from degenerate perturbation theory *are* parabolic eigenfunctions. That is, using

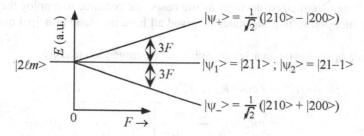

FIGURE 8.5. Schematic diagram showing the level splitting of the $n = 2$ energy level of hydrogen under the influence of an external electric field F. The kets represent spherical eigenfunctions.

the results of Problem 6.6, *viz.*

$$|2001\rangle_{par} = |211\rangle_{sph}$$

$$|2100\rangle_{par} = \frac{1}{\sqrt{2}}(|210\rangle_{sph} + |200\rangle_{sph})$$

$$|2010\rangle_{par} = \frac{1}{\sqrt{2}}(|210\rangle_{sph} - |200\rangle_{sph})$$

$$|211-1\rangle_{par} = |21-1\rangle_{sph} \qquad (8.62)$$

it is clear that the kets in Equation (8.61) are related to the parabolic kets in Equation (8.62) according to

$$|\psi_+\rangle_{sp} = |2010\rangle_{par}$$

$$|\psi_-\rangle_{sp} = |2100\rangle_{par} \qquad (8.63)$$

Of course, there is also a one-to-one correspondence between the remaining two eigenfunctions. This suggests that it may be simpler to treat the first-order Stark effect in hydrogen using parabolic coordinates instead of spherical coordinates.

To see how this solution follows immediately if parabolic eigenfunctions are used we must write the Stark Hamiltonian in terms of the Lenz vector. We suspect from the classical treatment of Chapter 5 that $\hat{p}_z \propto \hat{A}_z$. Our approach shows that, in fact, the quantum mechanical operator $\hat{p}_z = (3/2)\,n^2\hat{A}_z$ in atomic units as was found classically. This can be demonstrated by showing that the matrices representing these operators are proportional using, of course, the same basis set. Because $\hat{A}_z$ is one of the commuting operators used in the separation of the Schrödinger equation in parabolic coordinates the corresponding matrix using parabolic eigenfunctions as the basis functions is, by definition, diagonal. We would then, however, have to compute the matrix elements of z using parabolic coordinates. An alternate method is to obtain the matrices of $\hat{p}_z$ and $\hat{A}_z$ using spherical hydrogen atom eigenfunctions as the basis. Because we have already derived the action of $\hat{A}_z$ on a spherical eigenfunction we elect to use spherical hydrogen atom eigenfunctions as the basis. We continue to employ the weak field assumption so that the states involved all have the same principal quantum number n.

Now, z operating on an arbitrary spherical eigenfunction leads to

$$zR_{n\ell}(r)Y_\ell^m = r\cos\theta\, R_{n\ell}(r)Y_\ell^m(\theta,\phi)$$

$$= rR_{n\ell}(r)\sqrt{\frac{(\ell+m+1)(\ell-m+1)}{(2\ell+1)(2\ell+3)}}\, Y_{\ell+1}^m(\theta,\phi)$$

$$+ rR_{n\ell}(r)\sqrt{\frac{(\ell+m)(\ell-m)}{(2\ell+1)(2\ell-1)}}\, Y_{\ell-1}^m(\theta,\phi) \qquad (8.64)$$

where we have used the relation

$$\cos\theta Y_\ell^m(\theta,\phi) = \sqrt{\frac{(\ell+m+1)(\ell-m+1)}{(2\ell+1)(2\ell+3)}}Y_{\ell+1}^m(\theta,\phi)$$

$$+ \sqrt{\frac{(\ell+m)(\ell-m)}{(2\ell+1)(2\ell-1)}}Y_{\ell-1}^m(\theta,\phi) \qquad (8.65)$$

Thus, the diagonal elements vanish and the only nonzero elements of the z-matrix are those for which $\ell' = \ell \pm 1$ and $m = m'$; the nonzero matrix elements are

$$\langle n\ell m|z|n(\ell-1)m\rangle = \sqrt{\frac{(\ell-m)(\ell+m)}{(2\ell-1)(2\ell+1)}}R_{n\ell}^{n(\ell-1)} \qquad (8.66)$$

and

$$\langle n\ell m|z|n(\ell+1)m\rangle = \sqrt{\frac{(\ell-m+1)(\ell+m+1)}{(2\ell+1)(2\ell+3)}}R_{n\ell}^{n(\ell+1)} \qquad (8.67)$$

where

$$R_{n\ell}^{n\ell'} = \int r^3 R_{n\ell}(r) R_{n\ell'}(r)dr \qquad (8.68)$$

The $R_{n\ell}^{n(\ell\pm1)}$ are evaluated in Problem 4.6. They are

$$R_{nl}^{n(\ell-)1} = \frac{3}{2}n\sqrt{n^2-\ell^2}$$

$$R_{nl}^{n(\ell+)1} = \frac{3}{2}n\sqrt{n^2-(\ell+1)^2} \qquad (8.69)$$

The final results for the nonvanishing z-matrix elements are therefore

$$\langle n\ell m|z|n(\ell-1)m\rangle = \frac{3}{2}n\sqrt{\frac{(\ell-m)(\ell+m)(n^2-\ell^2)}{(2\ell-1)(2\ell+1)}} \qquad (8.70)$$

and

$$\langle n\ell m|z|n(\ell+1)m\rangle = \frac{3}{2}n\sqrt{\frac{(\ell-m+1)(\ell+m+1)[n^2-(\ell+1)^2]}{(2\ell+1)(2\ell+3)}} \qquad (8.71)$$

From Equation (6.59) we have

$$\hat{A}_z R_{n\ell}(r)Y_\ell^m(\theta,\phi) = \frac{1}{n}\sqrt{\frac{(\ell-m)(\ell+m)(n^2-\ell^2)}{(2\ell-1)(2\ell+1)}}R_{n(\ell-1)}(r)Y_{\ell-1}^m(\theta,\phi)$$

$$+ \frac{1}{n}\sqrt{\frac{(\ell-m+1)(\ell+m+1)[n^2-(\ell+1)^2]}{(2\ell+1)(2\ell+3)}}R_{n(\ell+1)}(r)Y_{\ell+1}^m(\theta,\phi)$$

so it is clear that the same matrix elements that are nonvanishing for $\hat{A}_z$ are identical to those for z. Moreover, these matrix elements are

$$\langle n\ell m|\hat{A}_z|n(\ell-1)m\rangle = \frac{1}{n}\sqrt{\frac{(\ell-m)(\ell+m)(n^2-\ell^2)}{(2\ell-1)(2\ell+1)}} \qquad (8.72)$$

and

$$\langle n\ell m|\hat{A}_z|n(\ell+1)m\rangle = \frac{1}{n}\sqrt{\frac{(\ell-m+1)(\ell+m+1)[n^2-(\ell+1)^2]}{(2\ell+1)(2\ell+3)}} \qquad (8.73)$$

It is clear then that

$$\langle n\ell m|z|n(\ell\pm1)m\rangle = \tfrac{3}{2}n^2\langle n\ell m|\hat{A}_z|n(\ell\pm1)m\rangle \qquad (8.74)$$

from which we infer that $\hat{p}_z = (3/2)n^2\hat{A}_z$ which is identical to the classical result.

Returning now to the Stark effect, the complete Hamiltonian is

$$\begin{aligned}
\hat{H} &= \hat{H}_0 + \hat{H}_S \\
&= \hat{H}_0 - \hat{p}_z F \\
&= \hat{H}_0 - \frac{3}{2}n^2 F\hat{A}_z
\end{aligned} \qquad (8.75)$$

The parabolic eigenfunctions are, however, eigenfunctions of both $\hat{H}_0$ and $\hat{A}_z$ so we may immediately write the eigenvalues. This is reminiscent of the strong field Zeeman effect for which the spherical eigenfunctions are eigenfunctions of both $\hat{H}_0$ and $\hat{H}_B$. In Chapter 6 we found that the eigenvalues of $\hat{A}_z$ are $(n_2-n_1)/n$ so the total energy shift from the degenerate Bohr levels are

$$E_S(n,n_1,n_2) = \tfrac{3}{2}n(n_1-n_2)F \qquad (8.76)$$

Notice that when the quantum numbers corresponding to $n=2$ are inserted in Equation (8.76) the energies are the same as those obtained using degenerate perturbation theory and the spherical eigenfunctions.

It is convenient to define another parabolic quantum number, the electric quantum number q. (This quantum number is often designated k, but we use q here to avoid confusion with the quantum number associated with $\hat{K}^2$.) Of course, it is dependent upon the quantum numbers that arise naturally from the solution to Schrödinger's equation in parabolic coordinates. We define

$$q \equiv n_1 - n_2 \qquad (8.77)$$

Because of the constraints on n_1 and n_2, q changes in steps of two for fixed m. In terms of the quantum number q the Stark energy is

$$E_{nq}^{(1)} = \tfrac{3}{2}nqF \qquad (8.78)$$

TABLE 8.7. Stark effect parameters in
parabolic coordinates for the $n = 2$
state of hydrogen in atomic units.

n_1	n_2	m	q	$E_{2q}^{(1)}$	p_z
1	0	0	1	$+3F$	$+3$
0	1	0	-1	$-3F$	-3
0	0	1	0	0	0
0	0	-1	0	0	0

We may summarize the results for $n = 2$ in TABLE 8.7 which includes the
parabolic quantum numbers, the Stark energy, and the permanent electric dipole
moment for that state.

The $n = 2$ results were relatively easily obtained using spherical eigenfunctions,
however, the situation for $n = 3$ is not so simple. To perform the calculation using
degenerate perturbation theory it would be necessary to diagonalize a 9×9 matrix
(actually only a submatrix). It is, however, a simple matter to obtain the Stark
energies using parabolic eigenfunctions. From Equation (8.78) we immediately
obtain the values listed in TABLE 8.8.

Using the operator formalism developed in Chapter 6 it is also possible to obtain
the wave functions corresponding to a particular Stark energy in terms of either the
parabolic or spherical basis set. Because the parabolic quantum numbers for each
energy are known, the corresponding parabolic eigenfunctions will be obvious. It
requires somewhat more work to obtain these eigenfunctions in terms of the
spherical basis set as shown below.

Because the operators $\hat{I}$ and $\hat{K}$ are angular momenta it is easiest to begin by
using the parabolic eigenfunctions designated by the set of quantum numbers that
correspond to $\hat{I}$ and $\hat{K}$, that is $(n\ i\ m_i\ m_k)$. It is not necessary to include k in this
list because $i = k$. We can, of course, convert the set of quantum numbers that
designate a particular wave function to the commonly used parabolic quantum
numbers $(n\ n_1\ n_2\ m)$ using Equation (6.88).

TABLE 8.8. Stark effect parameters in
parabolic coordinates for the $n = 3$ state of
hydrogen in atomic units.

n_1	n_2	m	q	$E_{3q}^{(1)}$	p_z
2	0	0	2	9	$+9$
0	2	0	-2	-9	-9
1	1	0	0	0	0
1	0	1	1	$+(9/2)$	$+9/2$
1	0	-1	1	$+(9/2)$	$+9/2$
0	1	1	-1	$-(9/2)$	$-9/2$
0	1	-1	-1	$-(9/2)$	$-9/2$
0	0	2	0	0	0

TABLE 8.9. Eigenfunctions that are the same in parabolic and spherical coordinates with the different sets of quantum numbers.

Spherical $|n \, \ell \, m\rangle$	Parabolic $|i \, m_i \, m_k\rangle$	Parabolic $|n \, n_1 \, n_2 \, m\rangle$	Parabolic $|n \, q \, m\rangle$
$|n(n-1)(n-1)\rangle_{sp}$	$\left\| \dfrac{(n-1)}{2} \dfrac{(n-1)}{2} \dfrac{(n-1)}{2} \right\rangle_{ik}$	$|n00(n-1)\rangle_{par}$	$|n0(n-1)\rangle_q$

Crucial to the development of this method for obtaining the Stark wave functions in terms of the spherical basis is the fact that there are always states for which a parabolic eigenfunction is exactly equal to a spherical eigenfunction. These are the parabolic states for which $q = 0$, that is, states for which $n_1 = n_2$, and for which $|m|$ has its maximum value; that is, $|m|_{max} = n - 1$. For example, in our treatment of the Stark effect for the $n = 2$ state of hydrogen it was found that $|2001\rangle_{par} = |211\rangle_{sp}$ and $|211 - 1\rangle_{par} = |21 - 1\rangle_{sp}$. The reason for this identity is that $|m|_{max} = n - 1$ can occur for only one value of ℓ; viz. $\ell = (n - 1)$. Because m is a good quantum number in each coordinate system there is only one component in the expansion of the parabolic ket with $m = (n - 1)$ on the spherical basis set, viz. the spherical ket with both and m and ℓ equal to $(n - 1)$. A similar argument holds for the parabolic ket with $m = -(n - 1)$. These are, of course, the top and bottom of the ladder states because the parabolic eigenstates are the uncoupled set whereas the spherical are the coupled.

We can use the raising and lowering operators to generate the set of Stark states using the formulas developed in Chapter 6. The states for which the parabolic and spherical eigenfunctions are identical are summarized in TABLE 8.9 for all different sets of quantum numbers. Of course, the m quantum number ($m = m_i + m_k$) is the same for all of these designations because it is a good quantum number in both spherical and parabolic coordinates. Note also that, for a given n, $n_1 = n_2$ means that $m_i = m_k$, Equation (6.88). Note that the Stark energy, Equation (8.76), vanishes for these states; that is, there is no Stark shift when F is applied. In the remainder of this chapter we attach subscripts to each ket to clarify which set of quantum numbers is being used.

From the definition of $\hat{I}$ in Chapter 6

$$\hat{I} = \left(\frac{1}{2}\right)(\hat{L} + \hat{A}') \tag{8.79}$$

and, because $\hat{A}' = n\hat{A}$,

$$\hat{I} = \frac{1}{2}(\hat{L} + n\hat{A}) \tag{8.80}$$

Therefore,

$$\hat{I}_\pm = \frac{1}{2}(\hat{L}_\pm + n\hat{A}_\pm) \tag{8.81}$$

To obtain the Stark eigenfunctions in terms of the spherical basis set we operate on both sides of the equation for which the parabolic and spherical eigenfunctions are identical, as in the construction of the Clebsch–Gordan coefficients in Chapter 3. In fact, this is exactly what we are doing as discussed later. We choose to begin by applying $\hat{I}_-$ to the wave function for which $m = n - 1$ and $m_i = m_k$; that is, $m = n - 1$ and $n_1 = n_2$. The left side is quite simple because $\hat{I}$ is an angular momentum so

$$\hat{I}_- \left| \frac{(n-1)}{2} \frac{(n-1)}{2} \frac{(n-1)}{2} \right\rangle_{ik} = \sqrt{n-1} \left| \frac{(n-1)}{2} \frac{(n-3)}{2} \frac{(n-1)}{2} \right\rangle_{ik} \tag{8.82}$$

To obtain the action of $\hat{I}_\pm$ on the $|n\ell m\rangle_{sp}$ we use Equations (2.61), (2.62), and (6.60) to obtain

$$
\begin{aligned}
&\hat{I}_\pm |n\ell m\rangle_{sp} \\
&= \left(\tfrac{1}{2}\right)\left(\hat{L}_\pm + n\hat{A}_\pm\right)|n\ell m\rangle_{sp} \\
&= \left(\tfrac{1}{2}\right)\sqrt{(\ell \mp m)(\ell \pm m + 1)}\,|n\ell(m \pm 1)\rangle_{sp} \\
&\quad \mp \left(\tfrac{1}{2}\right)\sqrt{\frac{[n^2 - (\ell+1)^2](\ell \pm m + 1)(\ell \pm m + 2)}{(2\ell+1)(2\ell+3)}}\,|n(\ell+1)(m \pm 1)\rangle_{sp} \\
&\quad \pm \left(\tfrac{1}{2}\right)\sqrt{\frac{[n^2 - \ell^2](\ell \mp m)(\ell \mp m - 1)}{(2\ell-1)(2\ell+1)}}\,|n(\ell-1)(m \pm 1)\rangle_{sp}
\end{aligned} \tag{8.83}
$$

and

$$
\begin{aligned}
&\hat{I}_- |n(n-1)(n-1)\rangle_{sp} \\
&= \tfrac{1}{2}\sqrt{2(n-1)}\,|n(n-1)(n-2)\rangle_{sp} + 0 \\
&\quad - \tfrac{1}{2}\sqrt{2(n-1)}\,|n(n-2)(n-2)\rangle_{sp} \\
&= \sqrt{\frac{(n-1)}{2}}\left[|n(n-1)(n-2)\rangle_{sp} - |n(n-2)(n-2)\rangle_{sp}\right]
\end{aligned} \tag{8.84}
$$

Equating (8.82) and (8.84) we obtain

$$\left| \frac{(n-1)}{2} \frac{(n-3)}{2} \frac{(n-1)}{2} \right\rangle_{ik} = \frac{1}{\sqrt{2}}\left[|n(n-1)(n-2)\rangle_{sp} - |n(n-2)(n-2)\rangle_{sp}\right] \tag{8.85}$$

Converting the left-hand side to the common parabolic number designations $|n\,n_1\,n_2\,m\rangle$ using Equation (6.88)

$$
\begin{aligned}
n_1 &= \frac{(n-1)}{2} - m_i &\Rightarrow\quad m_i &= \frac{(n-1)}{2} - n_1 \\
n_2 &= \frac{(n-1)}{2} - m_k &\Rightarrow\quad m_k &= \frac{(n-1)}{2} - n_2 \\
n &= 2i + 1 &\Rightarrow\quad i &= \frac{n-1}{2}
\end{aligned} \tag{8.86}
$$

we find that

$$|n\,10\,(n-2)\rangle_{par} = \frac{1}{\sqrt{2}}[|n(n-1)\,(n-2)\rangle_{sp} - |n(n-2)\,(n-2)\rangle_{sp}] \qquad (8.87)$$

Notice that for $n = 2$ we retrieve $|\psi_+\rangle$ as determined using degenerate perturbation theory, Equation (8.61). Also, because $m = (n-2)$ in this state there are two spherical components in the expansion of the parabolic eigenfunctions on the spherical basis set, *viz.* the kets for which $n = \ell - 1$ and $n = \ell - 2$. The same will be true for $m = -(n-2)$. Clearly, application of $\hat{I}_-$ to $|n\,10\,(n-2)\rangle_{par}$ will produce an expansion for $|n\,20\,(n-3)\rangle_{par}$ that contains three spherical eigenfunctions. Of course, the coefficients are merely the Clebsch–Gordan coefficients because the parabolic kets are the uncoupled set and the spherical kets the coupled set.

We see then that successively lowering m_i by repeated application of $\hat{I}_-$ will in turn lower m and produce the Stark states of maximum energy *for each value of* m. To obtain the remaining states for a given m we judiciously apply $\hat{I}_\pm$ and $\hat{K}_\mp$ because such application does not change $m = m_i + m_k$. For example, starting with the parabolic ket above we have

$$\hat{I}_-\hat{K}_+|n\,10\,(n-2)\rangle_{par} = \hat{I}_-\sqrt{n(n+1)}|n\,11\,(n-1)\rangle_{par}$$
$$= \sqrt{n(n+1)}\sqrt{(n+1)n}|n\,01\,(n-2)\rangle_{par} \qquad (8.88)$$

Although this might seem a tedious task it not necessary to continue it because, as noted above, knowledge of the Clebsch–Gordan coefficients is adequate. For example, if we wish to solve for the Stark eigenfunctions for $n = 3$ we can use TABLE 3.4 which is reproduced as TABLE 8.10 with the notation changed to be consistent with the notation used in this chapter.

For $n = 3$, the general result of Equation (8.85) is

$$|101\rangle_{ik} = |3101\rangle_{par}$$
$$= \frac{1}{\sqrt{2}}[|321\rangle_{sp} - |311\rangle_{sp}] \qquad (8.89)$$

TABLE 8.10. Table of Clebsch–Gordan coefficients for $i = 1 = k$ in the notation of this chapter.

$i = 1$	$k = 1$			$\ell = 2$				$\ell = 1$		$\ell = 0$
m_i	m_k	$m = 2$	$m = 1$	$m = 0$	$m = -1$	$m = -2$	$m = 1$	$m = 0$	$m = -1$	$m = 0$
1	1	1								
1	0		$\sqrt{1/2}$				$\sqrt{1/2}$			
1	-1			$\sqrt{1/6}$				$\sqrt{1/2}$		$\sqrt{1/3}$
0	1		$\sqrt{1/2}$				$-\sqrt{1/2}$			
0	0			$\sqrt{2/3}$				0		$-\sqrt{1/3}$
0	-1				$\sqrt{1/2}$				$\sqrt{1/2}$	
-1	1			$\sqrt{1/6}$				$-\sqrt{1/2}$		$\sqrt{1/3}$
-1	0				$\sqrt{1/2}$				$-\sqrt{1/2}$	
-1	-1					1				

TABLE 8.11. Stark energies for $n = 3$ with appropriate quantum numbers.

$i = k$	m_i	m_k	n_1	n_2	q	m	$E_{3q}^{(1)}$
1	-1	-1	2	2	0	-2	0
1	-1	0	2	1	-1	-1	9/2
1	-1	1	2	0	-2	0	9
1	0	-1	1	2	1	-1	$-9/2$
1	0	0	1	1	0	0	0
1	0	1	1	0	-1	1	0/2
1	1	-1	0	2	2	0	-9
1	1	0	0	1	1	1	$-9/2$
1	1	1	0	0	0	2	0

which can be read across the row for which $m_i = 0$ and $m_k = 1$. Examining a Stark state that has three spherical components we see that, for example

$$|11 - 1\rangle_{ik} = |302\rangle_{par}$$
$$= \frac{1}{\sqrt{6}}|320\rangle_{sp} + \frac{1}{\sqrt{2}}|310\rangle_{sp} + \frac{1}{\sqrt{3}}|300\rangle_{sp} \qquad (8.90)$$

The Stark energy for this state is $-9F$. TABLE 8.11 contains Stark energies for $n = 3$ that are identical with TABLE 8.8 except that the i and k quantum numbers are included.

8.4. Weak Electric Field—The Classical Stark Effect

As we saw in our discussion of the Kepler problem, for a pure Coulomb potential the orbit is fixed in space due to the constant of the motion, A, the Lenz vector. The fact that this constant is a vector that points along the semi-major axis ensures that the orbit is fixed in space. As for any central potential problem, the angular momentum vector L is also conserved. Conservation of L ensures that the motion takes place in a plane that is perpendicular to L. Application of a constant electric field destroys the spatial symmetry that causes L to be conserved so the motion is no longer constrained to a plane. If this field is weak then the orbit will resemble an ellipse, but the orbital plane and the shape of the ellipse do not remain constant. To simplify the problem we note that under these circumstances the electron moves around the nearly elliptical orbit in a time short compared with the time required for changes in the plane and shape of the trajectory. Thus, the time dependencies of the shape and orientation of the orbit provide a classical picture of the hydrogen atom subjected to a weak constant electric field.

It has been shown that the z-component of the dipole moment of the hydrogen atom in terms of the Lenz vector is

$$\hat{p}_z = \frac{3}{2}n^2 \hat{A}_z \qquad (8.91)$$

If we identify the quantum mechanical operators with their classical counterparts we see that that the energy shift caused by the external electric field $F = F\hat{k}$, is

$$\Delta E = -p \cdot F$$
$$= -\tfrac{3}{2}n^2 A_z F \qquad (8.92)$$

which is consistent with the quantum mechanical result. Previously, we found that the quantum mechanical eigenvalues of $\hat{A}_z$ in terms of the parabolic quantum numbers are $(n_1 - n_2)/n$. Inserting this in the classical result yields

$$\Delta E = \tfrac{3}{2}n(n_1 - n_2)F \qquad (8.93)$$

which is identical with $E_n^{(1)}$ as obtained in Equation (8.76).

We wish to find the effect of the electric field on the Keplerian orbit of the electron. It should again be emphasized that the orbital motion is rapid with respect to changes in the shape of the orbit so we may consider the orbit as the dynamical entity. The time dependence of L can be obtained by using the classical relationship for the torque on an electric dipole. The torque is the time derivative of the angular momentum so

$$\dot{L} = p \times F$$
$$= \tfrac{3}{2}n^2(A \times F) \qquad (8.94)$$

The appearance of n (and ℓ) does not signify a quantal calculation. These quantities may be regarded as being continuously variable for this calculation. We note from Equation (8.94) that, because F is in the z-direction, $\dot{L}_z = 0$ and L_z is a constant of the motion when the field is present.

There is considerably more algebra involved in obtaining the equation for $\dot{A}$ so we do not perform this derivation here. The reader is referred to the original paper.[6] The result is

$$\dot{A} = \tfrac{3}{2}(L \times F) \qquad (8.95)$$

Notice again that, because F is in the z-direction, $\dot{A}_z = 0$ so A_z is a constant of the motion. That both L_z and A_z are constants of the classical motion indicates that their quantum mechanical operators commute with the Hamiltonian, thus permitting the problem of a hydrogen atom in an electric field, the Stark effect, to be separated in *some* coordinate system, in this case parabolic coordinates.

The coupled symmetric equations for $\dot{A}$ and $\dot{L}$ can easily be uncoupled by differentiating one and substituting into the other. For example, the equation of motion for A is given by

$$\ddot{A} = -\left(\tfrac{3}{2}n\right)^2 \{A(F \cdot F) - F(F \cdot A)\} \qquad (8.96)$$

Because $F = F\hat{k}$

$$\ddot{A}_z = -\left(\tfrac{3}{2}n\right)^2 \left[A_z F^2 - F\left(FA_z\right)\right]$$
$$= 0 \qquad (8.97)$$

so that

$$\ddot{A} = -\left(\tfrac{3}{2}n\right)^2 \left[A\left(F \cdot F\right) - F\left(F \cdot A\right)\right]$$
$$= -\left(\tfrac{3}{2}n\right)^2 A\left(F \cdot F\right)$$
$$= -\left(\tfrac{3}{2}n\right)^2 F^2 \left(A_x \hat{i} + A_y \hat{j}\right) \qquad (8.98)$$

The x- and y-components of $\ddot{A}$ are therefore given by

$$\ddot{A}_i(t) = -\left(\tfrac{3}{2}nF\right)^2 A_i(t) \qquad (8.99)$$

Taking $A_y(t = 0) = 0$ we obtain

$$[A_x(t)]^2 + [A_y(t)]^2 = A_{x0}^2 \cos^2(\omega_S t) + A_{y0}^2 \sin(\omega_S t) \qquad (8.100)$$

where the A_{i0} are the values of the x- and y-components at $t = 0$ and

$$\omega_S = (3/2)nF \qquad (8.101)$$

is the Stark frequency. These relationships for the components of A show that the Lenz vector describes a rotating ellipse in a plane perpendicular to the direction of the applied field and that the frequency of the motion is ω_S. Although the motion of A indicates that the shape of the elliptical orbit is changing, this change can be associated with changes of only the minor axis because the energy remains constant and the energy is given by a, the semi-major axis.

If the equations of motion for A and L are uncoupled to give an equation that describes the motion of L it is found that it also outlines an ellipse rotating with frequency ω_S, but in a plane perpendicular to the z-axis. FIGURE 8.6 shows the rotation of the vectors A and L about the z-axis. Clearly, these vectors and the orbit revolve rigidly about the field direction because A is always in the plane of the orbit and L is perpendicular to the plane of the orbit. The entire assembly rotates with frequency $\omega_S = (3/2)nF$, the Stark frequency.

The classical view of the Stark effect on hydrogen atoms is one of a pulsating ellipse, rotating about the electric field vector. The pulsation causes the semi-minor, but not the semi-major, axis to change as the plane of the orbit rotates. FIGURE 8.7 shows the trajectory of an electron in a classical hydrogen atom that is subjected to an external electric field F that has orbital parameters corresponding to the parabolic quantum numbers $n = 11$, $m = 1$, and $g = 4$. The trajectory was generated by numerical solution of Hamilton's equations of motion with $F = 2.917 \times 10^{-7}$ a.u. (1500 V/cm). The nearly Keplerian orbits of the electron are slightly distorted and rotate around the electric field vector as described above.

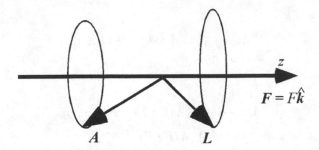

FIGURE 8.6. The rotation of L and A about the z-axis for an externally applied electric field in the z-direction.

Before leaving the subject of the classical hydrogen atom we examine the relation between the classical and quantal results. We expect the correspondence principle to manifest itself. To compare the classical and quantal effects we return to Equation (8.78) for the quantum mechanical Stark energy in parabolic coordinates. Because $q = (n_1 - n_2)$ changes by two for fixed values of m, the difference in energy between adjacent levels of the same m is given by

$$\Delta E = \tfrac{3}{2}n(2)F$$
$$= 3nF \qquad\qquad (8.102)$$

This is not, however, the energy separation between adjacent Stark states of different m. The Stark manifold of states having magnetic quantum number m is, in fact, interleaved with states having magnetic quantum number $(m + 1)$. The arrary corresponding to $(m + 2)$ coincides with the m-array, but with two fewer states, one at the top and one at the bottom. The interleaving of the arrays is illustrated in FIGURE 8.8 for adjacent values of m.

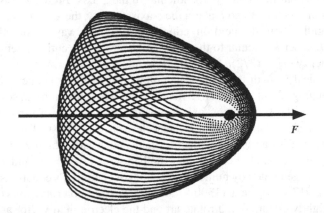

FIGURE 8.7. Classical trajectory of an electron having orbital parameters corresponding to the parabolic quantum numbers $n = 11$, $m = 1$, and $q = 4$. [From Reference 6.]

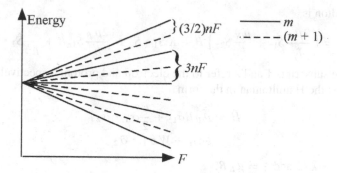

FIGURE 8.8. Stark energies for adjacent values of m.

From FIGURE 8.8 it is clear that the separation between adjacent Stark levels is $(3/2)nF$ which is precisely the Stark frequency ω_S given Equation (8.101). As suggested above, this is not an accident![6] According to the correspondence principle, radiation given off when a system undergoes a transition between levels separated by energy E is $E/\hbar$, which in this case is ω_S. The frequency of revolution of the orbit about F is also ω_S. The energy changes because it is the product of the z-component of the electric dipole moment and the magnitude of the field. Therefore, in our classical picture, different energy levels correspond to different orientations of the orbital plane with respect to the external field. Because the frequency of revolution remains constant, the energy levels at any given value of F, a vertical line in FIGURE 8.8, have constant separation.

Problems

8.1. Show that $\hat{\boldsymbol{L}} \cdot \hat{\boldsymbol{S}} = \hat{L}_z \hat{S}_z + \frac{1}{2}(\hat{L}_+ \hat{S}_- + \hat{L}_- \hat{S}_+)$.

8.2. Calculate the Zeeman splitting of each of the magnetic sublevels of hydrogen for $n = 2$ assuming that the electron has zero spin. Assume that the Zeeman energy and the fine-structure splitting are comparable.

8.3. Compute the energy shifts when a hydrogen atom in the $n = 2$ state is immersed in a very strong constant external magnetic field $\boldsymbol{B} = B\hat{\boldsymbol{k}}$. Sketch these energies as a function of $\mu_B B$ assuming that the (zero-field) fine-structure interval is so small compared with the Zeeman energies that it is not noticeable in the sketch. This is the Paschen–Bach effect. How do these energies compare with the asymptotic energies in Table 8.6?

8.4. Show that $E_Z^{(1)}$ for the strong and weak field cases, Equations (8.18) and (8.28), are the same for the top of the ladder state (or the bottom of the ladder state) and that they are the same as Equation (8.32).

8.5. Consider the effect of application of a constant magnetic field $\boldsymbol{B} = B\hat{\boldsymbol{k}}$ on the hyperfine levels of the ground state of hydrogen. Using the notation of Section 3.5

the Hamilton is

$$\hat{H} = \left(\frac{2\mu_B}{\hbar}\hat{S}_{1z} + \frac{\mu_N}{\hbar}\hat{S}_{2z}\right)B + K\hat{S}_1 \cdot \hat{S}_2 \approx \frac{2\mu_B}{\hbar}\hat{S}_{1z}B + \frac{2K}{\hbar^2}\hat{S}_1 \cdot \hat{S}_2$$

where the subscripts 1 and 2 refer to the electron and proton, respectively.
(a) Recast the Hamiltonian in the form

$$\hat{H} = \mu_B B\hat{\sigma}_{1z} + \frac{\kappa}{2}\hat{\sigma}_1 \cdot \hat{\sigma}_2$$
$$= \xi\hat{\sigma}_{1z} + W\hat{\sigma}_1 \cdot \hat{\sigma}_2$$

where $W = \kappa/2$ and $\xi = \mu_B B$.
(b) Find the exact values of the energies for this Hamiltonian. It is not necessary
to find the eigenkets.
(c) Correlate the energies that you found in part (b) with the eigenkets for $B = 0$
from which they emanate. These kets were found in Section 3.5. The eigenkets are
not needed for this part either. Draw a graph of the energy E versus B; E versus
ξ will do.

8.6. Consider an electron of mass m_e confined to an infinite one-dimensional
potential well of width L. Show that the polarizability of this "atom" in the ground
state is

$$\alpha = 256\frac{e^2 L^4 m_e}{\pi^6 \hbar^2}\sum_{q=1}^{\infty}\frac{4q^2}{\left(4q^2 - 1\right)^5}$$

Note that q is an index, not the electric quantum number. The following integral
will be helpful.

$$\int_0^{\pi} y \sin y \sin(ny)\,dy = \begin{cases} -4n/\left[(n+1)^2(n-1)^2\right], & n \text{ even} \\ 0, & n \text{ odd} \end{cases}$$

8.7. The energy levels in atomic units of a hydrogen atom in a constant electric
field are, to first-order,

$$E(n, n_1, n_2) = -\frac{1}{2n^2} + \frac{3}{2}n(n_1 - n_2)F$$

where n, n_1, and n_2 are parabolic quantum numbers.
(a) Show that the difference between adjacent levels having the same principal
quantum number is

$$\Delta E = \left(\frac{3}{2}\right)nF$$

(b) Show that the difference between the two extreme components for a given
principal quantum number and magnetic quantum number is

$$\Delta E = 3Fn(n-1)$$

8.8. It can be shown (see ref. 6) that the time rates of change of the classical
angular momentum and the Lenz vector, in atomic units, for an H-atom subjected

to a constant electric field $F = F\hat{k}$ are

$$\dot{L} = \left(\frac{3}{2}\right) n^2 (A \times F) \quad and \quad \dot{A} = \left(\frac{3}{2}\right)(L \times F)$$

Show that these equations lead to a picture of the Keplerian orbit of the electron rotating about the electric field vector with a frequency $\omega_S = (3/2)nF$.

8.9. (This solution to this problem is identical with that for Problem 6.6. The question is, however, phrased in the context of this chapter. If you did not work it in Chapter 6 it is recommended that you work it here.) Use the fact that the parabolic hydrogen atom eigenfunctions are also eigenfunctions in the presence of a constant electric field to derive the Stark effect wave functions for $n = 2$ in terms of the spherical eigenfunctions for a hydrogen atom in a constant electric field. Use the formalism of Chapter 6 by applying the operator $\hat{L}_- = \hat{I}_- + \hat{K}_-$ to the spherical eigenfunction $|n\ell m\rangle_{sp} = |211\rangle_{sp}$ which is identical to the parabolic eigenfunction $|nn_1n_2m\rangle_{par} = |2001\rangle_{par}$. Note that, although redundant, we use four quantum numbers in the parabolic kets.

References

1. D.J. Griffiths, *Introduction to Quantum Mechanics* (Prentice-Hall, Upper Saddle River, NJ, 1995).
2. P.A.M. Dirac, *Quantum Mechanics* (Oxford, London, 1970).
3. H.A. Bethe and E.E. Salpeter, *Quantum Mechanics of One- and Two-Electron Atoms* (Springer-Verlag, Berlin, 1957).
4. S. Borowitz, *Fundamentals of Quantum Mechanics* (Benjamin, New York, 1967).
5. S. Gasiorowicz, *Quantum Physics* (Wiley, New York, 2003).
6. T.P. Hezel, C.E. Burkhardt, M. Ciocca, et al., Am. J. Phys. **60**, 324 (1992).

9
The Helium Atom

9.1. Indistinguishable Particles

Our discussion has so far been concerned only with the hydrogen atom, a two-particle system. The transition to the study of multielectron atoms necessarily requires the use of approximation methods and consideration of the indistinguishability of electrons. We first consider the effects of indistinguishability and the fact that electrons are fermions. We write a ket that specifies a two-particle state as

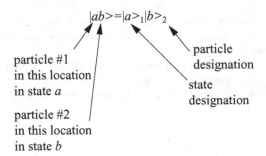

Now define the exchange operator $\hat{P}_{12}$ that exchanges the particles

$$\hat{P}_{12}|ab\rangle = |ba\rangle \tag{9.1}$$

or

$$\hat{P}_{12}|a\rangle_1|b\rangle_2 = |b\rangle_1|a\rangle_2 \tag{9.2}$$

For any acceptable state the "exchanged" state must be identical to the original state because the particles are indistinguishable. Therefore, we cannot tell if we have exchanged particles so that

$$\hat{P}_{12}|\psi\rangle = \lambda|\psi\rangle \tag{9.3}$$

where λ is the eigenvalue of the operator $\hat{P}_{12}$. In fact, the exchange operator must commute with the operator corresponding to any observable for a system of identical particles. In particular, we must have $[\hat{P}_{12}, \hat{H}] = 0$.

Because allowed states are eigenfunctions of $\hat{P}_{12}$ with eigenvalue λ, application of $\hat{P}_{12}$ twice returns the system to the original state; that is,

$$(\hat{P}_{12})^2|\psi\rangle = \lambda^2|\psi\rangle$$
$$= |\psi\rangle \qquad (9.4)$$

from which it is clear that $\lambda = \pm 1$. Now, if two identical particles are in the same state $|a\rangle$, they are in an eigenstate of $\hat{P}_{12}$ with $\lambda = +1$. That is,

$$\hat{P}_{12}|aa\rangle = |a\rangle|a\rangle \qquad (9.5)$$

which is symmetric under interchange of particles. On the other hand, a state $|ab\rangle$ where $b \neq a$ is not an eigenstate of $\hat{P}_{12}$. To make it one we must find linear combinations of the states $|ab\rangle$ and $|ba\rangle$ that are eigenstates of $\hat{P}_{12}$. The matrix representation of $\hat{P}_{12}$ using $|ab\rangle$ and $|ba\rangle$ as a basis is

$$\hat{P}_{12} \rightarrow \begin{pmatrix} \langle ab|\hat{P}_{12}|ab\rangle & \langle ab|\hat{P}_{12}|ba\rangle \\ \langle ba|\hat{P}_{12}|ab\rangle & \langle ba|\hat{P}_{12}|ba\rangle \end{pmatrix} \qquad (9.6)$$

Because $|ab\rangle$ and $|ba\rangle$ are orthonormal, the matrix elements are easily calculated. We have

$$\hat{P}_{12} \rightarrow \begin{pmatrix} 0 & 1 \\ 1 & 0 \end{pmatrix} \qquad (9.7)$$

The eigenvalue equation leads us to the secular equation

$$\begin{vmatrix} -\lambda & 1 \\ 1 & -\lambda \end{vmatrix} = 0 \Rightarrow \lambda^2 = 1 \Rightarrow \lambda = \pm 1 \qquad (9.8)$$

as was previously deduced. Now we find the ket corresponding to $\lambda = +1$. Calling the ket $|\psi_s\rangle$, where the s stands for "symmetric" (preempting the answer) we have

$$|\psi_s\rangle = c_1|ab\rangle + c_2|ba\rangle \qquad (9.9)$$

The eigenvalue equation with $\lambda = +1$ is then

$$\begin{pmatrix} 0 & 1 \\ 1 & 0 \end{pmatrix}\begin{pmatrix} c_1 \\ c_2 \end{pmatrix} = +1 \cdot \begin{pmatrix} c_1 \\ c_2 \end{pmatrix} \Rightarrow c_1 = c_2 \qquad (9.10)$$

and we have

$$|\psi\rangle_s = \frac{1}{\sqrt{2}}(|ab\rangle + |ba\rangle) \qquad (9.11)$$

Similarly, for $\lambda = -1$ we obtain

$$|\psi\rangle_a = \frac{1}{\sqrt{2}}(|ab\rangle - |ba\rangle) \qquad (9.12)$$

where the a stands for "antisymmetric".

Note that the identical particles must be in *either* $|\psi\rangle_s$ or $|\psi\rangle_a$, not a linear combination of them because the exchanging of particles must produce a state that differs from the original state by only a phase. That is,

$$\hat{P}_{12}(c_s|\psi_s\rangle + c_a|\psi_a\rangle) = c_s|\psi_s\rangle - c_a|\psi_a\rangle$$
$$\neq e^{I\delta}(c_s|\psi_s\rangle + c_a|\psi_a\rangle) \tag{9.13}$$

Now, there are two types of particles, bosons and fermions.

Bosons: Integral spin, only symmetric states, for example, deuterons
Fermions: 1/2-integral spin, only antisymmetric states, for example, electrons

Recall that if two identical particles are in the same state then the state is symmetric. This is the Pauli principle. That is, two electrons (fermions) cannot occupy the same state so the total wave function must be antisymmetric.

The Pauli principle is sometimes stated as "no two electrons in an atom can have the same set of quantum numbers." This is correct, but the statement above concerning an antisymmetric wave function is the basis for this statement.

9.2. The Total Energy of the Helium Atom

The helium atom is the prototype of a multielectron atom. The coordinates for the system are shown in FIGURE 9.1.

The Hamiltonian for this system in atomic units is

$$\hat{H} = \frac{p_1^2}{2} + \frac{p_2^2}{2} - \frac{Z}{|r_1|} - \frac{Z}{|r_2|} + \frac{1}{|r_1 - r_2|} \tag{9.14}$$

where $Z = 2$ for a helium atom, $Z = 3$ for once-ionized lithium, and so on. Retaining Z in the Hamiltonian (rather than replacing it by $Z = 2$ will turn out to be convenient).

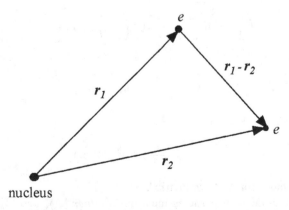

FIGURE 9.1. Coordinates used in the helium atom calculation.

The electron–electron repulsion enters the Hamiltonian in only the last term so we write $\hat{H}$ as the sum of three Hamiltonians

$$\hat{H}_{01} = \frac{p_1^2}{2} - \frac{Z}{|r_1|}; \quad \hat{H}_{02} = \frac{p_2^2}{2} - \frac{Z}{|r_2|}; \quad \hat{H}_1 = \frac{1}{|r_1 - r_2|} \qquad (9.15)$$

The unperturbed Hamiltonian is taken to be

$$\hat{H}_0 = \hat{H}_{01} + \hat{H}_{02} \qquad (9.16)$$

Because each of the terms in $\hat{H}_0$ involves the coordinates of only a single electron, $\hat{H}_0$ is separable. The eigenfunctions and eigenvalues are straightforward. The eigenfunctions are products of one-electron eigenfunctions with $Z = 2$ and the eigenvalues are the sums of the corresponding one-electron eigenvalues.

The separation of the unperturbed Hamiltonian into individual terms, each of which contains only coordinates of a single electron as in Equation (9.16), serves as the basis for describing all multielectron atoms. The "state" of a given electron is described by the one-electron quantum numbers that correspond to the one-electron eigenfunction that describes it. These one-electron states are referred to as "orbitals" and they are designated by the principal quantum number n and the lowercase letter corresponding to ℓ. Thus, for the ground state of helium both electrons occupy the $1s$ orbital. The listing of the orbitals is referred to as the electron configuration. The electron configuration of the ground state of helium is thus $1s^2$ where the superscript refers to the number of electrons in the designated orbital. If there is only one electron in an orbital the superscript is usually omitted.

The perturbation is taken to be $\hat{H}_1$. The eigenvalues of $\hat{H}_0$ are then the zeroth order approximation to the energy of the helium atom. Such an approximation is employed in the description of all multielectron atoms. The eigenvalues and eigenfunctions of $\hat{H}_0$ are

$$E_0 = E_{0n_1} + E_{0n_2} \qquad (9.17)$$

and

$$|\psi\rangle = |n_1 \ell_1 m_1\rangle \cdot |n_2 \ell_2 m_2\rangle \qquad (9.18)$$

where E_{0n_1} and E_{0n_2} are one-electron energies $-Z^2 R_H$. The kets $|n_1 \ell_1 m_1\rangle$ and $|n_2 \ell_2 m_2\rangle$ are the one-electron eigenfunctions where the subscripts refer to each electron. Note that $|\psi\rangle$ is only the spatial part of the wave function. Moreover, we have not yet imposed the symmetry requirements on the construction of $|\psi\rangle$.

The total wave function is the product of a spatial part and a spin part. The spatial wave function of the ground state of helium is

$$|1s\,1s\rangle_{space} = |100\rangle_1 \cdot |100\rangle_2 \qquad (9.19)$$

This ket is clearly symmetric under interchange of particles. Therefore, the spin wave function must be antisymmetric under interchange of particles so the total wave function is given by

$$|1s\,1s\rangle = |100\rangle_1 \cdot |100\rangle_2 \cdot \left[\frac{1}{\sqrt{2}} (|\alpha_1 \beta_2\rangle - |\alpha_2 \beta_1\rangle) \right] \qquad (9.20)$$

where the term in the brackets is the antisymmetric two-electron spin state (in the uncoupled representation). α and β are the usual "spin up" and "spin down" z-components of the spin and the subscripts designate each electron. Note that in the coupled representation this antisymmetric spin state is designated $|SM\rangle = |00\rangle$.

The requirement that the total wave function is antisymmetric dictates that the ground state of a two-electron system must have total spin zero (even though the Hamiltonian doesn't contain the spin). This is easily shown using the coupled representation because

$$\hat{S}^2|00\rangle = 0 \quad \text{and} \quad \hat{S}_z|00\rangle = 0 \tag{9.21}$$

In preparation for our discussion of the excited states of helium we note that there are three other possible spin states for two electrons. In the two representations they are

$$|\alpha_1\alpha_2\rangle = |11\rangle$$

$$\frac{1}{\sqrt{2}}(|\alpha_1\beta_2\rangle + |\alpha_2\beta_1\rangle) = |10\rangle$$

$$|\beta_1\beta_2\rangle = |1-1\rangle \tag{9.22}$$

These three states are all symmetric under interchange of particles and therefore do not qualify as constituents of the ground state because the space part of the ground state is manifestly symmetric. The ground state is therefore a "singlet".

Now, as discussed in more detail later, for most atoms the traditional designation for atomic states is "spectroscopic notation" which has the form $^{2S+1}L_J$ where S is the total spin, L is the total orbital angular momentum, and J is the total angular momentum, spin plus orbital angular momentum. The value of L is denoted by spectroscopic notation. That is, $L = 0, 1, 2, 3, \ldots$ correspond, respectively, to S, P, D, F (alphabetic order). Note that the "S" that represents $L = 0$ is not related to the same symbol that represents the total spin, that is, S in the coupled ket $|SM\rangle$. To confuse matters further, this scheme is referred to as LS notation. For the ground state of the helium atom the designation is

$$^{2S+1}L_J \rightarrow {}^1S_0$$

because the total spin, the total orbital angular momentum, and the total angular momentum are all zero. Note that the L-states are designated by capital letters and lowercase letters are used for designation of hydrogen atom states. The designation of hydrogen atom states is not, however, unique. Capital letters are frequently used. For multielectron atoms capital letters are always used.

The ground state energy of the helium atom is the minimum energy required to remove both electrons leaving behind a He^{++} ion. This minimum energy leads to isolated electrons and a He^{++} ion, all of which have zero kinetic energy. In the zeroth order approximation we ignore the interaction between the electrons, that is, the electron–electron Coulomb repulsion. Therefore, the zeroth order eigenfunctions are eigenfunctions of $\hat{H}_0$, products of the $|100\rangle$ one-electron wave functions for $Z = 2$ for each electron. The eigenvalue of $\hat{H}_0$ is the sum of the eigenvalues of

$\hat{H}_{01}$ and $\hat{H}_{02}$ for these states.

$$E_{1s1s}^{(0)} = \left(-\frac{1}{2}Z^2\right) + \left(-\frac{1}{2}Z^2\right) \tag{9.23}$$

For $Z = 2$

$$E_{1s1s}^{(0)} = -4$$
$$= -108.8\,\text{eV} \tag{9.24}$$

Because He$^+$ is indeed a one-electron atom, the ionization potential is correctly given as $4 \times R_H = 54.4\,\text{eV}$. Experimentally it is found that the ionization potential of neutral He (i.e., the energy required to remove a single electron from a helium atom in the ground state) is 24.6 eV. Thus, the correct total energy of the helium atom, the energy required to liberate both electrons is 79 eV. Clearly then the $e-e$ repulsion that was ignored in the zeroth order approximation that led to Equation (9.24) raises the energy by $\sim$30 eV. This is consistent with the fact that the $e-e$ repulsion is positive. It is, however, a significant fraction of 108 eV so we suspect that perturbation theory may not lead to a reliable answer. Nonetheless, it is clear that to obtain a more realistic value for the total energy of the helium atom than the zeroth approximation, 108.8 eV, we must include the correction due to $\hat{H}_1$, the $e-e$ repulsion.

9.3. Evaluation of the Ground State Energy of the Helium Atom Using Perturbation Theory

Although the correction due to the $e-e$ repulsion must be on the order of one-third of the unperturbed energy we begin by trying perturbation theory. As it turns out, we obtain a surprisingly good approximation to the true helium energy despite the magnitude of the correction.

Using first-order perturbation theory, we require the expectation value of the perturbing Hamiltonian using the unperturbed eigenfunctions so

$$E_0^{(1)} = \langle \psi_0 | \frac{1}{|r_1 - r_2|} | \psi_0 \rangle \tag{9.25}$$

Because the perturbation does not involve spin we may evaluate the above integral in coordinate space only so

$$E_0^{(1)} = {}_1\langle 100| \cdot {}_2\langle 100| \frac{1}{|r_1 - r_2|} |100\rangle_1 \cdot |100\rangle_2 \tag{9.26}$$

In terms of one-electron wave functions

$$E_0^{(1)} = \int dr_1 dr_2 |\psi_{1s}(r_1)|^2 \frac{1}{|r_1 - r_2|} |\psi_{1s}(r_2)|^2 \tag{9.27}$$

Note that the ground state wave function for the one-electron atom is independent of θ and ϕ so that $r_1 \rightarrow r_1$ (the spherical coordinate). The physical interpretation of this is the following.

$$|\psi_{1s}(r_1)|^2 = \text{probability of finding electron \#1 at } r_1$$

so that it is also the charge density due to electron #1 in atomic units. Similarly

$$|\psi_{1s}(r_2)|^2 = \text{charge density due to electron \#2 at } r_2$$

Therefore, the integral above is just the electrostatic interaction energy of two overlapping spherically symmetric charge distributions.

Inserting the specific ground state wave functions we have

$$E_0^{(1)} = \frac{Z^6}{\pi^2} \int dr_1 dr_2 \frac{1}{r_{12}} e^{-2Z(r_1+r_2)} \tag{9.28}$$

where

$$\frac{1}{r_{12}} = \frac{1}{|\mathbf{r_1} - \mathbf{r_2}|}$$

It is worthwhile to perform this calculation because it occurs many times in atomic physics as well as other branches of physics. The term $1/r_{12}$ may be expanded in terms of Legendre polynomials[1]

$$\frac{1}{r_{12}} = \frac{1}{r_1} \sum_{\ell=0}^{\infty} \left(\frac{r_2}{r_1} \right)^{\ell} P_{\ell}(\cos\theta) \quad r_1 > r_2$$

$$= \frac{1}{r_2} \sum_{\ell=0}^{\infty} \left(\frac{r_1}{r_2} \right)^{\ell} P_{\ell}(\cos\theta) \quad r_2 > r_1 \tag{9.29}$$

where θ is the angle between $\mathbf{r_1}$ and $\mathbf{r_2}$. The geometry of $\mathbf{r_1}$ and $\mathbf{r_2}$ is shown in FIGURE 9.2.

From simple trigonometry we have

$$\cos\theta = \cos\theta_1 \cos\theta_2 + \sin\theta_1 \sin\theta_2 \cos(\phi_1 - \phi_2) \tag{9.30}$$

For convenience, the expansion of $1/r_{12}$ is usually written more compactly as

$$\frac{1}{r_{12}} = \frac{1}{r_1} \sum_{\ell=0}^{\infty} \frac{(r_<)^{\ell}}{(r_>)^{\ell+1}} P_{\ell}(\cos\theta) \tag{9.31}$$

where $r_<$ is the smaller of r_1 and r_2 and $r_>$ is the larger. The addition theorem for spherical harmonics[1] is given by

$$P_{\ell}(\cos\theta) = \left(\frac{4\pi}{2\ell+1} \right) \sum_{m=-\ell}^{\ell} Y_{\ell m}^*(\theta_1, \phi_1) Y_{\ell m}(\theta_2, \phi_2) \tag{9.32}$$

from which

$$\frac{1}{r_{12}} = \sum_{\ell=0}^{\infty} \sum_{m=-\ell}^{\ell} \left(\frac{4\pi}{2\ell+1} \right) \cdot \frac{(r_<)^{\ell}}{(r_>)^{\ell+1}} Y_{\ell m}^*(\theta_1, \phi_1) Y_{\ell m}(\theta_2, \phi_2) \tag{9.33}$$

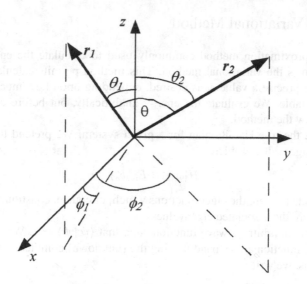

FIGURE 9.2. Geometry of the electron–electron repulsion calculation.

We may now insert this in Equation (9.28) and recall that the spherical harmonics are orthonormal to obtain

$$E_0^{(1)} = \frac{Z^6}{\pi^2} \sum_{\ell=0}^{\infty} \sum_{m=-\ell}^{\ell} \left(\frac{4\pi}{2\ell+1} \right) \int_0^{\infty} d^3 r_1 r_1^2 \int_0^{\infty} d^3 r_2 r_2^2 e^{-2Z(r_1+r_2)} \frac{(r_<)^\ell}{(r_>)^{\ell+1}} \delta_{\ell,0} \delta_{m,0}$$

(9.34)

All terms in the summations in Equation (9.34) vanish except those for which $\ell = m = 0$. The integrals are now straightforward and lead to

$$E_0^{(1)} = +\frac{5}{8} Z$$

(9.35)

Note that the positive sign indicates that the energy is indeed raised by the $e-e$ repulsion. To this level of approximation then the ground state energy is given by

$$E_0 = E_0^{(0)} + E_0^{(1)}$$
$$= -Z^2 + \frac{5}{8} Z$$
$$= -74.8 \, \text{eV}$$

(9.36)

This is surprisingly close to the experimentally determined value, 79 eV, especially in view of the fact that $\frac{5}{8} Z = \frac{5}{4}$ is a significant fraction of $Z^2 = 4$.

9.4. The Variational Method

Another approximation method commonly used to calculate the energy of the helium atom is the variational method. This method permits calculation of the energy to as precise a value as is desired, depending upon how much computer time is available. We evaluate the energy analytically, but before doing so we briefly review the method.

Let $\hat{H}$ be the true Hamiltonian for a given system. We pretend that we have solved the eigenvalue problem

$$\hat{H}|\phi_n\rangle = E_n|\phi_n\rangle \tag{9.37}$$

where the kets $|\phi_n\rangle$ are the eigenfunctions which, of course, constitute a complete set; the E_n are the associated eigenvalues.

Let $|\psi\rangle$ be an arbitrary wave function such that $\langle\psi|\psi\rangle = 1$. We use $|\psi\rangle$ as a "trial" wave function and expand it using the (unknown) complete set $|\phi_n\rangle$ as the basis. That is, we write

$$|\psi\rangle = \sum_{n=1}^{\infty} c_n|\phi_n\rangle \tag{9.38}$$

Using this trial wave function $|\psi\rangle$ we obtain the expectation value of the energy of the state that it represents

$$\langle E\rangle = \langle\psi|\hat{H}|\psi\rangle$$
$$= \sum_{n=1}^{\infty} |c_n|^2 E_n \tag{9.39}$$

Now, the lowest of the energy eigenvalues, E_0, is the ground state energy. For the helium case, E_0 is the lowest energy for which the process

$$He\left(1s^2\,{}^1S_0\right) \rightarrow He^{++} + 2e$$

can occur. In other words, E_0 is the energy that will twice ionize a ground state helium atom and leave the liberated particles with zero kinetic energy. If we replace energy E_n by E_0 in the summation in Equation (9.39) the result will be a number that is lower than the correct expectation value $\langle E\rangle$. This leads to the inequality

$$\langle E\rangle = \langle\psi|\hat{H}|\psi\rangle$$
$$= \sum_{n=1}^{\infty} |c_n|^2 E_n$$
$$\geq E_0 \sum_{n=1}^{\infty} |c_n|^2$$
$$\geq E_0 \tag{9.40}$$

Therefore, for *any* trial wave function $|\psi\rangle$ the true ground state energy E_0 will always be less than the expectation value $\langle E\rangle = \langle\hat{H}\rangle$. We may therefore select, perhaps by guessing, a $|\psi\rangle$ that includes parameters that may be varied to minimize $\langle E \rangle$. We leave these parameters in the trial wave function and evaluate $\langle \hat{H} \rangle$ in terms of them. We then minimize $\langle \hat{H} \rangle$ with respect to the parameters. This is the variational principle! According to this principle, the result will always be greater than the true energy. We see then that we may choose a complicated function, containing as many parameters as we please, and, if computer time is available, we can obtain an answer to any degree of precision. For our purposes, however, we use parameters that provide physical insight and for which the variational principle can be implemented analytically.

9.5. Application of the Variational Principle to the Ground State of Helium

We choose a trial wave function for which the integral $\langle \hat{H} \rangle$ may be readily computed. We use the product of two one-electron eigenfunctions with a parameter ζ that may be viewed as an *effective* nuclear charge. That is, we imagine that one electron screens the other electron from the effects of the full nuclear charge Z. Therefore, the effective charge ζ replaces Z in the usual one-electron wave function, but not in the Hamiltonian. It is the trial wave function that contains the variational parameters, not the Hamiltonian. We expect that ζ will be less than Z.

Explicitly writing ζ in the kets we have as our trial ket

$$|\psi\rangle = |100\zeta\rangle_1 |100\zeta\rangle_2 \tag{9.41}$$

The expectation value of the Hamiltonian is thus

$$\langle \hat{H} \rangle = {}_1\langle 100\zeta|_2\langle 100\zeta| \left(\hat{T}_1 + \hat{T}_2 - \frac{Z}{r_1} - \frac{Z}{r_2} + \frac{1}{r_{12}} \right) |100\zeta\rangle_1 |100\zeta\rangle_2 \tag{9.42}$$

where the $\hat{T}_1$ and $\hat{T}_2$ are the kinetic energies of the two electrons. Note that in Equation (9.42) the Hamiltonian explicitly contains Z; the variational parameter ζ occurs only in the trial wave function. In principle, we could replace Z by 2, but retaining it permits us to interpret the answer for any two-electron atom such as Li^+, Be^{++}, and so on. Moreover, we obtain ζ in terms of Z so the physical significance of it will be more apparent.

To evaluate $\langle \hat{H} \rangle$ we must evaluate five integrals, but symmetry reduces this number to three. We begin with the kinetic energy integrals which we evaluate using the virial theorem. For a potential energy of the form $V(r) \propto r^k$ the virial theorem states that $2\langle T \rangle = k\langle V \rangle$ so that for the Coulomb potential we have

$$\langle T \rangle = -\frac{1}{2}\langle V \rangle \tag{9.43}$$

Care must be taken, however, in applying Equation (9.43). This relation is valid only if the eigenfunctions used in the computation of the expectation value are eigenfunctions of the Hamiltonian. Because the eigenfunctions are trial eigenfunctions that contain ζ rather than Z the potential energy used in Equation (9.43) must be $-\zeta/r$. Thus, the viral theorem result provides the relation

$$\langle T \rangle = -\frac{1}{2}\left\langle -\frac{\zeta}{r} \right\rangle$$

$$= \frac{1}{2} \cdot \zeta^2 \qquad (9.44)$$

where we have used[2] $\langle 1/r \rangle = \zeta/n^2$. In this case $n = 1$. Note that in Equation (9.44) one ζ is from the potential energy and the other from the expectation value. By symmetry $\langle \hat{T}_1 \rangle = \langle \hat{T}_2 \rangle$.

Evaluation of the expectation values of the potential energies is straightforward. We have

$$\left\langle \frac{Z}{r} \right\rangle = Z \left\langle \frac{1}{r} \right\rangle$$

$$= Z\zeta \qquad (9.45)$$

where, again, we have used[2] $\langle 1/r \rangle = \zeta/n^2$ and $n = 1$.

We have already evaluated $\langle 1/r_{12} \rangle$ in terms of Z in the calculation of the ground state energy (in a.u.) of helium using perturbation theory. Letting $Z \rightarrow \zeta$ in Equation (9.35) we have

$$\left\langle \frac{1}{r_{12}} \right\rangle = \frac{5}{8}\zeta \qquad (9.46)$$

Putting this all together we have

$$\langle \hat{H} \rangle = \zeta^2 - 2Z\zeta + \frac{5}{8}\zeta \qquad (9.47)$$

which we must minimize by setting the derivative equal to zero. This yields

$$\zeta = Z - \frac{5}{16} \qquad (9.48)$$

As expected, $\zeta < Z$ so that we may regard the factor $5/16$ as a "screening constant". The interpretation of Equation (9.48) is that a fraction of the nuclear charge Z is obscured from one electron by the presence of the other. Inserting this value of ζ into $\langle \hat{H} \rangle$ we find

$$\langle \hat{H} \rangle = -\left(Z - \frac{5}{16} \right)^2$$

$$= \left(-Z^2 + \frac{5}{8}Z \right) - \frac{25}{256} \qquad (9.49)$$

The last line of Equation (9.49) is particularly illuminating because $(-Z^2 + 5/8) = -74.8$ eV is the total helium atom energy that was obtained using perturbation theory, Equation (9.36). We may therefore regard $25/256$ atomic units as a correction to the perturbation theory result. This correction lowers the perturbation theory result so that the answer from this simple variational treatment is

$$\langle \hat{H} \rangle = -74.8\,\text{eV} - \left(\frac{25}{256}\right) 27.2\,\text{eV}$$

$$= -77.5\,\text{eV} \tag{9.50}$$

Although -77.5 eV is still nearly two percent from the actual value of ~ 79 eV, it is clear that a more elaborate variational calculation would lead to a more precise answer, even if the variational parameters chosen have no obvious physical significance.

9.6. Excited States of Helium

In the case of the ground state we had two $1s$ electrons, so the space part of the wave function was necessarily symmetric. That is, it is an eigenfunction of $\hat{P}_{12}$ with eigenvalue $+1$. According to the Pauli principle then, the spin part of the ground state wave function is

$$|1s1s\rangle_{spin} = \frac{1}{\sqrt{2}}[|\alpha_1\beta_2\rangle - |\alpha_2\beta_1\rangle] \tag{9.51}$$

which is, necessarily, antisymmetric and an eigenfunction of $\hat{P}_{12}$ with eigenvalue -1. Recall also that $|1s1s\rangle_{spin}$ is, in the coupled representation, $|SM\rangle = |00\rangle$. There are three other possible spin states for two identical particles, all of which are symmetric under interchange of particles. They are

$$|11\rangle = |\alpha_1\alpha_2\rangle$$

$$|10\rangle = \frac{1}{\sqrt{2}}[|\alpha_2\beta_1\rangle + |\alpha_1\beta_2\rangle]$$

$$|1-1\rangle = |\beta_1\beta_2\rangle \tag{9.52}$$

Although the spin states in Equation (9.52) are prohibited for the ground state, they are acceptable for excited states because, in general, the electrons are not in the same spatial state. Because the singlet spin state $|00\rangle$ is an eigenstate of $\hat{P}_{12}$ with eigenvalue -1 and the triplet spin states $|1M\rangle$ are eigenfunctions of $\hat{P}_{12}$ with eigenvalue $+1$ the corresponding spatial wave functions must be symmetric and antisymmetric, respectively.

Consider now the possible electron configurations $1s2\ell$ of the helium atom, that is, excited states of helium for which one electron remains in the $1s$ (unperturbed) orbital and the other is excited to the $2s$ ($m_\ell = 0$) or $2p$ ($m_\ell = 0, \pm 1$) states. The unperturbed energy is that for which the electron–electron repulsion is ignored. It is given by

$$E_{1s\,n=2} = \left(-\frac{1}{2}\right) Z^2 \left(\frac{1}{1^2} + \frac{1}{2^2}\right)$$

$$= -\frac{5}{8} Z^2 \tag{9.53}$$

TABLE 9.1. Possible wave functions for $n = 2$.

$$\frac{1}{\sqrt{2}}\left[|100\rangle_1\,|2\ell m\rangle_2 + |2\ell m\rangle_1\,|100\rangle_2\right]|00\rangle \qquad \textbf{Singlets}$$

$$\frac{1}{\sqrt{2}}\left[|100\rangle_1\,|2\ell m\rangle_2 - |2\ell m\rangle_1\,|100\rangle_2\right]|1-1\rangle \qquad \textbf{Triplets}$$

$$\frac{1}{\sqrt{2}}\left[|100\rangle_1\,|2\ell m\rangle_2 - |2\ell m\rangle_1\,|100\rangle_2\right]|10\rangle$$

$$\frac{1}{\sqrt{2}}\left[|100\rangle_1\,|2\ell m\rangle_2 - |2\ell m\rangle_1\,|100\rangle_2\right]|11\rangle$$

The properly symmetrized possible spatial parts of the wave functions are

$$\frac{1}{\sqrt{2}}[|100\rangle_1|2\ell m\rangle_2 + |2\ell m\rangle_1|100\rangle_2] \quad \text{symmetric}$$

$$\frac{1}{\sqrt{2}}[|100\rangle_1|2\ell m\rangle_2 - |2\ell m\rangle_1|100\rangle_2] \quad \text{antisymmetric}$$

The subscript ℓ on the quantum number m has been dropped for convenience. The possible *total* wave functions including the spin $|SM\rangle$ are listed in TABLE 9.1.

Now we employ perturbation theory using the electron–electron repulsion $\hat{H}_1$ as the perturbation.

$$\hat{H}_1 = \frac{1}{r_{12}}$$

$$= \frac{1}{r_1}\sum_{\ell=0}^{\infty}\frac{(r_<)^\ell}{(r_>)^{\ell+1}}P_\ell(\cos\theta) \qquad (9.54)$$

Because this is independent of ϕ it commutes with $\hat{L}_z\ (= -\partial/\partial\phi)$ so the energy cannot depend upon the quantum number m and we may perform the calculation for $m = 0$. The calculation is more complicated for the excited states than it was for the ground state because each spatial part (the only part involved in the calculation) is a sum or difference of two integrals. The first-order correction to the energies may then be written

$$\left(E^{(1)}_{1s2\ell}\right)_\pm = J_{1s2\ell} \pm K_{1s2\ell} \qquad (9.55)$$

where J and K are integrals. These integrals depend on the quantum numbers assigned to each electron. $J_{1s2\ell}$ is given by

$$J_{1s2\ell} = {}_1\langle 100|_2\langle 2\ell m|\frac{1}{r_{12}}|100\rangle_1|2\ell m\rangle_2 \qquad (9.56)$$

It is called the Coulomb integral and it is manifestly positive. It represents electrostatic interactions because the integrals that it comprises involve the same electron in each state. That is, in Equation (9.56) electron 1 is in the $|100\rangle$ and electron 2 is in the $|2\ell m\rangle$ state. $J_{1s2\ell}$ would be the energy if the electrons were distinguishable.

In contrast to the Coulomb integral, the second integral $K_{1s2\ell}$ is given by

$$K_{1s2\ell} = {}_1\langle 100|_2\langle 2\ell m| \frac{1}{r_{12}} |2\ell m\rangle_1 |100\rangle_2 \tag{9.57}$$

This term consists of integrals with wave functions that are "exchanged", that is, bras and kets of the form $_i\langle 100|$ and $|2\ell m\rangle_i$ appear ($i = 1$ or 2). K is thus referred to as the exchange integral. It has no classical interpretation! It arises because the electrons are identical particles (a nonclassical concept). Although not as obvious as it is for J, K can also be shown to be manifestly positive. In the expression for the energy, Equation (9.55), the plus and minus signs refer to the energies associated with the symmetric and antisymmetric space wave functions, singlet, and triplet states, respectively. The singlet corrections to the unperturbed energy are therefore higher than the triplet corrections so that, for the same electron configuration, the triplet states are of lower energy. This can be rationalized by noting that the singlet wave functions for which the spins are in opposite directions permit the spatial coordinates of the electrons to be the same. This can most easily be seen by writing the probability of finding one electron at spherical coordinates (r_1, θ_1, ϕ_1) and the other at coordinates (r_2, θ_2, ϕ_2) in terms of the one-electron wave functions $\psi_{n\ell m}(r, \theta, \phi) = R_{n\ell}(r)Y_{\ell m}(\theta, \phi)$ (with, of course, $Z = 2$). This probability is given by

$$|\psi_{100}(r_1, \theta_1, \phi_1)\psi_{n\ell m}(r_2, \theta_2, \phi_2) \pm \psi_{n\ell m}(r_1, \theta_1, \phi_1)\psi_{100}(r_2, \theta_2, \phi_2)|^2$$

where the plus and minus signs represent the singlet and triplet states, respectively. We see that the probability of the electrons being close together (i.e., their spatial coordinates are nearly the same) is significant only for the singlet states. In fact, for the triplet states, the probability vanishes for $(r_1, \theta_1, \phi_1) \to (r_2, \theta_2, \phi_2)$. Thus, electrons in the singlet state tend to "attract" each other whereas in the triplet state they tend to "repel".

This "force" that causes the attraction and repulsion is not a force in the traditional sense. It is a consequence of the symmetry requirements imposed on the wave functions by the indistinguishability of identical particles. The electrons tend to avoid each other because of the symmetry requirements on the total wave function, including the spin. Because spin is not contained in the Hamiltonian this symmetry is not in the Hamiltonian. These forces are often referred to as "exchange forces" and the energy associated with them the "exchange energy". The exchange energy causes the singlet and triplet states to have different energies because the (positive) Coulomb repulsion must be greater for singlet states because the electrons tend to "attract" than it is for triplet states in which the electrons tend to avoid each other.

The J and K integrals can be evaluated exactly.[3] The energies $E_{1s2s}^{(1)}$ are

$$E_{1s2s}^{(1)} = J_{1s2s} \pm K_{1s2s}$$

$$= \frac{17}{81}Z \pm \frac{16}{729}Z$$

$$= 11.4\,\text{eV} \pm 1.2\,\text{eV} \tag{9.58}$$

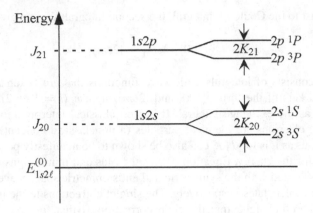

FIGURE 9.3. Schematic energy level diagram of the $n = 2$ states of helium. The solid line represents the energy if the interelectron repulsion is ignored. The Coulomb integral J raises the energy and the exchange integral K splits these levels.

and

$$E^{(1)}_{1s2p} = J_{1s2p} \pm K_{1s2p}$$

$$= \frac{59}{243}Z \pm \frac{112}{6561}Z$$

$$= 13.2\,\text{eV} \pm 0.9\,\text{eV} \qquad (9.59)$$

The relative positions of the energies of the four states of He($1sn\ell$) are shown in FIGURE 9.3. As discussed above, the singlet states are higher in energy than their corresponding triplet states. This is, in fact, a special case of one of the rules used to place the states in proper relative order for all multielectron atoms. These rules are referred to as Hund's rules and are discussed in detail in Chapter 10.

To put these energies in perspective relative to the ground state and the ionization limit FIGURE 9.4 shows the ground state and the $1s2\ell$ excited states together with the state of zero energy, that is, $He^{++}+ 2e$ in complete spectroscopic notation. The energy scale is electron volts.

9.7. Doubly Excited States of Helium: Autoionization

In addition to excited states having nominal electron configuration $1sn\ell$ there are also states of configuration $n\ell n'\ell'$ where $n, n' > 1$. Because states of configuration $1s2\ell$ have energies roughly 20 eV above the ground state and the ionization limit of He($1s^2$) is 24.6 eV, it is not surprising that these doubly excited states have energies in excess of the ionization limit. Thus, although they are bound states (because both electrons are bound to the He^{++} nucleus), they are degenerate with continuum states consisting of a He^+ ion and a free electron of the appropriate

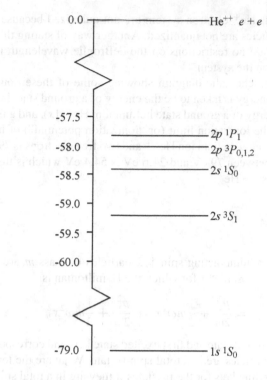

FIGURE 9.4. Energy level diagram of He showing the ground state, the first set of excited states, and the ionization limit.

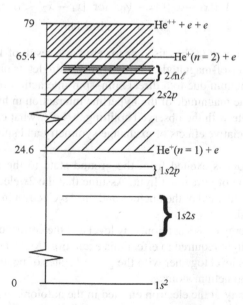

FIGURE 9.5. Schematic diagram showing some autoionizing states of He.

energy. States of this He^+-electron system are not quantized because the kinetic energies of free particles are not quantized. Another way of stating this is that for free particles there are no restrictions on the deBroglie wavelength that impose quantized energies on the system.

FIGURE 9.5 is a schematic diagram showing some of these doubly excited states. The zero of energy is taken to be the energy of a ground state helium atom. On this scale the energy of a ground state helium ion, $He^+(1s)$, and a free electron is 24.6 eV, that is, the ionization limit (or "ionization potential") of helium. The energy of a doubly charge helium ion He^{++} and two free electrons is 79 eV. Notice that the difference between 79 eV and 24.6 eV is 54.4 eV which is the ionization energy of a ground state He^+.

Problems

9.1. Two identical, noninteracting spin-1/2 particles of mass m are in the one-dimensional harmonic oscillator for which the Hamiltonian is

$$\hat{H} = \frac{p_{x_1}^2}{2m} + \frac{1}{2}m\omega^2 x_1^2 + \frac{p_{x_2}^2}{2m} + \frac{1}{2}m\omega^2 x_2^2 \tag{9.60}$$

(a) Determine the ground state and first excited state kets and corresponding energies when the two particles are in a total spin-0 state. What are the lowest energy states and corresponding kets for the particles if they are in a total spin-1 state?

(b) Suppose the two particles interact with a potential energy of interaction

$$V(|x_1 - x_2|) = -V_0 \quad \text{for } |x_1 - x_2| < a$$
$$= 0 \quad \text{elsewhere}$$

Will the energies of part (a) be raised or lowered as a result of $V(|x_1 - x_2|)$?

9.2. Make an order-of-magnitude estimate of the singlet–triplet splitting of the energy levels of helium due to a direct spin–spin interaction of the electrons by comparing with the magnitude of the hyperfine interaction in hydrogen. By comparing this estimate with the observed splitting, ~ 1 eV, what conclusions can be drawn about the relative effects of exchange symmetry and spin–spin interaction on the energy?

9.3. A helium atom is excited from the ground state to the autoionizing state $2s4p$ by absorption of ultraviolet light. Assume that the $2s$ electron moves in the unscreened Coulomb field of the nucleus and the $4p$ electron in the fully screened Coulomb potential.

(a) Obtain the energy of this autoionizing level and the corresponding wavelength of the ultraviolet light required to effect this excitation. Make an energy level diagram showing this level together with the ground states of neutral, singly ionized, and doubly ionized helium atoms.

(b) Find the velocity of the electron emitted in the autoionizing process in which the $2s4p$ state decays into a free electron and a He^+ ion in the ground state.

9.4. Use the variational principle to calculate the ground state energy of a hydrogen atom (in atomic units) using the normalized trial functions.

(a) $\psi(r) = \sqrt{\beta^3/\pi} \cdot \exp(-\beta r)$.

(b) $\psi(r) = 2\sqrt{2}(2\gamma^3/\pi)^{1/4} \cdot \exp(-\gamma r^2)$ where β and γ are parameters.

(c) Which gives the most accurate answer? Why?

9.5. A negative ion is formed when an electron attaches to an atom (or molecule), the result being that the nucleus of charge Z binds $(Z + 1)$ electrons. The binding energy of the electron to the neutral atom is referred to as the electron affinity of the atom. The electron affinity may also be thought of as the ionization potential of the negative ion. Not surprisingly, the halogen atoms form negative ions most readily. This means that the halogen atoms have the highest electron affinities of all atoms ($\sim$3 eV). Hydrogen atoms also form negative ions. The electron affinity of hydrogen is 0.75 eV. Use perturbation theory as it was applied to the helium atom to determine the total energy of the hydrogen atom negative ion. Compare the answer with the actual value. Compare the accuracy of the perturbation theory results for the hydrogen negative ion with the result for the helium atom. Why is perturbation theory more accurate for helium? Note that there are no "new" calculations necessary.

9.6. Assume the normalized trial wave function $|\psi\rangle = (1 + A^2)^{-1/2}[|100\rangle + A|210\rangle]$, where the kets on the right-hand side are spherical hydrogen atom eigenkets, represents a ground state hydrogen atom in a constant electric field F. This wave function represents a state that has "ground state character" and, assuming $A \ll 1$, a small amount of $|210\rangle$ character.

(a) Using this trial wave function show that, if terms in $\langle \hat{H} \rangle$ having powers greater than A^2 can be neglected, then second-order perturbation theory, as applied in Chapter 8, yields the same energy as the variational treatment. Note that if powers of A greater than two can be neglected in $\langle \hat{H} \rangle$ that powers of A greater than one may be neglected in $\partial \langle \hat{H} \rangle / \partial A$.

(b) Using this trial wave function estimate the dipole polarizability of ground state hydrogen and compare the answer with the exact answer given in Chapter 8.

References

1. G.B. Arfken and H.J. Weber, *Mathematical Methods for Physicists* (Harcourt, New York, 2001).
2. H.A. Bethe and E.E. Salpeter, *Quantum Mechanics of One- and Two-Electron Atoms* (Springer-Verlag, Berlin, 1957).
3. W. Kauzmann, *Quantum Chemistry* (Academic Press, New York, 1957).

10
Multielectron Atoms

10.1. Introduction

In our study of the helium atom we saw that the first approximation to the ground state energy was obtained by ignoring the electron–electron repulsion, thus making the Hamiltonian separable into the sum of two terms, each containing variables of a different electron. When a Hamiltonian is separable in this fashion the eigenfunctions of it are products of the eigenfunctions of each of the constituent Hamiltonians; the eigenvalues are the sums of the eigenvalues of the constituent Hamiltonians. Using this approximation, the eigenfunction for the ground state of helium is the product of two ground state one-electron wave functions with eigenvalue -4 a.u., -108.8 eV. This may be regarded as a zeroth-order approximation to the energy of the helium atom because the (ignored) electron–electron repulsion is substantial.

Although this level of approximation, sometimes referred to as the independent electron model, is not very elegant, it is used as the basis for describing the states of atoms over the entire periodic table, even though modern techniques in quantum chemistry permit calculation of atomic energies with almost arbitrary accuracy. Such calculations are not, however, a subject treated here. Rather, we present descriptions of two methods of describing atomic states, electron configuration and the quantum defect. In this chapter we concentrate on electron configuration.

10.2. Electron Configuration

The electron configuration of an atomic state is simply a listing of the one-electron states that constitute the eigenfunctions of the Hamiltonian excluding the electron–electron repulsion. It is assumed that each of these independent electrons is subjected to a central potential that depends upon the positions of the other electrons. As we have seen, the electron configuration of the ground state of helium is written $1s^2$ where the superscript 2, often (incorrectly) read as "squared", refers to the number of electrons in the atom that are in the $1s$ eigenstate. As noted in Chapter 9, such eigenstates (e.g., $1s$, $2p$, etc.) are frequently referred to as "orbitals" or "subshells", the term "shell" being reserved for the principal quantum number; the K-shell has $n = 1$, the L-shell $n = 2$, and so on. It is pointed out in introductory chemistry that the reason that two, and only two, electrons can occupy the $1s$ (or indeed the ns) orbital is that they have opposite spins. Therefore, the two electrons have different quantum numbers, n, ℓ, and m_ℓ being common to both, but $m_s = \pm 1/2$. In the language of quantum mechanics this situation is described by a wave function that is the product of a symmetric spatial function and an antisymmetric spin function.

After helium, the next atom in the periodic table is lithium ($Z = 3$) which has three electrons. The ground state configuration is therefore written as $1s^2 2s$. This procedure is continued and the subshells filled according to

$$1s^2 2s^2 2p^6 3s^2 3p^6 4s^x 3d^y$$

where a "new" subshell must be populated when all lower-lying subshells are full. The order in which these shells are filled is called the Aufbrau principle. Note that the "regularity" of the order of the shells is broken at $3d$ and $4s$, the energy of the latter lying lower than that of the $3d$ because s-electrons ($\ell = 0$) penetrate closer to the nucleus than do d-electrons ($\ell = 2$) electrons, thus making the Coulomb attraction between electron and nucleus (which lowers the energy) greater for s-electrons.

The above discussion is a familiar one first presented in introductory chemistry, without, of course, the emphasis on eigenfunctions and eigenvalues. A description of the independent electron model can also be framed in the context of quantum mechanics. Approximate though they may be, the independent electron eigenfunctions form a complete set. They may therefore be used as a basis set upon which any eigenfunction may be expanded. The independent electron model contains the assumption (actually, a first approximation) that the true eigenfunctions consist of only one (properly symmetrized) basis function, the one that is the product of *individual* electron eigenfunctions. As might be expected, this approximation breaks down in varying degrees, depending upon the particular atom and the state described. When the true eigenfunction is adequately described by a single basis function, it is referred to as a "pure" state. If the true eigenfunction requires two or more independent electron basis functions, that is, two or more "configurations", then it is said that these configurations interact. That is, configuration interaction occurs.

States are designated according to the scheme that most accurately describes the relationship of the orbital angular momentum of the electrons with the spin angular momentum of these electrons. In other words, how do the different angular momenta couple? The scheme that is appropriate for light atoms, and, indeed, most atoms, is called Russell Saunders or LS-coupling. Included in the group of atoms for which LS coupling is appropriate are certain heavy atoms, for example, alkali atoms. Although this coupling scheme may not accurately describe a given atom, states may be designated by their LS notation. In such cases, for example, mercury, the LS notation merely provides a name for the state.

10.3. The Designation of States—LS Coupling

To understand the LS coupling scheme we write the total Hamiltonian $\hat{H}_T$ for a multielectron atom as a sum of three terms

$$\hat{H}_T = \hat{H}_0 + \hat{H}_1 + \hat{H}_2 \tag{10.1}$$

where $\hat{H}_0$ represents the Coulomb interaction of each electron with the nucleus, $\hat{H}_1$ contains interelectron effects that include the electrostatic repulsion of the electrons, and $\hat{H}_2$ describes the spin-orbit interaction. Of course, the exchange energy is also present although not explicitly contained in the Hamiltonian. For the purpose of this discussion, however, we may regard it as being (symbolically) contained in $\hat{H}_1$. In fact, in LS-coupling the exchange force is greater than the electronic electrostatic repulsion, both of which are assumed to be greater than the spin-orbit coupling. That is, it is assumed that $\hat{H}_1 \gg \hat{H}_2$ so the unperturbed Hamiltonian is taken to be

$$\hat{H} = \hat{H}_0 + \hat{H}_1 \tag{10.2}$$

Because $\hat{H}$ does not contain the spin-orbit interaction (or even the spin) $\hat{J}$, $\hat{L}$, and $\hat{S}$ commute with the Hamiltonian and the eigenfunctions can be written as eigenfunctions of $\hat{L}^2$ and $\hat{S}^2$; that is, they have definite orbital angular momentum and definite spin angular momentum. It is assumed that there is a strong interaction between the orbital angular momentum of each electron with the other orbital angular momenta. Similarly, there is a strong interaction between the spin angular momentum of each electron with the other spins. The vector sum of the orbital angular momenta of all electrons is designated by the letter L, and S represents the vector sum of all the spin angular momenta. J designates the vector sum of L and S. Thus, the good quantum numbers are L, S, M_L, M_S, and, of course, J. States, sometimes referred to as "terms", are designated as follows.

$$^{2S+1}L_J$$

as was the case for helium. For reasons that are apparent shortly, the superscript $2S + 1$ is called the multiplicity.

Terms for Nonequivalent Electrons—Electrons Not in the Same Subshell

Once the electron configuration of a given atomic state is known, it is a relatively simple matter to find the possible terms. The case of two electrons serves as a guide. Extension to more than two electrons is obvious.

Example

Find all terms for two electrons of configuration $npn'd$. How many states are possible for this configuration?

Solution

The possible values of L for two electrons are given by

$$L = |\ell_1 + \ell_2|, \ |\ell_1 + \ell_2 - 1|, \ldots |\ell_1 - \ell_2|$$
$$= 3, 2, 1$$

so L can be F, D, or P. The possible values of S are

$$S = \frac{1}{2} \pm \frac{1}{2}$$
$$= 1, 0$$

Therefore, the states that can arise from an $npn'd$ configuration are

$$^3F_J, \ ^3D_J, \ ^3P_J, \ ^1F_J, \ ^1D_J, \ ^1P_J$$

The possible J-values that accompany each state depend upon the values of L and S for that state; that is,

$$J = |L + S|, \ |L + S - 1|, \ldots |L - S|$$

Thus, the states that result from configuration $npn'd$ are

$$^3F_4, \ ^3F_3, \ ^3F_2$$
$$^3D_3, \ ^3D_2, \ ^3D_1$$
$$^3P_2, \ ^3P_1, \ ^3P_0$$
$$^1F_3$$
$$^1D_2$$
$$^1P_1$$

There are three states, each of different J-value, for each-state for which $2S + 1 = 3$, but only one for those for which $2S + 1 = 1$. Therefore, the quantity $2S + 1$ is called the multiplicity and, usually (not always), corresponds to the number of J-values for a state of given L and S. An exception occurs (not in this example) when $L < S$. In such a case the number of J-values for each state is $2L + 1$. This is because the number of different vectors that result from the vector addition of L and S depends upon the smaller of L and S. Nevertheless, the antecedent superscript is always $2S + 1$ and it is usually called the multiplicity whether or not it corresponds to the number of J-values for each state of definite L and S. In this example, there are a total of twelve states: three each for the three different triplets and one each for each of the three singlets. Of course, the total number of states cannot depend upon the angular momentum coupling scheme.

To illustrate how the multiplicity, $2S + 1$, need not correspond to the total number of states for a multiplet we present another example.

Example

Find the terms for a $nsn's$ configuration. How many states are possible for this configuration?

Solution

Clearly $L = 0$ and $S = 0$ or 1. Therefore, the terms are $^3S_1$ and $^1S_0$. Although the multiplicity of the $^3S_1$ term suggests that this is a triplet, there is, in fact, only one value of J; that is, $J = 1 + 0 = 1$. Nonetheless, the term $^3S_1$ is usually read as "triplet S one". This configuration provides a total of two states.

Terms for Equivalent Electrons—Electrons in the Same Subshell

In the discussion of nonequivalent electrons we were justified in ignoring the implications of the Pauli principle. In the above examples either the principal quantum numbers of the electrons or the orbital angular momentum quantum numbers of the electrons were different so we did not have to consider violations of the Pauli principle in the form "no two electrons in an atom can have the same set of quantum numbers." For electrons that have different principal quantum numbers and/or different angular momentum quantum numbers the Pauli principle is satisfied a priori. If however, the electron configuration consists of two or more electrons having the same values n and ℓ, then care must be taken to ensure that some of the possible terms do not require two or more of the electrons to have the same quantum numbers, n, ℓ, m_ℓ, and m_s. Electrons that have the same n and ℓ quantum numbers are referred to as equivalent electrons. Again, a simple example serves to demonstrate the method of assigning terms to a given configuration.

Example

Consider two np electrons, that is, the np^2 configuration. Find all possible terms.

Solution

We first treat the electrons as though they are nonequivalent to obtain all possible designations of the states. We must then eliminate those that constitute a violation of the Pauli principle. If the electron configuration were $npn'p$ the possible states, ignoring J-values, would be

$$npn'p \rightarrow {}^3D, \ {}^3P, \ {}^3S, \ {}^1D, \ {}^1P, \ {}^1S$$

Now we must eliminate those terms for which the sets of four quantum numbers are identical. There are a few ways in which this may be done. The brute force method is to make a table of possible quantum numbers for each electron and to eliminate those terms for which the sets of quantum numbers are identical. This is, however, time consuming and unnecessary. A much simpler method is to use a table of equivalent electrons.[1] For comparison we first show a partial listing for *nonequivalent* electrons in TABLE 10.1.

This table is considerably reduced when the two electrons are equivalent as can be seen in TABLE 10.2.

TABLE 10.1. Terms for nonequivalent s, p, and d electron configurations.

ss	$^1S, {}^3S$
s, p	$^1P, {}^3P$
sd	$^1D, {}^3D$
pp	$^1S, {}^1P, {}^1D, {}^3S, {}^3P, {}^3D$
pd	$^1P, {}^1D, {}^1F, {}^3P, {}^3D, {}^3F$
dd	$^1S, {}^1P, {}^1D, {}^1F, {}^1G, {}^3S, {}^3P, {}^3D, {}^3F, {}^3G$

There are two configurations listed in the left-hand column of TABLE 10.2 because the same terms result when a given number of electrons are equal to the same number of "holes" in a subshell.

For the purpose of this example we use TABLE 10.2 and find that three of the terms for two *non*equivalent p-electrons are excluded if the electrons are equivalent. Thus,

$$np^2 \rightarrow {}^3P, {}^1D, {}^1S$$

There is an even easier way to find the possible terms if only two equivalent electrons are involved. For two equivalent electrons there is a rule that states

For two equivalent electrons the only states that are allowed are those for which the sum $(L + S)$ is even.

Inspection of the example of two equivalent p-electrons shows that this rule is obeyed as it is for two d-electrons (see TABLE 10.2).

After determining the possible terms for equivalent electrons, any other electrons not in a closed shell (valence electrons) may be considered.

Example

Determine the terms for the electron configuration $np^2n'p$.

Solution

We know that the np^2 configuration permits only the states ^3P, ^1D, and ^1S. We have then:

$$^1D + n'p : [L = 2, S = 0] + n'p \Rightarrow L = 3, 2, 1; S = 1/2$$
$$^3P + n'p : [L = 1, S = 1] + n'p \Rightarrow L = 2, 1, 0; S = 3/2, 1/2$$
$$^1S + n'p : [L = 0, S = 0] + n'p \Rightarrow L = 0; S = 1/2$$

TABLE 10.2. Terms for equivalent s, p, and d electron configurations.

p^1, p^5	^2P
p^2, p^4	$^3P, {}^1D, {}^1S$
p^3	$^4S, {}^2D, {}^2P$
d^1, d^9	^2D
d^2, d^8	$^3F, {}^3P, {}^1G, {}^1D, {}^1S$
d^3, d^7	$^4F, {}^4P, {}^2H, {}^2G, {}^2F, {}^2D(2), {}^2P$
d^4, d^6	$^5D, {}^3H, {}^3G, {}^3F(2), {}^3D, {}^3P(2), {}^1I, {}^1G(2), {}^1F, {}^1D(2), {}^1S(2)$
d^5	$^6S, {}^4G, {}^4F, {}^4D, {}^4P, {}^2I, {}^2H, {}^2G(2), {}^2F(2), {}^2D(3), {}^2P, {}^2S$

The combinations involving the ^3P two electron state give the following terms,

$$^4D_{7/2,5/2,3/2,1/2}; {}^4P_{5/2,3/2,1/2}; {}^4S_{3/2}; {}^2D_{5/2,3/2}; {}^2P_{3/2,1/2}; {}^2S_{1/2}$$

Notice that in this example we have quartets and doublets, but there are four and two states, respectively, only in those cases for which $L > S$. When $L < S$ the number of states is given by $2L + 1$.

Ordering of the States—Hund's Rules

This coupling scheme leads to two rules, referred to as Hund's rules, which determine the ordering of the states. If these rules do not strictly apply, then LS-coupling does not accurately describe the atom.

Hund's Rules
1. For states having the same electron configuration the state having the highest total spin S lies lowest.
2. For a given value of the total spin S the state with highest total orbital angular momentum L lies lowest.

The origin of these rules may be understood qualitatively using symmetry and a simple picture of the atom. For simplicity we discuss them for the helium atom; extension to multielectron atoms presents no conceptual problem.

In essence, the ordering of the possible states is determined by the electron–electron repulsion in the atom, that is, the term in the helium Hamiltonian

$$\hat{H}_{ee} = \left(\frac{1}{4\pi\varepsilon_0}\right)\frac{(-e)^2}{|r_1 - r_2|} \tag{10.3}$$

This term is manifestly positive. We see, therefore, that when the valence electrons are, on average, far apart then $|r_1 - r_2|$ is large and the positive contribution of this term to the energy will be small. On the other hand, when $|r_1 - r_2|$ is small so that, on average, the electrons are close together, this term will provide a large positive contribution to the energy. We may see this and also gain an appreciation for the consequences of indistinguishability and the Pauli principle by computing the average interelectron separation $\langle|r_1 - r_2|\rangle$ for two noninteracting electrons. Because the electrons are assumed to be noninteracting, we use the wave functions appropriate to the independent electron model. For the triplet state we have

$$|\text{triplet}\rangle = \frac{1}{\sqrt{2}}\left(|100\rangle_1 |n\ell m\rangle_2 - |n\ell m\rangle_1 |100\rangle_2\right)|1M_s\rangle \tag{10.4}$$

where we have used the coupled representation for the spin; M_s denotes the z-component of the *total* spin. Notice that for the triplet state, the spin part of the wave function is symmetric, so the space part is antisymmetric. For the singlet state we have

$$|\text{singlet}\rangle = \frac{1}{\sqrt{2}}\left(|100\rangle_1 |n\ell m\rangle_2 + |n\ell m\rangle_1 |100\rangle_2\right)|00\rangle \tag{10.5}$$

Because $|r_1 - r_2|$ does not contain spin, the spin part of the wave function does not enter into the computation. Because it is normalized, it merely multiplies the spatial integral by unity. Also, we simplify the notation by replacing the ground state and excited state kets by $|g(i)\rangle$ and $|e(i)\rangle$, respectively, where i denotes one or the other electron; it is, of course, either 1 or 2. The spatial part of the wave functions may then be written as

$$|\text{spatial}\rangle = \frac{1}{\sqrt{2}}\left[|g(1)\rangle|e(2)\rangle \pm |g(2)\rangle|e(1)\rangle\right] \qquad (10.6)$$

where the upper sign corresponds to the singlet state and the lower to the triplet states. For the computation we write $|r_1 - r_2|$ in terms of the individual vectors r_1 and r_2 and, for simplicity, consider the quantity $|r_1 - r_2|^2$,

$$\langle|r_1 - r_2|^2\rangle = \langle r_{12}^2\rangle$$
$$= \langle r_1^2 + r_2^2 - 2r_1 \cdot r_2\rangle \qquad (10.7)$$

Then, ignoring normalization, we have

$$\langle r_{12}^2\rangle = \frac{1}{2}\,\langle g(1)|\,r_1^2\,|g(1)\rangle + \langle e(1)|\,r_1^2\,|e(1)\rangle$$
$$+ \langle g(2)|\,r_2^2\,|g(2)\rangle + \langle e(2)|\,r_2^2\,|e(2)\rangle$$
$$- 2\,\langle g(1)|\,r_1\,|g(1)\rangle \cdot \langle e(2)|\,r_2\,|e(2)\rangle$$
$$- 2\,\langle e(1)|\,r_1\,|e(1)\rangle \cdot \langle g(2)|\,r_2\,|g(2)\rangle$$
$$\mp 2\left[\begin{array}{c}\langle g(1)|\,r_1\,|e(1)\rangle \cdot \langle g(2)|\,r_2\,|e(2)\rangle \\ -2\,\langle e(1)|\,r_1\,|g(1)\rangle \cdot \langle e(2)|\,r_2\,|g(2)\rangle\end{array}\right] \qquad (10.8)$$

Symmetry demands that certain of the integrals above be identical. We may also further simplify the notation letting

$$\langle g(1)|\,r_1^2\,|g(1)\rangle = \langle g(2)|\,r_2^2\,|g(2)\rangle = \langle r^2\rangle_g$$
$$\langle e(1)|\,r_1^2\,|e(1)\rangle = \langle e(2)|\,r_2^2\,|e(2)\rangle = \langle r^2\rangle_e$$
$$\langle g(1)|\,r_1\,|g(1)\rangle \cdot \langle e(2)|\,r_2\,|e(2)\rangle = \langle e(1)|\,r_1\,|e(1)\rangle \cdot \langle g(2)|\,r_2\,|g(2)\rangle$$
$$= \langle r\rangle_g \cdot \langle r\rangle_e$$
$$\langle g(1)|\,r_1\,|e(1)\rangle \cdot \langle g(2)|\,r_2\,|e(2)\rangle = \langle e(1)|\,r_1\,|g(1)\rangle \cdot \langle e(2)|\,r_2\,|g(2)\rangle$$
$$= \left(\langle r\rangle_{exchange}\right)^2 \qquad (10.9)$$

The expression for $\langle r_{12}\rangle$ becomes

$$\langle r_{12}^2\rangle = \langle r^2\rangle_g + \langle r^2\rangle_e - 2\,\langle r\rangle_g \cdot \langle r\rangle_e \mp 2\left[\langle r\rangle_{exchange}\right]^2 \qquad (10.10)$$

Now, if the electrons were distinguishable only the first three terms would appear. The term containing $\langle r\rangle_{exchange}$ arises from the necessary condition that the wave function, upon interchange of electrons, must transform into itself with a change of sign. That is, the *total* wave function must be antisymmetric. The upper sign corresponds to a symmetric spatial wave function and the lower to an antisymmetric spatial wave function. Because $\langle r \rangle_{exchange}$ is manifestly positive, the average inter-electron separation is smaller for singlet states (symmetric spatial) than it would be if the electrons were distinguishable. Correspondingly, the average interelectron separation for triplet states (antisymmetric spatial) is greater than it would be if the electrons were distinguishable.

The effect of the Pauli principle on electrons may be summarized as follows. Electrons having parallel spins tend to repel each other whereas electrons having opposite spins tend to attract. The repulsion and the attraction are due only to the symmetry requirements of the wave function. It is referred to as the *exchange force* because it arises from the indistinguishability of the electrons and the symmetry requirement of the Pauli principle. It is not caused by an electrical (Coulomb) repulsion.

The Coulomb repulsion of the electrons produces a positive contribution to the total energy, thus lowering the binding energy. Therefore, for triplet states, for which the exchange force requires the electrons to avoid each other, this positive contribution is lower than for singlet states for which the electrons can, in a sense, occupy the same volume. It is clear then that, for electrons of the same configuration, the state having the highest value of the total spin S will be the lowest lying state. This is, in fact, Hund's first rule.

To get a feeling for Hund's second rule we must envision the valence electrons in orbit about the ionic core. (In quantum mechanics we are not supposed to do this, but physicists do it anyway.) Suppose the valence electrons have nonzero angular momenta. Suppose further that the total orbital angular momentum L, the vector sum of the individual angular momenta, has the lowest possible value. This means that the individual electronic angular momenta vectors point in essentially the opposite directions, indicating that the electrons are revolving about the core in opposite directions. As such, they pass each other and the interelectron distance can be quite small. This leads to a large positive contribution to the total energy. At the other extreme, if L has its maximum value the electrons revolve in essentially the same direction and can maintain large interelectron separations. In this case the positive contribution to the energy is minimal. Such an argument provides, at least, a qualitative rationalization of Hund's second rule.

Using our knowledge of the labeling of LS-states and Hund's rules to order them we may construct an energy-level diagram for a given electron configuration. Before doing so, however, we must address the ordering of the J-states. The group of states having the same electron configuration and the same values of S and L, but different values of J is called a multiplet. The ordering of these different J-states within a multiplet is given by a separate rule.[1,2]

For a given multiplicity and value of L, the state having the lowest J lies lowest for subshells that are less than one-half full. If the subshell is more than half full then the state having the highest value of J lies lowest.

When subshells are less than half full the multiplets are referred to as being "regular"; when subshells are more than half full they are "inverted".

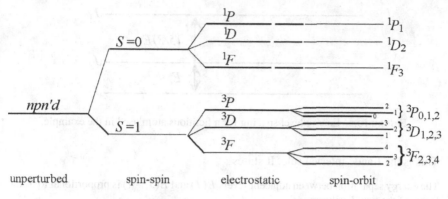

FIGURE 10.1. Schematic diagram of the energy splittings under *LS*-coupling of two electrons of *npn′d* configuration. [From Reference 3.]

As an example, we construct a schematic diagram showing the ordering of the *LS*-multiplets for two valence electrons in the *npn′d* configuration.[3] The complete diagram is shown in FIGURE 10.1.

The Landé Interval Rule for Multiplets

The relative separations between the states of a multiplet are given by the Landé interval rule. The splitting into the levels of the multiplet is caused by the spin-orbit interaction; the Hamiltonian $\hat{H}_2$ is

$$\hat{H}_2 = A\hat{L}\cdot\hat{S} \tag{10.11}$$

where A is a constant. Now $\hat{L}\cdot\hat{S}$ is given by

$$\hat{L}\cdot\hat{S} = \frac{1}{2}\left(\hat{J}^2 - \hat{L}^2 - \hat{S}^2\right) \tag{10.12}$$

so that eigenfunctions of $\hat{J}^2$, $\hat{L}^2$, and $\hat{S}^2$ are also eigenfunctions of $\hat{H}_2$. Thus, the eigenvalues of the operator $\hat{H}_2$ are

$$E_2(J, L, S) = \frac{1}{2}[J(J+1) - L(L+1) - S(S+1)] \tag{10.13}$$

Within a given multiplet we may compute the difference in the energies of adjacent J-levels. For adjacent levels of a multiplet the values of L and S are the same and we designate $J_{upper} = J$ so that $J_{lower} = J - 1$. We have then

$$E_2(J) - E_2(J-1) = \frac{1}{2}A[J(J+1) - L(L+1) - S(S+1)]$$
$$- \frac{1}{2}A[(J-1)J - L(L+1) - S(S+1)]$$
$$= AJ \tag{10.14}$$

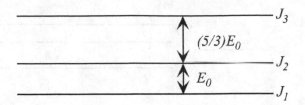

FIGURE 10.2. Energy level spacing for a fictitious atom used in the example.

This is the Landé interval rule. It states

The energy separation between adjacent levels $E(J)$ and $E(J-1)$ is proportional to J, the larger of the two J-values.

Example

The energy separations between the three levels of a triplet state of a fictitious atom that obeys LS coupling are observed to be as shown in FIGURE 10.2.

Find the correct LS designation for the state, that is, the correct term symbol, ^{2S+1}L.

Solution

According to the Landé interval rule, the energy separation between two levels of a multiplet is proportional to the J-value of the higher state. Using the notation shown in FIGURE 10.2 we have

$$\left(\frac{5}{3}\right)E_0 = KJ_3 \quad \text{and} \quad E_0 = KJ_2 \tag{10.15}$$

where K is a constant. Note that K is the same constant in both of these equations. Because $J_3 = J_2 + 1$

$$\left(\frac{5}{3}\right)E_0 = K(J_2 + 1) \quad \text{and} \quad E_0 = KJ_2 \tag{10.16}$$

Dividing the equations in (10.16), we have

$$\frac{5}{3} = \frac{J_2 + 1}{J_2} \tag{10.17}$$

or $J_2 = 3/2$, from which $J_3 = 5/2$ and $J_1 = 1/2$. J is related to L and S according to

$$J = |L + S|, |L + S - 1|, |L + S - 2|, \ldots, |L - S| \tag{10.18}$$

Now, assume that $L > S$, so we have $L + S = 5/2$ and $L - S = 1/2$. Adding these two equations we get $2L = 3$ or $L = 3/2$.

But this cannot be correct because L must be an integer. Because our arithmetic is correct, our assumption that $L > S$ must be incorrect. If $S > L$ then

$S - L = 1/2$. Adding this to $L + S = 5/2$ (which must be correct) we obtain $2S = 3$ or $S = 3/2$ from which we obtain $L = 1$. Because L is an integer, the assumption that $S > L$ must be correct and we are dealing with a P-state. The "multiplicity" is then $2S + 1 = 4$ and we have a quartet P-state. Specifically, we have the three states

$$^4P_{5/2}, \; ^4P_{3/2}, \; ^4P_{1/2}$$

Note that, although the multiplicity is 4, there are only three levels because $L < S$.

10.4. The Designation of States—*jj* Coupling

Although *LS* coupling is satisfactory for describing most atoms, the heaviest atoms require different considerations. *jj*-coupling is the antithesis of *LS*-coupling. In this scheme it is assumed that the spin-orbit interaction is much larger than interelectron effects. Therefore, the unperturbed Hamiltonian is taken to be

$$\hat{H} = \hat{H}_0 + \hat{H}_2 \tag{10.19}$$

and $\hat{H}_1$ is assumed to be the perturbation. The orbital angular momentum of each electron interacts strongly with its own spin angular momentum. Each electron is treated exactly as the spin-orbit coupling was treated for the hydrogen atom.

In Chapter 7 it was shown that the fine-structure correction is proportional to Z^4, and in Chapter 9 it was shown that the interelectron energy is proportional to Z. Therefore, the spin-orbit energy is expected to dominate the interelectron energy for high-Z atoms for which *jj*-coupling is most appropriate. Moreover, because the relativistic correction is also proportional to Z^4 we expect *jj*-coupling to be most appropriate for highly ionized atoms in which the electronic velocities are high.

For a single electron, *jj*-coupling states are designated by the quantum numbers n, ℓ, j, m where m is m_j. Using the letter designation for the quantum number ℓ, states of a single electron are labeled ℓ_j, so that possible one-electron states are $s_{1/2}, \, p_{1/2}, \, p_{3/2}, \, d_{3/2}, \, d_{5/2}, \ldots$ and so on. States are designated by first adding the orbital and spin angular momenta for each electron and then combining these angular momenta to form a total J. States are labeled $(j_1 \, j_2 \ldots)_J$ where the j_i are the total angular momenta for the ith electron. For nonequivalent electrons the values of these quantities are determined by the rules for addition of angular momenta.

Example

Consider the electron configuration $ndn'd$, two nonequivalent d-electrons. Find all terms for two electrons of this configuration. How many states are there?

Solution

The possible values of j_1 and j_2 are the same for each electron, 3/2 and 5/2. The total angular momentum can then range from $J = 0$ to $J = 5$. The states are designated

$$\left(\frac{3\,3}{2\,2}\right)_3 ; \left(\frac{3\,3}{2\,2}\right)_2 ; \left(\frac{3\,3}{2\,2}\right)_1 ; \left(\frac{3\,3}{2\,2}\right)_0 ;$$

$$\left(\frac{5\,5}{2\,2}\right)_5 ; \left(\frac{5\,5}{2\,2}\right)_4 ; \left(\frac{5\,5}{2\,2}\right)_3 ; \left(\frac{5\,5}{2\,2}\right)_2 ; \left(\frac{5\,5}{2\,2}\right)_1 ; \left(\frac{5\,5}{2\,2}\right)_0 ;$$

$$\left(\frac{3\,5}{2\,2}\right)_4 ; \left(\frac{3\,5}{2\,2}\right)_3 ; \left(\frac{3\,5}{2\,2}\right)_2 ; \left(\frac{3\,5}{2\,2}\right)_1 ;$$

$$\left(\frac{5\,3}{2\,2}\right)_4 ; \left(\frac{5\,3}{2\,2}\right)_3 ; \left(\frac{5\,3}{2\,2}\right)_2 ; \left(\frac{5\,3}{2\,2}\right)_1$$

for a total of 18 states. Note that this is the same number of states that is obtained for two nonequivalent d-electrons using LS coupling because the coupling scheme does not determine the number of states (see TABLE 10.1).

As for LS coupling, the number of states is reduced if there are equivalent electrons in the configuration. For two equivalent electrons it is relatively simple to eliminate those terms that violate the Pauli principle. There are two rules.

1. If $j_1 \neq j_2$ then all states from nonequivalent electrons are allowed, but, because the electrons are equivalent, the states $(j_1, j_2)_J$ are identical with states $(j_2, j_1)_J$ so retain only one set, for example, those for which $j_1 < j_2$.
2. If $j_1 = j_2$ then the allowed values of J are $J = (2j - 1)$, $(2j - 3)$, $(2j - 5) \ldots$ until J becomes negative. Values of J for which $J = (2j)$, $(2j - 2)$, $(2j - 4) \ldots$ are forbidden.

If the two d-electrons in the example above are now considered to be equivalent, then, of the eight terms for which $j_1 \neq j_2$ only four are distinct. We retain those for which $j_1 < j_2$.

$$\left(\frac{3\,5}{2\,2}\right)_4 ; \left(\frac{3\,5}{2\,2}\right)_3 ; \left(\frac{3\,5}{2\,2}\right)_2 ; \left(\frac{3\,5}{2\,2}\right)_1$$

The remaining terms must also be reduced. We see that of the two terms with $j_1 = j_2 = 3/2$ the ones for which $J = 0$, 2 are allowed. For the states for which $j_1 = j_2 = 5/2$ those that are allowed are those for which $J = 4$, 2, 0. Therefore, the allowed states for two equivalent d-electrons are

$$\left(\frac{3\,3}{2\,2}\right)_2 ; \left(\frac{3\,3}{2\,2}\right)_0$$

$$\left(\frac{5\,5}{2\,2}\right)_4 ; \left(\frac{5\,5}{2\,2}\right)_2 ; \left(\frac{5\,5}{2\,2}\right)_0$$

$$\left(\frac{3\,5}{2\,2}\right)_4 ; \left(\frac{3\,5}{2\,2}\right)_3 ; \left(\frac{3\,5}{2\,2}\right)_2 ; \left(\frac{3\,5}{2\,2}\right)_1$$

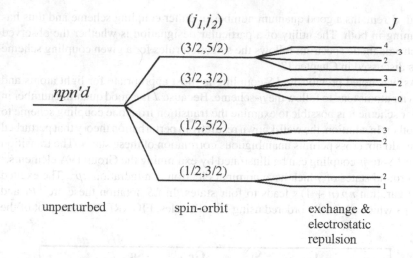

FIGURE 10.3. Schematic diagram of the energy splittings under *jj*-coupling of two electrons of *npn'd* configuration. [From Reference 3.]

We see that there are a total number of nine states from the $(nd)^2$ configuration as there are in the *LS* coupling scheme (see TABLE 10.2). It should be remarked that when the valence electrons are equivalent their electrostatic interaction is large as a result of their proximity. Thus, *jj*-coupling is more appropriate for *non*equivalent electrons.

To illustrate the splitting of the levels, FIGURE 10.3 shows the different *jj*-coupling states for the *npn'd* configuration, the same configuration used to illustrate *LS*-states in FIGURE 10.1. To order these states we note from the discussion of hydrogen fine-structure in Chapter 7 that, for a single electron, the energy due to the spin-orbit interaction depends upon the alignment of the orbital and spin angular momenta. It was found that

$$E_{SO}^{(1)} = -\frac{1}{2}\alpha^2 E_n^{(0)} \frac{1}{n\left(\ell+\frac{1}{2}\right)(\ell+1)} \qquad \text{for } j = \ell + \frac{1}{2}$$

$$= \frac{1}{2}\alpha^2 E_n^{(0)} \frac{1}{n\ell\left(\ell+\frac{1}{2}\right)} \qquad \text{for } j = \ell - \frac{1}{2} \qquad (10.20)$$

Because $E_n^{(0)}$ is an intrinsically negative quantity, the spin-orbit energy for $j = (\ell - 1/2)$ is lower than the spin-orbit energy for $j = (\ell + 1/2)$. This indicates that the lowest states in *jj*-coupling will be those for which the individual *j*-values are lowest.

It is important to note that the *LS* and *jj*-designations are merely names for the states. The efficacy of any name lies in the reliability of the parameters that are contained in them. Therefore, to designate a state using, for example, its *LS* name is not very helpful if *L* and *S* are not good quantum numbers. On the other hand, J remains a good quantum number in either coupling scheme and thus has meaning in both. The utility of a particular designation is whether the observed electromagnetic spectrum follows the selection rules for a given coupling scheme as is discussed in Chapter 13.

As was noted previously, LS-coupling is most appropriate for light atoms and heavier atoms tend to follow the jj-scheme. Because J is a good quantum number in either scheme it is possible to examine the transition from one coupling scheme to the other. In addition, the well-known result from perturbation theory that perturbed states do not cross permits unambiguous correlation of these states. The transition from LS- to jj-coupling can be illustrated by examining the Group IVA elements,[2] the ground states of which have nominal electron configuration np^2. The excited configuration $np\,(n+1)\,s$ leads to four states. In LS notation these are $^1P_1$ and $^3P_{0,1,2}$ which are easily ordered using Hund's rules. FIGURE 10.4 is a plot of the

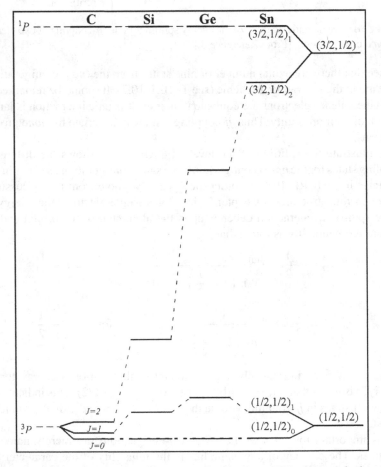

FIGURE 10.4. The transition from LS- to jj-coupling for Group IVA elements in the excited configuration $np(n+1)s$.

relative separations of these states as they evolve from C to Sn. Shown at the extreme left and right, respectively, are the pure LS and jj-levels before having been split according to their J-values. Because J is a good quantum number, the correlations $^3P_0 \rightarrow \left(\frac{1}{2}\,\frac{1}{2}\right)_0$ and $^3P_2 \rightarrow \left(\frac{3}{2}\,\frac{1}{2}\right)_2$ are obvious. Using the same arguments [see Equations (10.20)] that led to ordering of the states in FIGURE 10.3, it is clear that $^3P_1 \rightarrow \left(\frac{1}{2}\,\frac{1}{2}\right)_1$ because this jj-designated state has smaller j-values than $\left(\frac{3}{2}\,\frac{1}{2}\right)_1$.

For the purpose of illustration the separations between the lowest and highest levels of each element have been scaled to be the same in FIGURE 10.4. It can be seen that for C the levels are arranged in accord with Hund's rules. The singlet state is higher than the triplet states by roughly 50 times the separations between successive triplet states. Moreover, the triplet states are ordered in increasing values of J because the subshell is less than half full. Si exhibits a near-LS spacing of the analogous levels, but substantial deviation occurs for Sn where the highest triplet state actually lies closer to the state that evolved from $^1P_1$ than the state that evolved from $^3P_1$. For Sn the states have regrouped according to (j_1, j_2) and, as can be seen by comparison with the unperturbed jj-states on the right, this atom exhibits nearly pure jj-coupling. The data in FIGURE 10.4 show clearly that the coupling scheme that is appropriate for many atoms is a mixture of the two extreme coupling schemes discussed here. An additional point of interest is that, ignoring the spacing between energy levels, the correlations pertain to states of *any* atoms (not just group IVA atoms) for which the valence electrons have sp configuration.

Although high-Z atoms generally exhibit jj-coupling, high-Z does not necessarily mean that jj-coupling is appropriate. This is illustrated in the alkaline earth elements, group IIA. The ground states of the group IIA elements have nominal electron configuration ns^2 so they have relatively low-lying excited states of configuration $ns\,(n+1)\,p$. Notice that this sp configuration will lead to identically labeled states as those we encountered in the case of the group IVA elements. The difference is that in the group IVA elements the p-electron remains with the "core" of the atom. It is the s-electron that occupies the higher orbital. The situation is reversed for the group IIA elements. The p-electron occupies the higher orbital. FIGURE 10.5 shows the energy levels of the group IIA elements having $ns\,(n+1)p$ electron configuration. Note that, even at high Z, the states group, more or less, in accord with what is expected for LS-coupling. This is in sharp contrast to the situation for the group IVA elements as was seen in FIGURE 10.4 despite the same nominal electron configurations.

The difference between these two cases lies in the relative magnitudes of n and n'. If $n > n'$, as for the group IVA elements, then a transition from LS- to jj-coupling occurs. If, however, $n < n'$, the energy level structure remains nearly that of LS-coupling. For the s-electron, $\ell = 0$ so there is no spin-orbit interaction. It remains then to understand the spin-orbit interaction for the p-electron in each case. From Equation (10.20) it is seen that the magnitude of the spin-orbit energy is proportional to $1/n^3$. Thus, this excited state configuration for the group IIA elements features the p-electron in a higher shell (not subshell) than the comparable state of the group IVA elements. Therefore, in the group IIA elements the

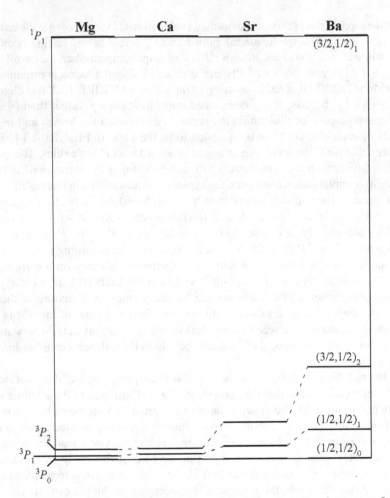

FIGURE 10.5. Energy levels of Group IIA elements having $ns(n + 1)p$ electron configuration illustrating that LS-coupling is valid even for the high Z elements. The energy scale has been adjusted to make the spacing between the lowest and highest levels the same in each atom.

spin-orbit energy in $ns(n + 1)p$ configuration will be considerably smaller than the spin-orbit energy in $np(n + 1)s$ configuration of the group IVA elements. Because jj-coupling is most pertinent when the spin-orbit coupling is maximum, the $nsn'p$ states of the group IIA elements exhibit near LS-coupling.

Problems

10.1. Find the LS terms that arise from the following configurations:
(a) $nsnp$; (b) $npnd$; (c) $(np)^2ns$.

10.2. Write the complete ground state term in LS notation, $^{2s+1}L_J$, for the elements in the first row of the periodic table, that is, Li through Ne.

10.3. The total number of states for given values of L and S is the sum of all M_J states for each possible value of J. Show that this number is $(2S + 1)(2L + 1)$ and verify that it is true for ^4D states and ^4P.

10.4. (a) Write all terms for the electron configuration $npn'p$ in both the LS- and jj-coupling notation.

(b) Make a diagram similar to FIGURE 10.3 for the jj-coupling states showing the effects of spin-orbit interaction and exchange and electrostatic repulsion. Put all terms in proper order.

10.5. (a) Write all terms for the electron configuration np^2 in both LS- and jj-coupling notation.

(b) Make a diagram similar to FIGURE 10.4 showing the transition from LS- to jj-coupling. Put all terms in proper order.

10.6. An excited configuration of the Ca atom is: $[Ar](3d)^2$.

(a) What are the allowed LS terms?

(b) A particular multiplet of a Ca atom having the above electron configuration is observed to have the energy spacing between adjacent J levels as follows.

$$E_J - E_{J-1} = \frac{4}{3}E_0 \quad \text{and} \quad E_{J-1} - E_{J-2} = E_0$$

where E_0 is a constant. What is J? What is the LS term designation ^{2S+1}L of the multiplet for which the energy levels are as shown?

References

1. G. Herzberg, *Atomic Spectra and Atomic Structure* (Dover, New York, 1957).
2. H.G. Kuhn, *Atomic Spectra* (Longmans, London and Harlow, 1962).
3. R.B. Leighton, *Principles of Modern Physics* (McGraw-Hill, New York, 1959).

11
The Quantum Defect

11.1. Introduction

In 1890 Rydberg showed[1] that for multielectron atoms the energy may be written in a form that resembles the now familiar $-1/2n^2$ for the hydrogen atom. (In this chapter we use atomic units.) For multielectron atoms there is no accidental degeneracy as for the hydrogen atom so the expression for the energy should depend on both n and ℓ. The form in which Rydberg wrote the energy is

$$E_{n\ell} = -\frac{1}{2(n - \delta_\ell)^2} \qquad (11.1)$$

where n is, as usual, the principal quantum number and δ_ℓ is called the quantum defect. The attraction of the quantum defect approximation is that it is found experimentally that for many atoms, in particular the alkali atoms, δ_ℓ is very nearly independent of n for a given value of ℓ. Thus, the energy-level diagram for such an atom can be broken into different series of states, each series corresponding to a different value of δ_ℓ and hence ℓ. FIGURE 11.1 is an energy-level diagram for hydrogen, lithium, and sodium with the common zero of energy taken to be the ionization limits of each atom. This figure shows that, as expected, all lithium and sodium levels approach those of the hydrogen atom as the principal quantum number increases. It also shows that, for a given principal quantum number, the higher angular momentum states are more nearly hydrogenic than the lower angular momentum states.

The concept of the quantum defect extends beyond being merely a numerical factor that leads to the correct energies. It is, in fact, a measure of the phase shift of the wave function for the multielectron atom from that for a hydrogen atom. In this picture the outermost or "valence" electron experiences a pure Coulomb potential when it is beyond some critical distance from the nucleus, but is "scattered" when it encounters the ionic core composed of the nucleus and the remaining electrons. Although the valence electron is bound, this is analogous to the scattering problem of an unbound electron scattering off the ionic core. In fact, when extrapolated to positive energies the phase shift $\pi\delta_\ell$ provides a link to the cross section for scattering of the ℓth partial wave.[2] In this book, however, we concentrate on the

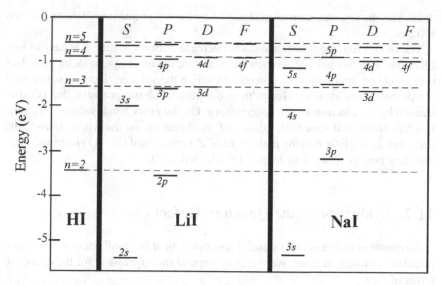

FIGURE 11.1. Partial energy-level diagrams of hydrogen, lithium, and sodium showing that the levels approximate those of hydrogen for high principal quantum numbers and high angular momentum quantum numbers.

consequences of the quantum defect on the bound state energy levels and their behavior under the influence of perturbations, notably an external electric field.

Experimentally it is found that Equation (11.1) works particularly well for alkali atoms, atoms that have one electron outside a closed shell (rare gas) configuration. Thus, an alkali atom may be regarded a "one-electron" atom, the electron residing in a potential provided by the nucleus and the rare gas core of electrons. The nucleus and these core electrons are referred to as the ionic core. In accord with the observation that the higher angular momentum states are more nearly hydrogenic in energy than the lower angular momentum states it is expected that the quantum defects δ_ℓ for the higher angular momentum states should be very small compared with those of the lower angular momentum states. TABLE 11.1 is a listing of the quantum defects for the first few angular momentum states of the alkali atoms. This table clearly shows the dramatic decrease in δ_ℓ with increasing angular momentum. As expected, the characterization of these atoms as one-electron atoms is best for

TABLE 11.1. Quantum defects of the alkali atoms for $\ell = 0 - 3$.

ℓ	Li	Na	K	Rb	Cs
s	0.40	1.35	2.19	3.13	4.06
p	0.04	0.85	1.71	2.66	3.59
d	0.00	0.01	0.25	1.34	2.46
f	0.00	0.00	0.00	0.01	0.02

those with the fewest electrons as signified by the lower quantum defects for even s-states.

The reason that the quantum defects decrease with increasing ℓ is easy to see. Electrons with low angular momentum penetrate the ionic core making interelectron repulsion important. The potential to which the valence electron is subjected is therefore substantially different from the pure Coulomb potential that is experienced by the electron in a hydrogen atom. On the other hand, valence electrons in alkali atoms that have high values of ℓ must remain, on the whole, outside the core, thus essentially moving in the field of Z protons and $(Z-1)$ electrons. They therefore behave more as hydrogen, for which $\delta_\ell \equiv 0$.

11.2. Evaluation of the Quantum Defect

If the quantum defect is to be a useful concept it must be small compared with the principal quantum number. We therefore expand the expression for the energy in terms of δ_ℓ/n

$$
\begin{aligned}
E_{n\ell} &= -\frac{1}{2} \cdot \frac{1}{(n-\delta_\ell)^2} \\
&= -\frac{1}{2n^2} \cdot \frac{1}{\left(1-\dfrac{\delta_\ell}{n}\right)^2} \\
&= -\frac{1}{2n^2}\left(1+2\frac{\delta_\ell}{n}+\cdots\right) \\
&\approx -\frac{1}{2n^2} - \frac{\delta_\ell}{n^3} \\
&= E_n - \frac{\delta_\ell}{n^3}
\end{aligned}
\tag{11.2}
$$

According to Equation (11.2) the nonhydrogenic contribution to the energy is given by $-\delta_\ell/n^3$ which is the average value of the non-Coulombic part of the potential energy. Because this non-Coulombic term arises because of the core electrons we refer to this potential as $V_c(r)$. Thus, this energy may be computed clasically by averaging over a single orbit. We have

$$
-\frac{\delta_\ell}{n^3} = \frac{1}{(2\pi n^3)} \int_{orbit} V_c(r)dt
\tag{11.3}
$$

because the period is $2\pi n^3$ (see Chapter 1). Solving for δ_ℓ,

$$
\delta_\ell = -(1/2\pi) \int V_c(r)\,dt
\tag{11.4}
$$

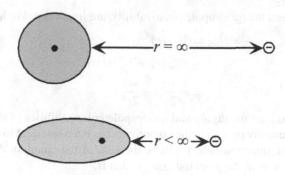

FIGURE 11.2. Schematic diagram showing the polarization of the ionic core by the valence electron.

If the $V_c(r)$ consists of several terms then δ_ℓ is a simple sum

$$\delta_\ell = \sum \delta_{i_\ell} \tag{11.5}$$

In general $V_c(r)$ will be given by

$$V_c(r) = V_{pol}(r) + V_{pen}(r) + V_{rel}(r) \tag{11.6}$$

$V_{pol}(r)$ describes the effects of polarization of the ionic core by the valence electron, $V_{pen}(r)$ is due to core penetration, and $V_{rel}(r)$ is due to relativistic effects such as spin-orbit coupling. The average values over an orbit of all three of these potentials scale as $1/n^3$. We, however, concentrate on the case for which $\ell > \ell_{core}$, where ℓ_{core} is the maximum angular momentum of a core electron. For example, for sodium $\ell_{core} = 2$ and it is found that $\delta_\ell \ll 1$ for electrons for which $\ell > \ell_{core}$ (see TABLE 11.1). Under these circumstances the largest contribution to the quantum defect comes from $V_{pol}(r)$.

By concentrating on the effects of $V_{pol}(r)$ we are considering the atom to consist of the ionic core and a valence electron (sometimes referred to as a Rydberg electron because such atoms are sometimes called Rydberg atoms). The picture is then one of the Rydberg electron polarizing the charge distribution that is the ionic core. That is, the electron distorts the cloud consisting of $(Z - 1)$ electrons that surrounds the nuclear point charge $+Z$ as illustrated in FIGURE 11.2. In Chapter 12 we show, however, that this simple model of core polarization does not always lead to a useful physical picture.

The total Hamiltonian may be written as

$$\hat{H} = \hat{H}_{Coul} + \hat{H}_{pol} \tag{11.7}$$

where $\hat{H}_{Coul}$ is the hydrogen atom Hamiltonian and $\hat{H}_{pol}$ contains all the potential energy terms resulting from polarization of the ionic core by the Rydberg electron. That is, $\hat{H}_{pol}$ consists of all terms of the multipole expansion of the potential energy of the core. In fact, however, only the first and perhaps the second terms, the dipole polarizability and the quadrupole polarizability, are important. We have then

$$\hat{H}_{pol} = V_{pol}(r)$$
$$= V_{pol}^d(r) + V_{pol}^q(r) + \cdots$$
$$\cong -\frac{\alpha_d}{2r^4} - \frac{\alpha_q}{2r^6} \tag{11.8}$$

where α_d and α_q are the dipole and quadrupole polarizabilities of the ionic core. (The e^2 that is usually present in the numerators has been set equal to one in atomic units.) Now, the quantum defect will be the sum of the quantum defects due to each of these terms in the potential energy; that is,

$$\delta_\ell = \delta_\ell^d + \delta_\ell^q \tag{11.9}$$

Our goal is to evaluate δ_ℓ^d in terms of α_d and the angular momentum quantum number ℓ. Evaluation of δ_ℓ^q is left as an exercise at the end of this chapter (see Problem 11.5). To evaluate δ_ℓ^d we compute the shift in energy (from E_n) resulting from $V_{pol}^d(r)$. This shift may be evaluated using quantum mechanical perturbation theory for stationary states.[3] This shift may also be evaluated using classical mechanics.[4] The answers, to the same level of approximation, are the same. We elect to use classical mechanics.

The central nature of this potential energy assures us that Kepler's second law pertains. In accordance with Kepler's second law, the electron spends a major fraction of the Keplerian period at large values of r for which the $1/r^4$ is negligible. Consequently, during a single revolution of the Rydberg electron this electron executes a very nearly Keplerian orbit. We therefore compute the energy of the Rydberg electron by averaging over an orbital period. Of course, the average of the Coulombic part of the potential energy is just the hydrogenic energy E_n. We evaluate the energy shift by averaging the expression for $V_{pol}^d(r)$ over a single orbit of the Rydberg electron assuming that each individual orbit is Keplerian. Note that this uses a hydrogenic feature, a Keplerian orbit, to evaluate a decidedly nonhydrogenic quantity $V_{pol}^d(r)$. This is analogous to the technique that would be employed if quantum mechanical perturbation theory were employed. That is, the correction to the unperturbed energy is obtained by computing the expectation value of the perturbing Hamiltonian using unperturbed wave functions which, in this case, are hydrogen atom wave functions.

Now, the average value of $V_{pol}^d(r)$ is

$$\langle V_{pol}^d(r) \rangle = -\frac{\alpha_d}{2} \cdot \left\langle \frac{1}{r^4} \right\rangle \tag{11.10}$$

so the task is to evaluate $\langle r^{-4} \rangle$ over one Keplerian orbit. For generality, we evaluate the quantity $\langle r^m \rangle$.

$$\langle r^{-m} \rangle = \frac{1}{\tau} \int_0^\tau r^{-m} dt$$
$$= \left(\frac{1}{2\pi n^3} \right) \int_0^\tau r^{-m} dt \tag{11.11}$$

where τ is the Keplerian period equal to $2\pi n^3$. To evaluate the integral we convert it to an integral over ϕ by replacing dt by $d\phi/\dot{\phi}$ and, using the equation of the Keplerian orbit of eccentricity ε (see Chapter 5),

$$\frac{1}{r} = \frac{1 + \varepsilon \cos \phi}{\ell^2} \tag{11.12}$$

to eliminate r. Using angular momentum conservation we have

$$r^2 \dot{\phi} = \ell \tag{11.13}$$

where ℓ is the (constant) angular momentum. Then $\langle r^{-m} \rangle$ becomes

$$\langle r^{-m} \rangle = \left(\frac{1}{2\pi n^3} \right) \cdot \frac{1}{\ell} \cdot (\ell^2)^{2-m} \int_0^{2\pi} (1 + \varepsilon \cos \phi)^{m-2} \, d\phi \tag{11.14}$$

provided, of course, that $m \geq 2$. Now, expanding the integrand using the binomial theorem we have

$$(1 + \varepsilon \cos \phi)^{m-2} = 1 + \frac{(m-2)}{1!} \varepsilon \cos \phi + \frac{(m-2)(m-3)}{2!} \varepsilon^2 \cos^2 \phi$$
$$+ \frac{(m-2)(m-3)(m-4)}{3!} \varepsilon^3 \cos^3 \phi + \cdots \tag{11.15}$$

The integrals of the odd powers of the cosine vanish so we have

$$\langle r^{-m} \rangle = \frac{\ell^{3-2m}}{n^3} \left[1 + \frac{(m-2)(m-3)}{2!} \varepsilon^2 \left(\frac{1}{2} \right) \right]$$
$$+ \frac{\ell^{3-2m}}{n^3} \left[\frac{(m-2)(m-3)(m-4)(m-5)}{4!} \varepsilon^4 \left(\frac{3}{8} \right) + \cdots \right] \tag{11.16}$$

For $m = 4$ we have

$$\langle r^{-4} \rangle = \left(\frac{1}{n^3} \right) \cdot \ell^{-5} \left(1 + \frac{1}{2} \varepsilon^2 \right) \tag{11.17}$$

The eccentricity is

$$\varepsilon^2 = 1 - \frac{\ell^2}{n^2} \tag{11.18}$$

so that

$$\langle r^{-4} \rangle = \left(\frac{\ell^{-5}}{n^3} \right) \left[1 + \frac{1}{2} \left(1 - \frac{\ell^2}{n^2} \right) \right]$$
$$= \left(\frac{\ell^{-5}}{n^3} \right) \left(\frac{3}{2} - \frac{\ell^2}{2n^2} \right) \tag{11.19}$$

and, retaining only terms to order $1/n^3$ ($\varepsilon^2 \approx 1$ in this approximation), we have

$$\langle r^{-4} \rangle = \left(\frac{3}{2} \right) \left(\frac{\ell^{-5}}{n^3} \right) \tag{11.20}$$

Then for $\langle V_{pol}(r) \rangle$, we have

$$\langle V_{pol}^d(r) \rangle = -\frac{\alpha_d}{2} \left\langle \frac{1}{r^4} \right\rangle$$

$$= -\frac{3}{4} \left(\alpha_d \ell^{-5} \right) \frac{1}{n^3} \tag{11.21}$$

The total energy is given by

$$\langle E \rangle = -\frac{1}{2n^2} - \frac{3}{4} \left(\alpha_d \ell^{-5} \right) \frac{1}{n^3} \tag{11.22}$$

and we identify the coefficient of the $1/n^3$ term as δ_ℓ so that

$$\delta_\ell^d = \frac{3}{4} \left(\alpha_d \ell^{-5} \right) \tag{11.23}$$

As noted previously, this is the same result as that obtained using quantum mechanical perturbation theory.[3] The result is also identical to that obtained using classical perturbation theory.

11.3. Classical Formulation of the Quantum Defect and the Correspondence Principle

Although seemingly self-contradictory, it is possible to relate the quantum defect to the classical precession frequency. Assuming that dipole polarizability of the core is the only non-Coulomb part of the potential, the total potential in atomic units is

$$V(r) = -\frac{1}{r} - \frac{\alpha_d}{2r^4} \tag{11.24}$$

Inasmuch as this is a central potential the classical orbit is confined to a plane because angular momentum is conserved. If we also assume that

$$\frac{\alpha_d}{2r^4} \ll \frac{1}{r} \tag{11.25}$$

then the classical orbit of the Rydberg electron is very nearly a Keplerian ellipse. As discussed in Chapter 5, the non-Coulombic term in $V(r)$ destroys the conservation of the Lenz vector A that would exist if the potential were purely Coulombic. Because the non-Coulombic term is assumed small compared with the Coulombic term the orbit may be envisioned as a Keplerian ellipse precessing about the force center as shown in Figure 5.4. This is, of course, analogous to the celebrated

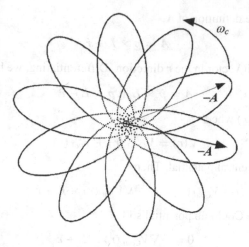

FIGURE 11.3. Precession of the Lenz vector and, hence, the Keplerian ellipse caused by a small non-Coulombic term in the potential.

precession of the perihelion of the planet Mercury that is caused by the relativistic correction to the attractive Newtonian $1/r$ potential. Precession of the Keplerian ellipse is illustrated in FIGURE 11.3.

It is interesting to contrast FIGURE 11.3 with Figure 5.5. In Chapter 5 the permanent electric dipole moment that is unique to hydrogen was understood classically by appealing to Kepler's third law. In a Keplerian orbit the electronic motion produces an asymmetric charge distribution because it moves more slowly at apocenter than at pericenter. The presence of the small non-Coulombic term causes the (nearly) Keplerian orbit to precess about the force center thus symmetrizing the charge distribution and destroying the permanent electric dipole moment of the pure Keplerian orbit. This classical precession is equivalent to destruction of the quantum mechanical accidental degeneracy.

It is seen then that in multielectron atoms, deviations from hydrogenic energies are due to the non-Coulombic part of the potential. These deviations may be viewed as manifestations of the quantum defect. Therefore, the quantum defect must be related to the classical precession frequency ω_c. If ω_c is small then δ_ℓ is expected to be small.

We seek the relationship between the classical precession frequency and the quantum defect. The precession rate is ω_c, which, when the non-Coulombic term is small compared to the Coulomb potential, is simply the "frequency" of the revolution of A about the force center. Strictly speaking, of course, this is not a frequency because the classical motion is not periodic. Within the approximation, however, it is assumed that each revolution of the particle from $\phi = 0$ to $\phi = 2\pi$ is a Keplerian orbit and that A is fixed over one such revolution. The precession frequency may then be identified as the time rate of change of A from one revolution to the next. We therefore compute $\langle \dot{A} \rangle_t$, the time average of $\dot{A}$ over one "period".

We begin with the definition of A.

$$A = p \times L - \hat{r} \tag{11.26}$$

where $\hat{r}$ is the unit vector in the r direction. Differentiating, we have

$$\dot{A} = \dot{p} \times L + p \times \dot{L} - \dot{\hat{r}} \tag{11.27}$$

Now, $\dot{p} = -\nabla V(r)$ where

$$V(r) = V_{Coul}(r) + V_{pol}(r) \tag{11.28}$$

and $\dot{L} = 0$ for a central potential. Then

$$\dot{A} = -\nabla V_{Coul}(r) \times L - \nabla V_{pol}(r) \times L - \dot{\hat{r}} \tag{11.29}$$

Now, $\dot{A} \equiv 0$ for a Coulomb potential so

$$0 = -\nabla V_{Coul}(r) \times L - \dot{\hat{r}} \tag{11.30}$$

Thus

$$-\nabla V_{Coul}(r) \times L = \dot{\hat{r}} \tag{11.31}$$

and

$$\dot{A} = -\nabla V_{pol}(r) \times L \tag{11.32}$$

Also,

$$\nabla V_{pol}(r) = \hat{r} \frac{dV_{pol}(r)}{dr}$$
$$= \frac{r}{r} \cdot \frac{dV_{pol}(r)}{dr} \tag{11.33}$$

so that

$$\dot{A} = \frac{1}{r} \frac{dV_{pol}(r)}{dr} (L \times r)$$
$$= \frac{1}{r} \left(4 \frac{\alpha_d}{2r^5} \right) (L \times r)$$
$$= \frac{2\alpha_d}{r^6} (L \times r) \tag{11.34}$$

We must now compute $\langle \dot{A} \rangle_t$ which is given by

$$\langle \dot{A} \rangle = 2\alpha_d \cdot L \times \left\langle \frac{r}{r^6} \right\rangle_t \tag{11.35}$$

because L is constant throughout. Also, because A is very nearly constant over a single revolution the direction of the vector r is along the major axis, the direction of A, and its magnitude is $r\cos\phi$. Therefore,

$$\dot{A} = 2\alpha_d \left\langle \frac{\cos\phi}{r^5} \right\rangle_t \left(L \times \frac{A}{|A|} \right) \tag{11.36}$$

To compute the average value of $\left\langle \dfrac{\cos\phi}{r^5} \right\rangle_t$ we first compute the average for the more general quantity $\left\langle \dfrac{\cos\phi}{r^m} \right\rangle_t$.

$$\left\langle \frac{\cos\phi}{r^m} \right\rangle_t = \frac{1}{\tau} \int_0^{2\pi} \frac{\cos\phi}{r^m} dt \qquad (11.37)$$

This time average is most easily computed by converting to an average over ϕ using Kepler's second law $r^2\dot{\phi} = \ell$. Also, because the orbit is assumed Keplerian over one revolution the period is $\tau = 2\pi n^3$. We obtain

$$\left\langle \frac{\cos\phi}{r^m} \right\rangle_\phi = \frac{1}{2\pi n^3} \cdot \frac{1}{\ell} \int_0^{2\pi} \frac{\cos\phi}{r^m} d\phi$$

$$= \frac{1}{2\pi n^3} \cdot \frac{1}{\ell} \cdot \left(\frac{1}{\ell^2}\right)^{m-2} \int_0^{2\pi} \cos\phi\, (1 + \varepsilon\cos\phi)^{m-2} d\phi$$

$$= \frac{1}{2\pi n^3} \cdot \frac{1}{\ell} \cdot \left(\frac{1}{\ell^2}\right)^{m-2} \int_0^{2\pi} \cos\phi\Big[1 + \frac{(m-2)}{1!}\varepsilon\cos\phi$$

$$+ \frac{(m-2)(m-3)}{2!}\varepsilon^2 \cos^2\phi$$

$$+ \frac{(m-2)(m-3)(m-4)}{3!}\varepsilon^3 \cos^3\phi$$

$$+ \frac{(m-2)(m-3)(m-4)(m-5)}{4!}\varepsilon^4 \cos^4\phi + \cdots]d\phi \qquad (11.38)$$

The integrals of the odd powers of $\cos\phi$ vanish and we have

$$\left\langle \frac{\cos\phi}{r^m} \right\rangle_\phi = \frac{1}{2\pi n^3} \cdot \frac{1}{\ell} \left(\frac{1}{\ell^2}\right)^{m-2}$$

$$\times \left[(m-2)\varepsilon\pi + \frac{(m-2)(m-3)(m-4)}{3!}\varepsilon^3 \frac{3}{4}\pi + \cdots \right] \qquad (11.39)$$

Now, for the dipole polarizability we require $m = 5$ giving

$$\left\langle \frac{\cos\phi}{r^m} \right\rangle_\phi = \frac{1}{2n^3} \cdot \frac{1}{\ell} \left(\frac{1}{\ell^2}\right)^{m-2} 3\varepsilon \left[1 + \frac{3}{4}\varepsilon^2\right] \qquad (11.40)$$

To cast Equation (11.40) in a suitable form we keep only terms to first power of (ℓ/n). Thus, $\varepsilon^2 = 1 - \ell^2/n^2 \approx 1$, but, from Chapter 5, $\varepsilon = |A|$ so we have

$$\left\langle \frac{\cos\phi}{r^5} \right\rangle_\phi = \frac{15}{8} \frac{1\cdot}{n^3\ell^7} \cdot |A| \qquad (11.41)$$

Equation (11.36) becomes

$$\dot{A} = \frac{15}{4} \cdot \frac{\alpha_d}{\ell^7 n^3} (L \times A)$$

$$= \omega_c \left(\frac{L}{\ell} \times A \right) \tag{11.42}$$

where

$$\omega_c = \frac{15}{4} \cdot \frac{\alpha_d}{\ell^6 n^3}$$

$$= \frac{5}{n^3 \ell} \delta_\ell \tag{11.43}$$

where the last expression follows from Equation (11.23).

We may show that this is indeed the frequency of revolution of the vector A about the angular momentum unit vector L/ℓ as follows. Suppose that L is in the z-direction so that $\hat{k} = L/\ell$ and that A rotates about it with frequency ω so that

$$A = A \left(\cos \omega t \hat{i} + \sin \omega t \hat{j} \right) \tag{11.44}$$

and

$$\dot{A} = -\omega A \left(\sin \omega t \hat{i} - \cos \omega t \hat{j} \right)$$

$$= \omega \hat{k} \times A \tag{11.45}$$

The relationship between ω_c and δ_ℓ in Equation (11.43) shows explicitly that, from a classical point of view, the nonhydrogenic nature of Rydberg atoms as manifested in the quantum defect is a result of the non-Keplerian part of the potential energy. It is this non-Keplerian portion of the potential energy that causes the precession and, consequently, the deviations from hydrogenic energies. Thus, δ_ℓ or ω_c may each be regarded as a consequence of the other. The precession frequency and the quantum defect are both very strong functions of $1/\ell$ showing that, as expected, low angular momentum orbits penetrate the ionic core and are thus more non-Keplerian than higher angular momentum orbits for which the electron is forced to avoid the core.

We may gain further insight into the relationship between δ_ℓ or ω_c by invoking the correspondence principle. According to the correspondence principle, the energy between adjacent states should be equal to $\hbar$ times the frequency of the classical motion. Radiation given off in transitions between these adjacent states should be of frequency equal to the frequency of the classical motion. Our interest is in the difference in energy between angular momentum states of the same Keplerian energy so the appropriate frequency here is not the Keplerian frequency, but, rather, the precessional frequency. In atomic units $\hbar = 1$ so, using Equation (11.2), we may write

$$\omega_c = E_{n\ell} - E_{n(\ell+1)}$$

$$= \frac{1}{n^3} (\delta_\ell - \delta_{\ell+1}) \tag{11.46}$$

Inserting the value of δ_ℓ from Equation (11.23) we have

$$\omega_c = \frac{1}{n^3} \cdot \frac{3\alpha_d}{4} \left[\frac{1}{\ell^5} - \frac{1}{(1+\ell)^5} \right]$$

$$= \frac{1}{n^3} \cdot \frac{3\alpha_d}{4} \frac{\left(5\ell + 10\ell^2 + 10\ell^3 + 5\ell^4\right)}{(1+\ell)^5 \, \ell^5}$$

$$= \frac{15}{4} \cdot \frac{\alpha_d}{\ell^6 n^3} + \text{terms of higher order in } \frac{1}{\ell} \qquad (11.47)$$

The leading term in this expression goes as ℓ^{-6}, as does the precession frequency as calculated previously. In fact, retaining only the leading term we arrive at exactly the same expression for the precession frequency that we derived previously, Equation (11.43).

This treatment is, of course, identical to treating ℓ as a continuous variable (as it is classically). Thus, Equation (11.46) may be written as

$$\omega_c = -\frac{1}{n^3} \frac{\partial \delta_\ell}{\partial \ell} \qquad (11.48)$$

Differentiating Equation (11.23) and inserting it in Equation (11.48) we have

$$\omega_c = \frac{1}{n^3} \left(\frac{15}{4} \right) \left(\alpha_d \ell^{-5} \right) \qquad (11.49)$$

which is identical with the result of Equation (11.43). This treatment employing the correspondence principle makes it clear that it is the rate of change of the quantum defect with angular momentum that determines the precession rate of the near Keplerian orbit and not the actual values of δ_ℓ.

11.4. The Connection Between the Quantum Defect and the Radial Wave Function

The success of the quantum defect description of the energy levels of multielectron atoms in terms of those of hydrogen, stems from the fact that many core potentials are central in nature and have a shorter range than the Coulomb potential. The central nature of the core potential guarantees the angular dependence of the wave function is given by the spherical harmonics, just as in hydrogen. Thus, the core potential only affects the radial wave function.

To understand the role that the quantum defect plays in the radial wave function of multielectron atoms, we consider the WKB approximation which provides a good approximation to the radial wave function between the turning points[5]

$$r_{1,2} = n^2 \pm n\sqrt{n^2 - (\ell + \frac{1}{2})^2} \qquad (11.50)$$

where, as is common in the WKB approximation, $\ell(\ell + 1)$ has been replaced by $(\ell + \frac{1}{2})^2$. The radial wave function for $R(r)$ is given by the solution to

Equation (4.2). Using atomic units we have

$$\frac{d^2u(r)}{dr^2} + \left[2E + 2V(r) - \frac{(\ell + 1/2)^2}{r^2} \right] u(r) = 0 \tag{11.51}$$

where $u(r) = r R(r)$. This has the approximate solution[5] for $R(r)$

$$R(r) = \frac{C}{r} \left(\left[2V(r) + 2E - \frac{(\ell + 1/2)^2}{r^2} \right]^{-1/2} \right)$$

$$\times \cos \left[\int_{r_1}^{r} \sqrt{2V(r) + 2E - \frac{(\ell + 1/2)^2}{r^2}} \, dr - \frac{\pi}{4} \right] \tag{11.52}$$

where C is a constant.

Although Equation (11.52) does indeed follow from the WKB approximation to the radial Schrodinger equation, it is easily interpreted as a deBroglie wave with a wavelength given by the "local" kinetic energy and momentum of the electron.

The addition of a small core potential $V_c(r)$ to the Coulomb potential will have the greatest effect on the argument of the cosine function, which will become

$$\int_{r_1}^{r} \sqrt{\frac{2}{r} - 2V_c(r) - \frac{1}{n^2} - \frac{(\ell + 1/2)^2}{r^2}} \, dr - \frac{\pi}{4} \tag{11.53}$$

Expanding (11.53) in a Taylor series about $V_c(r) = 0$ gives

$$\int_{r_1}^{r} \sqrt{\frac{2}{r} - \frac{1}{n^2} - \frac{(\ell + 1/2)^2}{r^2}} \, dr - \int_{r_1}^{r} \frac{V_c(r)}{\sqrt{\frac{2}{r} - \frac{1}{n^2} - \frac{(\ell + 1/2)^2}{r^2}}} \, dr - \frac{\pi}{4} \tag{11.54}$$

Although, in general, we cannot evaluate the second integral, we already know the answer as r becomes large (but not larger than r_2). Appealing to the semiclassical picture presented in Section 11.3 we can see that the denominator is dr/dt because it is $\sqrt{2[E - V(r)]}$. For large r we may extend the upper limit to r_2 so that the argument of the cosine in Equation (11.52) approaches

$$\int_{r_1}^{r_2} \sqrt{\frac{2}{r} - \frac{1}{n^2} - \frac{(\ell + 1/2)^2}{r^2}} \, dr - \int_{r_1}^{r_2} V_c(r) dt - \frac{\pi}{4} \tag{11.55}$$

for large r.

Equation (11.4) shows that the quantum defect is given by $\delta_\ell = -(1/2\pi) \int_{orbit} V_c(r) dt$ where the integral is evaluated over the entire Keplerian ellipse. The integral from the inner turning point to the outer is over one half the ellipse so we see that the core term in the radial wave function approaches $\pi \delta_\ell$ for large r.

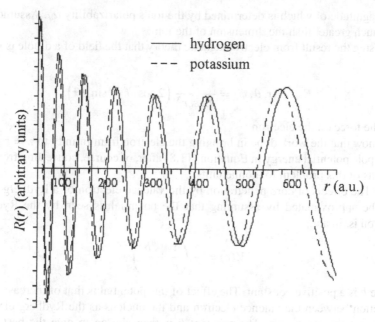

FIGURE 11.4. WKB radial wave functions of hydrogen and potassium in the classically allowed region illustrating the shift in phase caused by the ionic core of potassium.

The effect of a core potential such as the polarization potentials discussed in this chapter is to cause a phase shift from radial functions of hydrogen. This phase shift "accumulates" throughout the classically allowed region and approaches $\pi \delta_\ell$. FIGURE 11.4 shows a graph of the radial wave functions of d-states of hydrogen and potassium for $n = 20$ as given by Equation (11.52). Potassium was chosen because its d-state quantum defect is 0.25, large enough to make the phase shift observable in the graph. It is assumed that the quantum defect is produced by only the dipole polarization of the core and using Equation (11.23) to compute α_d so that $V_c(r)$ is given by $-\alpha_d/2r^4$. The solid line is hydrogen. The expected phase shift approaching 0.25π (see TABLE 11.1) is indicated.

Problems

11.1. Using quantum defects calculate the ionization potentials of lithium and sodium in electron volts. The ionization potential is simply the ground state energy of the atom.

11.2. A spherically symmetric singly charged positive ion (the "ionic core") is situated at the origin of coordinates. An electron is assumed to be at a fixed distance z along the z-axis. This electron induces in the ion a dipole moment, the magnitude of which is determined by the ion's polarizability α_d. Assume that z is much greater than the dimension of the ion.

(a) Using the result from electromagnetic theory that the field of a dipole is given by

$$E_{dipole}(r, \theta, \phi) = \frac{p}{4\pi\varepsilon_0 r^3}\left(2\cos\theta\hat{r} + \sin\theta\hat{\theta}\right)$$

find the force on the electron.

(b) Show that the work done in bringing the electron from infinity to $z = r$ gives the dipole potential energy in Equation (11.8) after, of course, converting from the SI units of this problem to atomic units.

11.3. The effects of core penetration by the valence electron of a Rydberg atom may be approximated by assuming that the potential as seen by the Rydberg electron is, in a.u.,

$$V(r) = -\left(\frac{1}{r} + \frac{b}{r^2}\right)$$

where b is a positive constant. The effect of this potential is that of increasing the attraction between the valence electron and the nucleus as the Rydberg electron penetrates the ionic core. The constant b is then chosen to give the best fit to spectral data. By averaging the $1/r^2$ term in the potential energy over a Keplerian orbit find the quantum defect in terms of b and the angular momentum.

11.4. The quantum defect for the potential of the previous problem,

$$V(r) = -\left(\frac{1}{r} + \frac{b}{r^2}\right)$$

where b is a positive constant, may also be obtained from the quantum mechanical solution of the radial part of the Schrödinger equation because the $1/r^2$ term may be combined with the centrifugal term. Show that the quantum defect is given by $\delta_\ell \approx b/(\ell + 1/2)$. Can you account for the difference between this answer and the previous answer? How do these energy states vary for a given n with ℓ?

11.5. In addition to the dipole term in the Hamiltonian that led to the quantum defect

$$\delta_\ell^d = \frac{3}{4}\left(\alpha_d\ell^{-5}\right)$$

there is also a potential energy due to quadrupole polarizability of the ionic core that can be included. This quadrupole term is $V_{pol}^q(r) = -\alpha_q/2r^6$. Use the same technique as that used in the text to show that the portion of the quantum defect that is due to quadrupole polarizability is

$$\delta_\ell^q = \frac{35}{16}\left(\alpha_q\ell^{-9}\right)$$

References

1. J.R. Rydberg, K. Svenska Vetenskaps Akad. Handlinger **23**, 1 (1890).
2. E.W. McDaniel, *Atomic Collisions* (John Wiley, New York, 1989).
3. R.R. Freeman and D. Kleppner, Phys. Rev. **A14**, 1614 (1976).
4. T.P. Hezel, C.E. Burkhardt, M. Ciocca, et al., Am. J. Phys. **60**, 329 (1992).
5. H.A. Bethe and E.E. Salpeter, *Quantum Mechanics of One- and Two-Electron Atoms* (Springer-Verlag, Berlin, 1957).

12
Multielectron Atoms in External Fields

12.1. The Stark Effect

In Chapter 8 we saw that the quadratic Stark effect on the ground state of hydrogen is a consequence of the nondegenerate nature of this state. The linear Stark effect that is characteristic of the excited states of hydrogen is a direct consequence of the accidental degeneracy. On the other hand, excited states of multielectron atoms are nondegenerate and should therefore exhibit the quadratic Stark effect. In fact, this is true, but in some instances excited states of multielectron atoms behave in a hydrogenic fashion. We therefore make a distinction between hydrogenic states (of multielectron atoms) and states of hydrogen. The existence of hydrogenic states is most often observed in alkali metal atoms when they are subjected to an external electric field. As might be expected from the discussion of Chapter 11, high angular momentum states, those having the smallest quantum defects, are most nearly hydrogenic.

In our discussion of the Stark effect in multielectron atoms we use sodium as our example because it exhibits features not present in the lighter lithium. Certain states behave as if they have a negative polarizability. That is, in terms of the static polarization picture represented in Chapter 11 (see Figure 11.2), the field seems to polarize the atomic charge distribution "backwards", thus leading to a negative polarizability. This concept is difficult to understand in terms of the simple static picture. We emphasize, however, that in this chapter we are discussing polarization of the entire atom, not the ionic core as was discussed in Chapter 11. We merely compare the effects of the external field with those of the field of the valence electron in the static model presented in Chapter 11. We return to the concept of a negative atomic polarizability later in this chapter and find that the classical model[1] offers an alternative explanation.

FIGURE 12.1 is a schematic diagram of the energy levels of the sodium atom when an electric field F is applied. This diagram, also referred to as a Stark map, can be understood in terms of the quantum defects for sodium.

To understand this diagram we first examine the ordering of the levels with $F = 0$. Because the quantum defects of the states having $\ell \geq 2$ are very small, their

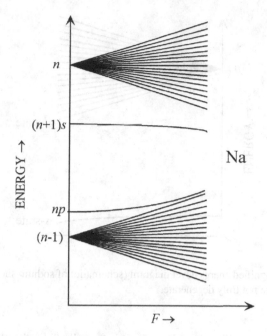

FIGURE 12.1. Schematic energy-level diagram of sodium showing the behavior of the levels when an external electric field F is applied. [From Reference 1.]

zero-field energies are very nearly the same as those of hydrogen. The groups of states emanating from n and $(n-1)$ in FIGURE 12.1 are referred to as hydrogenic manifolds of states. These manifolds comprise only those states for which the zero-field energies are very nearly those of hydrogen. On the other hand, the quantum defects of s- and p-states of sodium are 1.34 and 0.85, respectively (see Table 11.1), so that the $(n+1)s$ and np states are isolated from the hydrogenic manifolds. Examination of this table indicates that for lithium only the s-states would be isolated from the hydrogenic manifolds.

In fact, the placement of states having substantial quantum defects with respect to the hydrogenic manifolds depends only on the fractional part of the quantum defect. Because the quantum defect of the p-states is greater than one-half, np-states lie nearer to the $(n-1)$ hydrogenic manifold than to the n manifold. This is easily seen by noting that if the quantum defect were exactly unity then the energy would be coincident with the hydrogen energy of principal quantum number $(n-1)$. In contrast, the quantum defect of sodium s-states is 1.35. The fact that $\delta_s > 1$ means that a given s-state lies below the next lowest hydrogenic manifold as shown in FIGURE 12.1. In this sense, the quantum defects of the s- and p-states may be regarded as being 0.34 and -0.15, respectively.

If now, we turn on the electric field, the energies of the nondegenerate s- and (p)-states decrease (increase) quadratically. From quantum mechanics we know

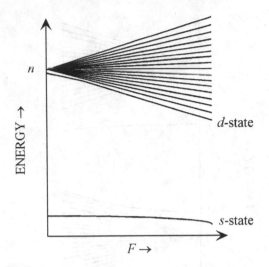

FIGURE 12.2. Magnified energy-level diagram (schematic) of sodium showing that the hydrogenic states are not truly degenerate.

that the s-state energy decreases because it is "repelled" by the closest state to it, the lowest state of the n hydrogenic manifold above it. The p-state is, however, different. It too is repelled by its nearest neighbor state, but this is the highest state of the $(n-1)$ manifold so its energy must increase as shown. This is the negative atomic polarizability mentioned earlier in this chapter.

As for the hydrogenic manifold, although the quantum defects for $\ell \geq 2$ are very small, they are not zero. Therefore, sufficiently magnified, there is a separation between states, even at zero field. Moreover, at very small fields these states must behave quadratically because they are not states of hydrogen. FIGURE 12.2 illustrates this point. It shows the hydrogenic manifold corresponding to principal quantum number n, but magnified from that in FIGURE 12.1 to show the separation between the d-state and the remainder of the hydrogenic manifold. Of course, magnified further, the f-state and higher angular momentum states would also separate from the nominal hydrogen energy. As the field increases these separated states behave quadratically, but, with further increase in the field these states "join the hydrogenic manifold" and behave as linear Stark states.

The Stark effect was studied early in the twentieth century by Stark and Lo Surdo,[2] but, because the high field required to affect atoms in low-lying states was substantial, experiments were difficult. With the availability of lasers as a laboratory tool, however, there was renewed interest in the Stark effect.[3] Lasers made it possible to excite atoms to specific highly excited states which, as we saw in Chapter 8, are much more sensitive to the electric field because the Stark energy increases dramatically with increasing principal quantum number [see Equation (8.76)]. This, of course, makes sense because in highly excited states the intra-atomic electric field to which the valence electron is subjected is small because of the remoteness of the ionic core.

Because highly excited atoms are so sensitive to external electric fields it is possible to gain insight into the Stark effect by again appealing to a classical model. We saw in Chapter 8 that application of the electric field causes the Keplerian orbit of a hydrogen atom to rotate about the electric field vector. From Chapter 11 we know that the non-Coulombic term in the potential energy causes the near-Keplerian orbit to precess about the force center in the plane of the orbit. The behavior of the atom under this influence is clearly determined by the relative magnitudes of the rotation and precession frequencies. These frequencies are given by

$$\omega_S = (3/2)\, n F$$
$$\omega_c = \frac{5}{n^3 \ell} \delta_\ell \tag{12.1}$$

so that at low fields when $\omega_S \ll \omega_c$ the Keplerian orbit rapidly spins about the force center, symmetrizing the charge distribution and destroying the permanent electric dipole moment extant for a fixed Keplerian orbit. The result is that there is no linear Stark effect. The atom does, however, respond quadratically to the electric field. At the other extreme, at high fields where $\omega_S \gg \omega_c$, the plane of the orbit rotates rapidly about F. In effect this freezes the Keplerian orbit so the atom behaves as though it had a permanent electric dipole moment and responds linearly to the electric field.

To obtain an approximate value for the magnitude of the electric field at which hydrogenic behavior is expected, we may equate ω_S and ω_c. This yields

$$F = \frac{10}{3} \frac{1}{n^4 \ell} \delta_\ell \tag{12.2}$$

which shows the strong dependence on n, ℓ, and the quantum defect which itself is strongly dependent upon ℓ. In fact, because we know that $\delta_\ell \ll 1$ for $\ell > \ell_{core}$ it is clear that the term "hydrogenic manifold" is quite descriptive for $\ell > \ell_{core}$.

We may use the classical model applied to multielectron atoms to understand the negative polarizabilities of certain atomic states, for example, p-states of sodium discussed earlier in this chapter. It was seen in Chapter 11, Equation (11.48), that the field-free precession frequency ω_c may be written as

$$\omega_c = -\frac{1}{n^3} \frac{\partial \delta_\ell}{\partial \ell} \tag{12.3}$$

As usual, we treat the orbit as the dynamical entity. When the field is present, the potential is no longer central and angular momentum is not conserved. There will then be a torque on the orbit that causes it to rotate about the applied field. We assume low fields so that this rotation rate is slow compared with the precession rate ω_c. The orbit may therefore be assumed to be frozen for the purpose of this derivation. We designate the plane of this frozen orbit as the yz-plane and the field

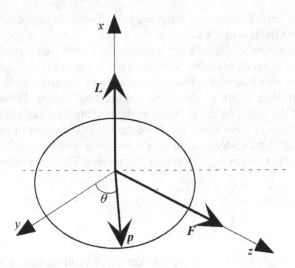

FIGURE 12.3. Geometry used in the derivation of the classical expression for the atomic polarizability.

to be in the z-direction so that $F = F\hat{k}$ $L = \ell\hat{i}$. The coordinates are shown in FIGURE 12.3.

The torque on the orbit is

$$\dot{L} = p \times F \tag{12.4}$$

where p is the instantaneous electric dipole moment and θ the angular displacement of p from the y-axis.

Because ω_c is a function of the changing angular momentum ℓ we find $\omega_c(\ell)$. Then, using $\omega_c(\ell)$, we average the instantaneous dipole moment over one precessional period to find the effective atomic dipole moment. This effective dipole moment then leads to the polarizability of the atom. We use the symbol α for the polarizability in this chapter. Note that this is not the core polarizability α_d of Chapter 11.

From FIGURE 12.3 and Equation (12.4), the magnitude of the torque is

$$\begin{aligned}
|\dot{L}| &= \frac{d\ell}{dt} \\
&= p_y F
\end{aligned} \tag{12.5}$$

Also,

$$\begin{aligned}
\frac{d\ell}{dt} &= \frac{d\ell}{d\theta}\dot{\theta} \\
&= \frac{d\ell}{d\theta}\omega_c(\ell)
\end{aligned} \tag{12.6}$$

Therefore,

$$\frac{d\ell}{d\theta}\omega_c\,(\ell) = p_y F \tag{12.7}$$

Defining

$$\omega_0 = \omega_c\,(\ell = \ell_0) \tag{12.8}$$

and using Equation (11.43) which is

$$\omega_c = \frac{5}{n^3\ell}\delta_\ell \tag{12.9}$$

we have

$$\omega_c\,(\ell) = \omega_0\left(\frac{\ell_0}{\ell}\right)^6 \tag{12.10}$$

Equation (12.7) becomes

$$\omega_0\left(\frac{\ell_0}{\ell}\right)^6 d\ell = p\cos\theta\,F d\theta \tag{12.11}$$

which integrates to

$$\frac{\omega_0\ell_0^6}{5}\left(\frac{1}{\ell_0^5} - \frac{1}{\ell^5}\right) = pF\sin\theta \tag{12.12}$$

If ℓ deviates from ℓ_0 by only a small amount Δ we may write

$$\frac{\omega_0\ell_0^6}{5}\left(\frac{1}{\ell_0^5} - \frac{1}{\ell^5}\right) = \frac{\omega_0\ell_0^6}{5}\left[\frac{1}{\ell_0^5} - \frac{1}{(\ell_0 + \Delta)^5}\right]$$

$$\cong \frac{\omega_0\ell_0^6}{5}\left[\frac{1}{\ell_0^5} - \frac{1}{\ell_0^5}\left(1 - 5\frac{\Delta}{\ell_0}\right)\right]$$

$$= \omega_0\Delta$$

$$= \omega_0\,(\ell - \ell_0) \tag{12.13}$$

Solving for ℓ we have

$$\ell = \frac{pF\sin\theta}{\omega_0} + \ell_0$$

$$= \ell_0\,(1 + \eta\sin\theta) \tag{12.14}$$

where $\eta = pF/\ell_0\omega_0$. The precessional frequency as a function of the angular momentum is then

$$\omega_c\,(\ell) = \frac{\omega_0}{(1 + \eta\sin\theta)^6} \tag{12.15}$$

from which we may obtain the precessional period τ_c.

$$
\begin{aligned}
\tau_c &= \int_0^{\tau_c} dt \\
&= \int_0^{2\pi} \frac{d\theta}{\dot{\theta}} \\
&= \int_0^{2\pi} \left[\left(\frac{1 + \eta \sin \theta}{\omega_0} \right) \right]^6 d\theta \\
&\cong \frac{2\pi}{\omega_0} \left(1 + \frac{15}{2} \eta^2 \right)
\end{aligned}
\tag{12.16}
$$

The average dipole moment is the average of the z-component of the dipole moment. By symmetry $\langle p_y \rangle = 0$. The average value of the z-component is

$$
\begin{aligned}
\langle p_z \rangle &= \frac{1}{\tau_c} \int_0^{\tau_c} p_z(\theta) \, dt \\
&= \frac{1}{\tau_c} \int_0^{\tau_c} p \sin \theta \frac{d\theta}{\dot{\theta}} \\
&= \frac{1}{\tau_c} \int_0^{2\pi} p \frac{\sin \theta (1 + \eta \sin \theta)^6}{\omega_0} d\theta
\end{aligned}
\tag{12.17}
$$

For the weak fields under consideration we retain only linear powers of the field, and therefore linear powers of η. We have

$$
\begin{aligned}
\langle p_z \rangle &= \frac{\omega_0}{2\pi} \int_0^{2\pi} \frac{p}{\omega_0} \left(\sin \theta + 6\eta \sin^2 \theta \right) d\theta \\
&= 3p\eta \\
&= 3 \frac{p^2}{\ell_0 \omega_0} F
\end{aligned}
\tag{12.18}
$$

From the results of Problem 5.3 we know that

$$
p^2 = \frac{9}{4} n^4 \left(1 - \ell^2/n^2 \right)
\tag{12.19}
$$

so that

$$
\langle p_z \rangle = \frac{27}{4} \frac{n^4 \left(1 - \ell^2/n^2 \right)}{\ell_0 \omega_0} F
\tag{12.20}
$$

Because the energy is given by

$$
\begin{aligned}
E &= -\frac{1}{2} \langle p_z \rangle F \\
&= -\frac{27}{8} \frac{n^4 \left(1 - \ell^2/n^2 \right)}{\ell_0 \omega_0} F^2 \\
&= -\frac{1}{2} \alpha F^2
\end{aligned}
\tag{12.21}
$$

we see that the polarizability is

$$\alpha = -\frac{27}{4}\frac{n^4\left(1 - \ell^2/n^2\right)}{\ell_0\omega_0} \tag{12.22}$$

We wish, however, to cast this in terms of the quantum defect. We therefore replace ω_0 using Equation (12.9) with $\ell \to \ell_0$ and obtain

$$\alpha = -\frac{27}{20}\left(1 - \ell^2/n^2\right)\frac{n^7}{\delta_\ell} \tag{12.23}$$

The polarizability as given in Equation (12.23) is consistent with the concept of a negative quantum defect assigned to sodium p-states to account for their positive Stark shifts.

Thus, according to this formulation, any state for which the fractional part of the quantum defect, call it f, is such that $f > 1/2$ should be assigned a quantum defect $\delta = f - 1$, a negative number. All such states will exhibit negative polarizabilities. Equation (12.23) was derived using purely classical methods. The appearance of the "quantum numbers" n and ℓ and the quantum defect δ_ℓ in no way suggests a quantum mechanical origin of this expression. Indeed, n and ℓ are simply the energy and angular momentum, respectively, and are continuous variables. It is interesting that the expression obtained for the polarization using classical methods, Equation (12.23), is exactly the same expression obtained using second-order perturbation theory.[4]

Positive and negative polarizabilities are illustrated in FIGURE 12.4. The figure shows precessing near-Keplerian orbits with a variable precessional rate caused by application of an external field. The parameters used were identical except the signs of the quantum defects were reversed.

In FIGURE 12.4(A) there is clearly a buildup of negative charge on the downfield side of the nucleus as is expected from the charge polarization model. In this case,

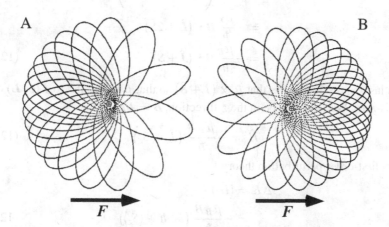

FIGURE 12.4. Precession of the nearly Keplerian orbit of an electron due to the charge-induced polarization potential. In (A) the polarizability is taken to be positive whereas in (B) it is negative. [From Reference 1.]

application of the field leads to a negative shift in energy as would occur if the isolated state were closer to the hydrogenic manifold above. On the other hand, if the state were closer to the manifold below then $\partial \delta_\ell / \partial \ell > 0$ causing the orbit to precess in the opposite direction. This leads to a buildup of negative charge on the upfield side of the nucleus and a negative polarizability as shown in FIGURE 12.4(B).

12.2. The Zeeman Effect

In Chapter 8 we discussed at length the Zeeman effect in the hydrogen atom. A major difference between the hydrogen atom treatment and the multielectron case is that there is no limit on the total electronic spin for multielectron atoms. In hydrogen the electronic spin is necessarily one-half. Moreover, we have to deal with coupling between the spins and the total orbital angular momenta.

We consider the weak field Zeeman effect and define a "weak" magnetic field B to be one for which the energy associated with the atom in this field, $\hat{\mu} \cdot B$, is comparable with the spin-orbit term. Thus

$$\hat{\mu} \cdot B \sim \xi(r) \hat{L} \cdot \hat{S}$$

We take the unperturbed Hamiltonian to contain all intra-atomic effects not associated with the external field

$$\hat{H}_0 = \hat{H}_{Coulomb} + \xi(r) \hat{L} \cdot \hat{S} \qquad (12.24)$$

so that the unperturbed wave functions are eigenfunctions of $\hat{L}^2$, $\hat{S}^2$, $\hat{J}^2$, and $\hat{J}_z$; that is, the eigenfunctions are the coupled wave functions. Recall that these are not eigenfunctions of either $\hat{L}_z$ or $\hat{S}_z$. The perturbing Hamiltonian is taken to be

$$\hat{H}' = -B \cdot \hat{\mu}$$
$$= -B \cdot \left(\frac{\mu_B}{\hbar} \hat{L} + 2\frac{\mu_B}{\hbar} \hat{S} \right)$$
$$= -\frac{\mu_B}{\hbar} B \cdot (\hat{L} + 2\hat{S})$$
$$= -\frac{\mu_B}{\hbar} B \cdot (\hat{J} + \hat{S}) \qquad (12.25)$$

This last expression shows that $\hat{\mu} \propto (\hat{J} + \hat{S})$ so that $\hat{\mu}$ is parallel to $\hat{J}$ (or $\hat{L}$) only if $\hat{S}$ is zero. We take B to be in the z-direction, $B = B\hat{k}$, so that

$$\hat{H}' = -\frac{\mu_B B}{\hbar} (\hat{J}_z + \hat{S}_z) \qquad (12.26)$$

From first-order perturbation theory

$$\Delta E = \langle \hat{H}' \rangle$$
$$= -\frac{\mu_B B}{\hbar} (M_J \hbar + \langle \hat{S}_z \rangle) \qquad (12.27)$$

But, this expression for ΔE does not lend itself to easy evaluation because, although $\hat{J}_z$ is well-defined on the coupled eigenstates of $\hat{H}_0$, $\hat{S}_z$ is not. It would be more

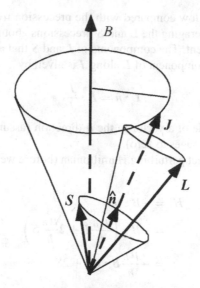

FIGURE 12.5. Diagram showing the relationship of the various angular momenta.

useful if the Hamiltonian could be written in the form

$$\hat{H}' = -g_J \left(\frac{\mu_B}{\hbar} \right) B \hat{J}_z \tag{12.28}$$

because $\hat{J}_z$ is well defined (with eigenvalues $M_J \hbar$). This procedure is equivalent to replacing the magnetic dipole moment with an "effective" magnetic dipole moment

$$\mu_{eff} = g_J \left(\frac{\mu_B}{\hbar} \right) \hat{J}_z \tag{12.29}$$

The constant g_J, the Landé g-factor, is to be determined such that μ_{eff} is a good approximation to the true magnetic dipole moment, $\left(\frac{\mu_B}{\hbar} \right) (\hat{J}_z + \hat{S}_z)$. The Landé g-factor in Equation (12.29) will be found to be a generalization of the Landé g-factor that was obtained for the hydrogen atom in Chapter 8.

Consider a general direction for the field. Figure 12.5 illustrates the relationship of the three angular momenta under consideration; $\hat{J} = J\hat{n}$, where $\hat{n}$ is a unit vector in the direction of $\hat{J}$.

There are three precessional motions to consider:

1. S about J
2. L about J
3. J about B

Classically the magnetic moment μ precesses about the magnetic field B with the Larmor frequency ω_L given by

$$\omega_L = \frac{\mu_B}{\hbar} B \tag{12.30}$$

If B is weak, ω_L is low compared with the precession frequencies of L and S about J. Therefore, averaging the L and S precessions about J gives the effective magnetic dipole moment. The components of L and S that are perpendicular to J average to zero. The component of L along J is given by

$$L \cdot \hat{n} = L \cdot \frac{J}{J} \tag{12.31}$$

which is the magnitude of a vector in the $\hat{n}$ direction (assuming that the perpendicular component averages to zero).

Returning to the exact perturbing Hamiltonian (before we replaced J by $L + S$) we have

$$\hat{H}' = -B \cdot \mu$$
$$= -B \cdot \left(\frac{\mu_B}{\hbar} \hat{L} + 2 \frac{\mu_B}{\hbar} \hat{S} \right)$$
$$= -\frac{\mu_B}{\hbar} B \cdot (\hat{L} + 2\hat{S}) \tag{12.32}$$

which shows that we must evaluate $B \cdot L$ and $B \cdot S$ with our approximation. We have

$$B \cdot L \rightarrow B \cdot \langle L \rangle$$
$$= B \cdot |\langle L \rangle| \hat{n}$$
$$= B \cdot \left(\frac{L \cdot J}{J} \right) \hat{n}$$
$$= B \cdot \left(\frac{L \cdot J}{J} \right) \frac{J}{J}$$
$$= \left(\frac{L \cdot J}{J^2} \right) (B \cdot J) \tag{12.33}$$

Similarly, we have

$$B \cdot S \rightarrow B \cdot \langle S \rangle$$
$$= \left(\frac{S \cdot J}{J^2} \right) (B \cdot J) \tag{12.34}$$

Now we may insert Equations (12.33) and (12.34) in Equation (12.32) for $\hat{H}'$ to obtain

$$\hat{H}' = -\frac{\mu_B}{\hbar} B \cdot (\hat{L} + 2\hat{S})$$
$$= -\frac{\mu_B}{\hbar} \left[\left(\frac{\hat{L} \cdot \hat{J}}{J^2} \right) (B \cdot \hat{J}) + 2 \left(\frac{\hat{S} \cdot \hat{J}}{J^2} \right) (B \cdot \hat{J}) \right]$$
$$= -\frac{\mu_B}{\hbar} (\hat{L} \cdot \hat{J} + 2\hat{S} \cdot \hat{J}) \left(\frac{B \cdot \hat{J}}{J^2} \right) \tag{12.35}$$

Now, $\hat{J} = \hat{L} + \hat{S}$ so

$$\begin{aligned}
\hat{L} \cdot \hat{J} &= \hat{L}^2 + \hat{L} \cdot \hat{S} \\
&= \hat{L}^2 + \frac{1}{2}\left(\hat{J}^2 - \hat{L}^2 - \hat{S}^2\right) \\
&= \frac{1}{2}\left(\hat{J}^2 + \hat{L}^2 - \hat{S}^2\right)
\end{aligned} \tag{12.36}$$

Similarly

$$2\hat{S} \cdot \hat{J} = \left(\hat{J}^2 + \hat{S}^2 - \hat{L}^2\right) \tag{12.37}$$

and we have

$$\begin{aligned}
\hat{H}' &= -\frac{\mu_B}{\hbar}\left(\frac{1}{2}\hat{J}^2 + \frac{1}{2}\hat{L}^2 - \frac{1}{2}\hat{S}^2 + \hat{J}^2 + \hat{S}^2 - \hat{L}^2\right)\frac{B \cdot \hat{J}}{J^2} \\
&= -\frac{\mu_B}{\hbar}\left[\hat{J}^2 + \frac{1}{2}\left(\hat{J}^2 + \hat{S}^2 - \hat{L}^2\right)\right]\frac{B \cdot \hat{J}}{J^2}
\end{aligned} \tag{12.38}$$

Moreover, we have judiciously chosen the direction of B (i.e., $B = B\hat{k}$) so that

$$B \cdot \hat{J} = B\hat{J}_z$$

and the Hamiltonian becomes

$$\hat{H}' = -\frac{\mu_B}{\hbar}\left[\hat{J}^2 + \frac{1}{2}\left(\hat{J}^2 + \hat{S}^2 - \hat{L}^2\right)\right]\frac{B}{J^2}\hat{J}_z \tag{12.39}$$

If it is assumed that $\hat{H}'$ operates on coupled eigenfunctions, we may write $\hat{H}'$ as

$$\hat{H}' = -\frac{\mu_B}{\hbar}\left(1 + \frac{J(J+1) + S(S+1) - L(L+1)}{2J(J+1)}\right)B\hat{J}_z$$

Comparing this equation with the desired form of $\hat{H}'$, Equation (12.28), we make the identification

$$g_J = 1 + \frac{J(J+1) + S(S+1) - L(L+1)}{2J(J+1)} \tag{12.40}$$

which is the Landé g-factor for multielectron atoms. The Zeeman energy is thus

$$\begin{aligned}
\Delta E &= \langle \hat{H}' \rangle \\
&= \frac{\mu_B B}{\hbar}g_J M_J \hbar \\
&= \mu_B B g_J M_J
\end{aligned} \tag{12.41}$$

Thus, in the presence of a B-field, a given state will split into $2J + 1$ magnetic sublevels, one for each value of the quantum number M_J.

The Normal Zeeman Effect

For singlet states we have $S = 0$ so $J = L$. Therefore, $g_J = 1$ for all states for which the total spin is zero. As a consequence, the splitting of all such states will be

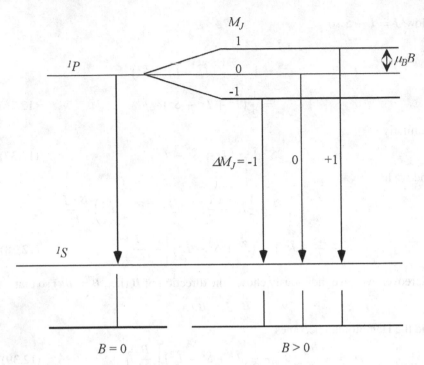

FIGURE 12.6. Energy-level diagram showing the level splitting for the normal Zeeman effect. Also shown are the transitions that would be observed between levels.

the same [see Equation (12.41)] and the "normal Zeeman effect" will be observed. In fact, the normal Zeeman effect is really abnormal because states for which $S = 0$ are the exception rather than the rule. The normal Zeeman effect can, however, be understood in terms of orbiting electrons, thus the appellation "normal". The anomalous Zeeman effect was only understood after it was realized that electrons had intrinsic magnetic moments, spin.

To understand the nature of the spectrum when a spin zero atom is immersed in a constant B-field consider transitions between an upper ^1P level and a lower ^1S level. As noted above, the Landé g-factor is unity for both of these states. There are three magnetic sublevels of the P-state and none of the S-state. Thus, there will be only three transitions between these states as shown in Figure 12.6.

A somewhat more complicated case is encountered for transitions between an upper ^1D level and a lower ^1P level which have five and three sublevels, respectively. Although there will be a number of possible combinations between these sublevels, only three spectral lines will be observed because of the selection rule $\Delta M_J = 0, \pm 1$ as will be discussed in Chapter 13. The Zeeman levels and the transitions between them are shown in Figure 12.7. Notice that, although there are nine allowed transitions, only three distinct lines would be observed in the emission spectrum because the energy differences between some of the different transitions are the same.

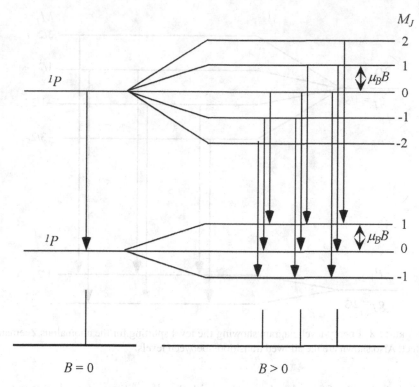

FIGURE 12.7. Energy-level diagram showing the level splitting for the normal Zeeman effect. Also shown are the transitions that would be observed between levels.

The Anomalous Zeeman Effect

This effect is the more common one because most systems do not have $S = 0$. In fact, it is the anomalous Zeeman effect that is observed in hydrogen because all states are necessarily doublets. As an example of the anomalous Zeeman effect we consider the $^2D_{3/2} \rightarrow {}^2P_{1/2}$ transition. The Landé g-factors for the two states are

$$(g_J)_D = 1 + \frac{J(J+1) + S(S+1) - L(L+1)}{2J(J+1)}$$

$$= \frac{4}{5} \tag{12.42}$$

and

$$(g_J)_P = \frac{2}{3} \tag{12.43}$$

The different g-factors cause the Zeeman splitting of these states to be different thus complicating the spectrum. This is illustrated in Figure 12.8 which shows the

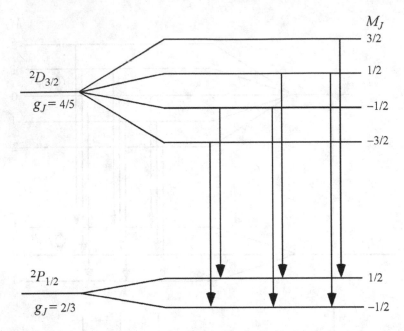

FIGURE 12.8. Energy-level diagram showing the level splitting for the anomalous Zeeman effect. Also shown are the allowed transitions between levels.

splitting of each of these levels together with the allowed transitions between the levels.

Problems

12.1. The quantum defects of the potassium atom are:

$$\delta_0 = 2.19$$
$$\delta_1 = 1.71$$
$$\delta_2 = 0.25$$
$$\delta_\ell = 0.00 \text{ for } \ell \geq 3$$

Sketch a diagram similar to Figure 12.1 indicating the positions of the field-free $n\ell$ states (i.e., ns, np, etc.). To avoid cluttering the diagram, show only the states having principal quantum number n, not $(n + 1)$, $(n − 1)$, and so on. Indicate the locations of hydrogenic manifolds. Sketch the Stark states when the dc electric field, F, is turned on. Be sure to indicate details such as which states exhibit quadratic Stark effects and which exhibit linear Stark effects. For clarity, show the linear Stark effect on only the n and $n − 2$ hydrogenic manifolds. Be sure to indicate the correct curvature of the quadratic Stark states. To start you off, the positions of several hydrogenic states are shown. Use atomic units.

$$E_{n\ell} = -\frac{1}{2\,(n-\delta_\ell)^2} = -\frac{1}{2n^2}\left(1 - \frac{\delta_\ell}{n}\right)^{-2} \approx -\frac{1}{2n^2}\left(1 + 2\frac{\delta_\ell}{n} + \cdots\right)$$

12.2. Consider the Bohr atom to be a rigid rotor in the xy-plane with a proton at the center and an electron circling it at radius $r_n = n^2 a_0$ where n is the principal quantum number. Find the polarizability of this atom for any state of principal quantum number n. Assume that the field is weak enough so there is no distortion of the circular orbit by the field. Note that the rotation of this dipole symmetrizes the charge distribution, thus negating the electric dipole moment of the proton/electron combination. Show that the first-order correction to the energy vanishes so the first nonzero correction to the energy is proportional to the square of the applied electric field strength.

12.3. Show that there is no weak field Zeeman splitting for an atom having a $^4D_{1/2}$ term.

12.4. (a) Make a diagram of the energy levels for the weak field Zeeman effect on $^1F \rightarrow {}^1D$ transitions. How many lines will be observed in the emission spectrum? (b) Make a diagram of the energy levels for the weak field Zeeman effect on $^3P \rightarrow {}^3S$ transitions. How many lines will be observed in the emission spectrum?

12.5. Make a diagram showing all energy levels resulting from application of a weak magnetic field to an excited sodium atom having electron configuration [Ne] $3p$.

References

1. T.P. Hezel, C.E. Burkhardt, M. Ciocca, et al., Am. J. Phys. **60**, 329 (1992).
2. Ryde, *Atoms and Molecules in Electric Fields* (Almqvist & Wiksell International, Stockholm, 1976).
3. M.L. Zimmerman, M.G. Littman, M.M. Kash, et al., Phys. Rev. **20**, 2251 (1979).
4. J.F. Baugh, D.A. Edmonds, P.T. Nellesen, et al., Am. J. Phys. **65**, 602 (1997).

13
Interaction of Atoms with Radiation

13.1. Introduction

To this point we have considered only the properties of isolated atoms. We know, however, that atoms radiate and absorb electromagnetic radiation. Indeed, it was investigation of the wavelengths of these emissions and absorptions that led to the development of the quantum theory of matter.

Before delving into the details of the interaction of atoms with radiation we first address the question of how, within the framework of quantum mechanics, an atom radiates (or absorbs) electromagnetic energy. Having adopted the Rutherford model of the atom, electrons "orbiting" a point nucleus, the most important question to be asked is: why do these accelerating charges not radiate energy, lose their kinetic energy, and collide with the positively charged nucleus? Recall that, in order to formulate his model of the atom, Bohr simply postulated that atoms in "stationary states" do not radiate. We now know that these ad hoc stationary states are closely related to energy eigenstates.

Suppose we have an atom in an eigenstate. To be definite we use spherical eigenfunctions and assume that the Hamiltonian does not contain the time. Because the Hamiltonian does not contain time, the energy eigenfunctions are products of the spatial wave functions and exponential time factors. The total wave function including time is given by

$$\Psi_{n\ell m}(\boldsymbol{r}, t) = \psi_{n\ell m}(r, \theta, \phi) \cdot \exp\left[-i\left(\frac{E_{n\ell}}{\hbar}\right)t\right]$$
$$= \psi_k \exp[-i\omega_k t] \tag{13.1}$$

where we have represented the quantum numbers n, ℓ, and m by k. E_k is the energy eigenvalue corresponding to the eigenfunction of ψ_k and $\omega_k = E_k/\hbar$.

The electronic charge distribution associated with an atom in this eigenstate is

$$\rho(\boldsymbol{r}, t) = e\Psi_{n\ell m}^*(\boldsymbol{r}, t)\, \Psi_{n\ell m}(\boldsymbol{r}, t)$$
$$= e\psi_k^*(\boldsymbol{r})\, \psi_k(\boldsymbol{r})\left(e^{ik_n t} e^{-ik_n t}\right)$$
$$= e\psi_k^*(\boldsymbol{r})\, \psi_k(\boldsymbol{r}) \tag{13.2}$$

The important point here is that because the atom is in an eigenstate, the electronic charge distribution $\rho(r, t)$ is independent of time. That is,

$$\rho(r, t) = \rho(r) \tag{13.3}$$

The charge distribution associated with the atom is electro*static*. Classically such a distribution does not radiate because it is not changing in time; only accelerating charges radiate.

Now, suppose the atom in a stationary state is perturbed so that it emits radiation and decays to a lower (stationary) state, the wave function for which is

$$\Psi_{n'}(r, t) = \psi_{n'\ell'm'}(r, \theta, \phi) \cdot \exp\left[-i\left(\frac{E_q}{\hbar}\right)t\right]$$

$$= \psi_q e^{-iqt} \tag{13.4}$$

where q represents the primed quantum numbers. We may conclude that, after the perturbation, but before the atom is in the lower eigenstate, that is, during the transition, the atom exists in a superposition of the initial and final states. The wave function that describes this "intermediate" state is given by

$$\Psi(r, t) = a(t) \psi_k e^{-i\omega_k t} + b(t) \psi_q e^{-i\omega_q t} \tag{13.5}$$

where the expansion coefficients $a(t)$ and $b(t)$ are functions of time. Clearly $|a(t = 0)|^2 = 1$ and $|b(t = 0)|^2 = 0$ and after the transition $|a(t)|^2 = 0$ and $|b(t)|^2 = 1$. During the transition $|a(t)|^2 + |b(t)|^2 = 1$.

For an atom in such a superposition of states $\rho(r, t)$ is given by

$$\rho(r, t) = e\Psi^*(r, t) \Psi(r, t)$$
$$= \left|a\psi_k e^{-i\omega_k t} + b\psi_q e^{-i\omega_q t}\right|^2$$
$$= |a|^2 |\psi_k|^2 + |b|^2 |\psi_q|^2 + a^*b\psi_k^*\psi_q e^{i(\omega_n - \omega_{n'})t} + ab^*\psi_k\psi_q^* e^{-i(\omega_k - \omega_q)t}$$

$$\tag{13.6}$$

Although the first two terms are independent of time, the last two are not. In fact, they oscillate in time with a frequency $\omega = \omega_k - \omega_q = (E_k - E_q)/\hbar$. Thus, unlike the stationary state case, the charge distribution for an atom in a superposition of states is an oscillating function of time. Classically, such an oscillating distribution radiates! Moreover, it radiates at the frequency of oscillation of the charge distribution, in this case at the frequency ω. But, according to Bohr theory, ω is the frequency of the photon given off when the atom undergoes a transition from the state k to the state q. How then does one reconcile these two different frequencies? As discussed in Chapter 1, the answer is that, for large n, the energy separation $E_n - E_{n'}$ divided by $\hbar$ approaches the frequency of oscillation of the charge distribution. That is, the correspondence principle is at the heart of the matter.

13.2. Time Dependence of the Wave Function

Consider a Hamiltonian

$$\hat{H}(t) = \hat{H}_0 + \hat{H}'(t) \tag{13.7}$$

where $\hat{H}_0$ is independent of time. The Hamiltonian $\hat{H}(t)$ might represent an atom in an applied electromagnetic field with $\hat{H}_0$ being the time-independent atomic Hamiltonian and $\hat{H}'(t)$ the time-dependent function that represents the interaction of the atom with the field. It is assumed that the solutions of the time-independent Schrödinger equation are known and given by

$$\hat{H}_0 |u_n\rangle = E_n |u_n\rangle \tag{13.8}$$

where the $|u_n\rangle$ do not contain time. We use $|u_n\rangle$ to designate the stationary states of $\hat{H}_0$ to avoid confusion with our designation of the nonstationary state $|\Psi(r, t)\rangle$.

It is assumed that at $t = 0$ the system is in a state $|\Psi(r, t = 0)\rangle$ which is not a stationary state. The wave function for the nonstationary state $|\Psi(r, t)\rangle$ may be expanded on the basis set of stationary states of $\hat{H}_0$, but the time dependences of the stationary states (basis vectors) must be included. Therefore

$$|\Psi(r, t)\rangle = \sum_n c_n(t)|u_n\rangle \cdot \exp[-i(E_n/\hbar)t] \tag{13.9}$$

The time evolution of this wave function is governed by the time-dependent Schrödinger equation

$$\hat{H}(t) |\Psi(r, t)\rangle = i\hbar \frac{\partial}{\partial t} |\Psi(r, t)\rangle \tag{13.10}$$

The probability of finding the system in one of the eigenstates $|u_k\rangle$ at a time t is given by the square of the amplitude $|c_k(t)|^2$ for that state. This quantity is, of course, time dependent. To determine these amplitudes we substitute the expansion for $|\Psi(r, t)\rangle$ in the Schrödinger time-dependent wave equation and obtain

$$i\hbar \dot{c}_k(t) = \sum_n c_n(t) \hat{H}'_{kn}(t) e^{i\omega_{kn}t} \tag{13.11}$$

where

$$\omega_{kn} = \frac{E_k - E_n}{\hbar}$$

and

$$\hat{H}'_{kn}(t) = \langle u_k| \hat{H}'(t) |u_n\rangle$$

are the matrix elements of the perturbing Hamiltonian. These matrix elements are said to connect the initial and final states.

These equations are exact. They represent an infinite set of coupled equations for the expansion coefficients $c_n(t)$. For most cases it is not possible to solve for these expansion coefficients exactly, so it is necessary to make approximations to do so. Making approximations to solve these equations is called time-dependent perturbation theory.

13.3. Interaction of an Atom with a Sinusoidal Electromagnetic Field

The case that interests us here is that of a transition between initial and final stationary states $|u_i\rangle$ and $|u_f\rangle$ that is caused by application of an external electromagnetic field. The interaction energy of the atom with this field is the perturbing Hamiltonian. The atom may absorb energy from the field and thus the energy of the final state E_f will be higher than the energy of the initial state E_i. In this case the atom is said to have absorbed a quantum of electromagnetic energy from the field, or, more simply the atom has "absorbed a photon". If, on the other hand, E_i is higher than E_f then the atom can give up a quantum of electromagnetic radiation to the field and "emit a photon". Note that, under the present assumptions, the presence of the external field is required for both of these processes, absorption and emission, because without the field all matrix elements of the perturbing Hamiltonian vanish and there is no way to connect these states. We show in the next section that it is possible for an excited atom, that is, an atom for which $E_i > E_f$, to spontaneously emit a photon, but that situation is not covered by the present formulation. For this reason we use the term "stimulated" to describe both the emission and the absorption of a photon when the external field is present. In this sense, the terms "stimulated" and "spontaneous" are antitheses. Clearly, it is not necessary to refer to absorption as "stimulated absorption" because "spontaneous absorption" is impossible (even in quantum mechanics) because it violates conservation of energy.

To obtain the transition probabilities for interactions of atoms with electromagnetic radiation it is necessary to find the proper form of the interaction Hamiltonian $\hat{H}'(t)$ when an atom is immersed in an electromagnetic field. This field consists of time-varying electric and magnetic fields, $F(r_j, t)$ and $B(r_j, t)$, respectively. The force on the jth constituent charge of the atom, electrons, and the nucleus, due to the field is given by

$$f_j = q_j F(r_j, t) + q_j v_j \times B(r_j, t) \tag{13.12}$$

For electromagnetic waves, however, $|B| = (1/c)|F|$; moreover, the orbital velocities of electrons are also proportional to $1/c$. Therefore, we may, to a first approximation, ignore the interaction with the magnetic field. Additionally, we know that the energy separations between atomic energy levels are the order of electron volts for which the wavelength of radiation connecting them is $\sim 10^4$ Å. Because atomic dimensions are ~ 1 Å it is permissible to ignore the spatial variation of the field over the extent of the atom. We therefore regard the perturbation as being time dependent with no spatial dependence. It can have some distribution of frequencies of the electromagnetic waves that constitute the field. First, however, we consider the radiation field to be a plane wave with frequency ω and polarization vector $\hat{n}$ so that, with these approximations, we represent the electromagnetic field by

$$F(r, t) = F_0 \hat{n} \cos(\omega t) \tag{13.13}$$

We examine the transition probability as a function of the frequency of the monochromatic plane wave ω.

The *total* force on the atom due to the field is

$$
\begin{aligned}
f &= \sum_j f_j \\
&= \sum_j q_j F(r_j, t) \\
&= -\sum_j q_j \nabla \phi(r_j, t)
\end{aligned}
\tag{13.14}
$$

where $\phi(r_j, t)$ is the potential that gives rise to the field at the location of the jth charge in the atom. The energy due to the interaction of this field with the atom is $\hat{H}'(t)$ and is given by

$$\hat{H}'(t) = \sum_i q_j \phi(r_j, t) \tag{13.15}$$

We expand the potential in a Taylor's series[1] about $r = 0$ so $\hat{H}'(t)$ becomes

$$\hat{H}'(t) = \sum_i q_i \left\{ \phi(0, t) + (r_i \cdot \nabla) \phi(0, t) + \frac{1}{2}(r_i \cdot \nabla)^2 \phi(0, t) + \cdots \right\}$$

$$= Q\phi(0, t) + \left[\sum_i (q_i r_i) \right] \cdot \nabla \phi(0, t) + \frac{1}{2} \sum_i q_i (r_i \cdot \nabla)^2 \phi(0, t) + \cdots$$

$$\tag{13.16}$$

The first term includes the total charge on the atom Q. Usually, $Q = 0$ so this term does not contribute. Even if $Q \neq 0$, for example, if an electron is removed from the atom leaving a positive ion, this term does not contribute because the matrix element of $\hat{H}'(t)$ connecting two different states vanishes because of orthogonality of the eigenfunctions.

If we now take the origin of coordinates to be the position of the nucleus, the two remaining summations over the charges do not include the nucleus. The term in brackets is the electric dipole moment of the atom having Z electrons

$$\hat{p} = -e \sum_{j=1}^{Z} r_j \tag{13.17}$$

so the time-dependent Hamiltonian that represents the interaction of the atom with the field to first order is then

$$\hat{H}'(t) = -\hat{p} \cdot F(t)$$
$$= -F_0 (\hat{p} \cdot \hat{n}) \cos(\omega t) \qquad (13.18)$$

This will be the dominant term, the one that leads to "electric dipole radiation".

13.4. A Two-State System—The Rotating Wave Approximation

Although most problems of interest require treatment using perturbation theory, the "two-state atom" subjected to a sinusoidal electric field can be solved more or less exactly. The solution is only nearly exact because an approximation to be discussed below is required. We assume that the atom has two levels. The eigenstates are designated $|u_1\rangle$ and $|u_2\rangle$ with eigenenergies E_1 and E_2, respectively. We assume that $E_2 > E_1$. As was deduced in the previous section, the perturbing Hamiltonian may be written

$$\hat{H}'(t) = -\hat{p} \cdot F(t)$$
$$= -F_0 (\hat{p} \cdot \hat{n}) \cos(\omega t)$$
$$= \hat{V} \cos(\omega t) \qquad (13.19)$$

where $\hat{V}$ is a time-independent operator. From Equation (13.11) the equations for the expansion coefficients are then

$$i\hbar \dot{c}_1(t) = c_1(t) \hat{V}_{11} \cos(\omega t) + c_2(t) \hat{V}_{12} \cos(\omega t) e^{-i\omega_{21}t}$$
$$i\hbar \dot{c}_2(t) = c_1(t) \hat{V}_{21} \cos(\omega t) e^{i\omega_{21}t} + c_2(t) \hat{V}_{22} \cos(\omega t) \qquad (13.20)$$

where $\omega_{kn} = (E_k - E_n)/\hbar$ is the Bohr frequency and we have used $\omega_{12} = -\omega_{21}$.

Replacing the cosine coefficients of $e^{\pm i\omega_{21}t}$ with the exponential form we obtain

$$i\hbar \dot{c}_1(t) = c_1(t) \hat{V}_{11} \cos(\omega t)$$
$$+ c_2(t) \hat{V}_{12} \left(\frac{1}{2}\right) \{\exp[i(\omega - \omega_{21})t] + \exp[-i(\omega + \omega_{21})t]\}$$

$$i\hbar \dot{c}_2(t) = c_1(t) \hat{V}_{21} \left(\frac{1}{2}\right) \{\exp[i(\omega + \omega_{21})t] + \exp[-i(\omega - \omega_{21})t]\}$$
$$+ c_2 \hat{V}_{22} \cos(\omega t) \qquad (13.21)$$

These equations are exact, but intractable. We note, however, that near resonance, when $\omega \approx \omega_{21}$, the sinusoidal terms, for which the arguments are $(\omega + \omega_{21})$ and ω, vary much more rapidly than those of argument $(\omega - \omega_{21})$. These rapidly varying terms do not make significant contributions to the differential equations for long times. They therefore "average out". This is referred to as the "rotating wave approximation". The name originated with the description of the response of an electron spin to a magnetic induction field that is sinusoidally varying in time, but constant in direction. Such a field may be described as the superposition of two fields that are rotating in the opposite sense. The rotating wave approximation amounts to ignoring one of the rotating fields because it causes the rapidly varying term as described above. In essence, the field that is constant in direction is approximated by a rotating field.

For the rotating wave approximation we retain only terms containing $(\omega - \omega_{21})$. Letting $\delta = (\omega - \omega_{21})$ we obtain

$$i\hbar \dot{c}_1(t) = c_2(t) \hat{V}_{12} \left(\frac{1}{2} \right) e^{i\delta t}$$

$$i\hbar \dot{c}_2(t) = c_1(t) \hat{V}_{21} \left(\frac{1}{2} \right) e^{-i\delta t} \tag{13.22}$$

Note that δ is a measure of the amount by which the frequency of the applied field is "off-resonance" from the natural frequency of the system, the Bohr frequency.

When these equations are uncoupled they yield a linear second-order differential equation. If it is assumed that the atom is initially in the lower level, then these equations can be solved (see Problem 13.2). The results are

$$c_1(t) = e^{i\delta t/2} \cos(\omega_R t) - i \left(\frac{\delta}{2\omega_R} \right) \sin(\omega_R t)$$

$$c_2(t) = \frac{\hat{V}_{12}}{2i\hbar\omega_R} e^{-i\delta t/2} \sin(\omega_R t) \tag{13.23}$$

where

$$\omega_R = \frac{1}{2} \sqrt{\delta^2 + \frac{|\hat{V}_{12}|^2}{\hbar^2}} \tag{13.24}$$

so the probability that the atom undergoes a transition from state 1 to state 2 is

$$P_{1 \to 2} = |c_2(t)|^2$$

$$= \frac{1}{1 + (\hbar\delta)^2 / |\hat{V}_{12}|^2} \sin^2(\omega_R t) \tag{13.25}$$

and the probability of returning to state 1 from state 2 is

$$P_{2 \to 1} = |c_1(t)|^2$$

$$= \cos^2(\omega_R t) + \left(\frac{\delta}{2\omega_R} \right)^2 \sin^2(\omega_R t) \tag{13.26}$$

Clearly the atom oscillates between the two states with frequency ω_R which is

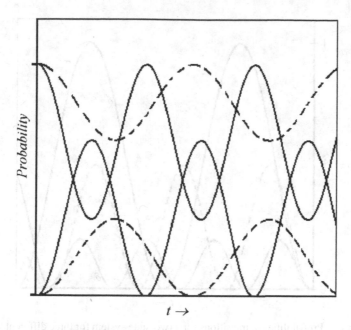

FIGURE 13.1. Probabilities of transitions for a two-state system for two different values of the matrix element, but the same value of the detuning δ.

referred to as the Rabi flopping frequency. Note that the "flopping" from state to state (including upper to lower) requires the presence of the field. It does not occur spontaneously.

If the frequency of the applied field is exactly resonant, $\omega = \omega_{kn}$ so $\delta = 0$, then the system oscillates between the two states with frequency $\hat{V}_{12}/2\hbar$. At resonance the probability of stimulated absorption or stimulated emission can be unity at certain times. When, however, the applied field is off resonant the amplitude for transitions is somewhat less than unity depending upon the value of the matrix element $\hat{V}_{12}$. FIGURE 13.1 shows plots of Equations (13.25) and (13.26) for two different values of the matrix element, but the same value of the detuning δ. The solid curves represent the case for which $\left|\hat{V}_{12}\right|^2/\hbar^2 = 2\delta^2$ and the dashed curves denote the case for which $\left|\hat{V}_{12}\right|^2/\hbar^2 = 0.5\delta^2$.

It is clear that the Rabi frequency is different for the two cases [see Equation (13.24)]. Notice also that, as expected, the stronger coupling (solid curve) leads to a higher probability of transition between the two states. This is clear if we examine the transition probability from lower to upper state $P_{1 \to 2}$ for a fixed value of the matrix element, but different values of δ as shown in FIGURE 13.2 where δ is in units of $\left|\hat{V}_{12}\right|^2/\hbar^2$.

For the resonant case, $\delta = 0$, the probability of excitation reaches unity, but as the detuning increases, the probability of excitation decreases. Moreover, the frequency of the transitions, the Rabi frequency, increases with the detuning.

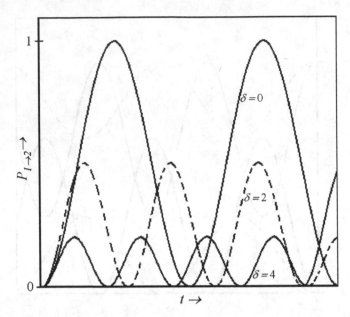

FIGURE 13.2. Probabilities of transitions for a two-state system for three different values of the detuning δ, but the same value of the matrix element.

13.5. Stimulated Absorption and Stimulated Emission

Having solved the two-state problem "exactly" we now turn our attention to the more general problem of stimulated emission and absorption in the presence of an external electromagnetic field. We use time-dependent perturbation theory and compute the transition rate when multistate atoms are subjected to an external electromagnetic field. As previously, we consider the perturbation to be a monochromatic sine wave of the form

$$\hat{H}'(t) = \hat{V}\cos(\omega t) \tag{13.27}$$

where

$$\hat{V} = -F_0(\hat{p} \cdot \hat{n}) \tag{13.28}$$

Assume that $\hat{H}'(t) \ll \hat{H}_0$. Further assume that at $t = 0$ the system is in an eigenstate $|u_i\rangle$ of $\hat{H}_0$ so that

$$c_n(t = 0) = \delta_{ni} \tag{13.29}$$

where the subscript i denotes the initial state. Now we expand the coefficients $c_n(t)$ power series in some arbitrary parameter ζ which we use to keep track of the order of the approximation retained.

$$c_k(t) = \sum_{j=0}^{\infty} \zeta^j c_k^{(j)} \tag{13.30}$$

Applying the initial condition on the $c_k(t)$, Equation (13.29), inserting the expansion, Equation (13.30), into the exact equation for the expansion coefficients, Equation (13.11), and retaining only terms to first order we have

$$i\hbar \frac{d}{dt} \left(\zeta^0 c_k^{(0)} + \zeta^1 c_k^{(1)} \cdots \right) = \sum_n \left(\zeta^0 c_n^{(0)} + \zeta^1 c_n^{(1)} \cdots \right) \left[\zeta \hat{H}'_{kn}(t) \right] e^{i\omega_{kn}t} \tag{13.31}$$

Notice that, in the spirit of the expansion of $c_k(t)$ in a power series of ζ we must replace the time-dependent perturbation Hamiltonian $\hat{H}'(t)$ with $\zeta \hat{H}'(t)$.

The zeroth-order term $dc_k^{(0)}/dt = 0$ in Equation (13.31). Thus, to zeroth order, nothing happens. On the other hand, the first-order correction provides the response of the atom to the perturbation. In view of the initial conditions of Equation (13.29), the first-order correction is

$$i\hbar \dot{c}_k^{(1)} = \sum_n c_n^{(0)} \hat{H}'_{kn}(t) e^{i\omega_{kn}t}$$

$$= \hat{H}'_{ki}(t) e^{i\omega_{ki}t} \tag{13.32}$$

This expression for $\dot{c}_k(t)$ holds for all k (including $k = i$). We now integrate this equation from $t = 0$ to some arbitrary time t and obtain

$$c_k(t) = \frac{1}{i\hbar} \int_0^t \hat{H}'_{ki}(t') e^{i\omega_{ki}t'} dt' \qquad \text{for } k \neq i \tag{13.33}$$

and

$$c_i(t) = 1 + \frac{1}{i\hbar} \int_0^t \hat{H}'_{ii}(t') dt' \qquad \text{for } k = i \tag{13.34}$$

Now, $|c_k(t)|^2$ is the probability that, at time t, the system has undergone a transition from the state $|u_i\rangle$ to the state $|u_k\rangle$. For perturbation theory to be applicable, this probability must be small, so that

$$|c_i(t)|^2 \approx 1 \tag{13.35}$$

and

$$1 - |c_i(t)|^2 = \sum_{q \neq i} |c_q(t)|^2$$

$$\ll 1 \tag{13.36}$$

Inserting the Hamiltonian in the expression for $c_k(t)$, letting $k = f$ to represent a final state, and integrating we obtain

$$c_f(t) = \frac{1}{2i\hbar} \hat{p}_{fi} \left\{ \frac{\exp\left[i\left(\omega_{fi} - \omega\right)t\right] - 1}{i\left(\omega_{fi} - \omega\right)} + \frac{\exp[i\left(\omega_{fi} + \omega\right)] + 1}{i\left(\omega_{fi} + \omega\right)} \right\} \tag{13.37}$$

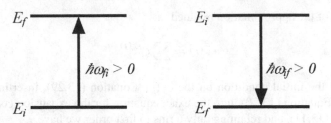

FIGURE 13.3. Schematic energy-level diagram illustrating the absorption and emission of a photon by the atom δ.

where

$$\hat{p}_{fi} = \langle u_f | \hat{\boldsymbol{p}} \cdot \hat{\boldsymbol{n}} | u_i \rangle \qquad (13.38)$$

The first term in the bracket in Equation (13.37) is significant only if $\omega \approx \omega_{fi}$ and the second is large only if $\omega \approx -\omega_{fi}$. There are therefore two different conditions under which the transition may be expected to occur, that is, conditions under which the transition probability, $|c_f(t)|^2$, would be large. This would occur when $\omega \approx \pm\omega_{fi}$. These two cases are illustrated in FIGURE 13.3.

For absorption of a photon, ω_{fi} is positive and the first term in the equation for $c_f(t)$ dominates. The probability of a transition from the lower state to the higher state after time t is therefore given by

$$P_{i \to f}(t) = |c_f(t)|^2$$

$$= \frac{1}{\hbar^2} |\hat{p}_{fi}|^2 F_0^2 \frac{\sin^2\left[\frac{1}{2}(\omega_{fi} - \omega) t\right]}{(\omega_{fi} - \omega)^2} \qquad (13.39)$$

For emission of a photon, ω_{fi} is negative so the second term in the equation for $c_f(t)$ dominates and the probability of a transition from the higher state to the lower state is given by

$$P_{i \to f}(t) = |c_f(t)|^2$$

$$= \frac{1}{\hbar^2} |\hat{p}_{fi}|^2 F_0^2 \frac{\sin^2\left[\frac{1}{2}(\omega_{fi} + \omega) t\right]}{(\omega_{fi} + \omega)^2} \qquad (13.40)$$

Before continuing discussion of this result it should be remarked that, within the given assumptions, this expression should apply to the two-state system of Section 13.3. This result is, however, more restrictive than that solution because it was assumed that the perturbation is "small". No such assumption was invoked in the treatment of the two-state problem. Nonetheless, the result from the two-state system should reduce to the present result if the interaction Hamiltonian is suitably small.

To check the compatibility of these two solutions we cast the two-state result in the notation of the more general solution. In the present notation the two-state result may be written

$$P_{i \to f} = \frac{\left|\hat{V}_{if}\right|^2}{4\hbar^2 \omega_R^2} \sin^2 (\omega_R t) \tag{13.41}$$

where ω_R is given by Equation (13.24). If, as required by the perturbation theory treatment, $\hat{V}_{12}$ is small, then ω_R may be approximated as

$$\omega_R = \frac{(\omega - \omega_{fi})}{2} \sqrt{1 + \frac{\left|\hat{V}_{12}\right|^2}{(\omega - \omega_{fi})^2 \hbar^2}}$$

$$\approx \frac{1}{2} (\omega - \omega_{fi}) \tag{13.42}$$

so that

$$P_{i \to f} = \frac{\left|\hat{V}_{if}\right|^2}{\hbar^2 (\omega - \omega_{fi})^2} \sin^2 \left[\frac{1}{2} (\omega - \omega_{fi}) t\right] \tag{13.43}$$

Because

$$\left|\hat{V}_{if}\right|^2 = F_0^2 \left|\hat{p}_{fi}\right|^2 \tag{13.44}$$

the result from the rotating wave approximation reduces to that obtained using perturbation theory.

The emission from an upper state to a lower state is not spontaneous. Rather, it is induced, or stimulated, by the presence of the field. In terms of photons, we may say that for the transition to occur, the atom must be stimulated by the presence of a photon of precisely the "correct" frequency, ω_{fi}, to emit a second photon of the same frequency. It is seen then that the term "stimulated emission" is the origin of the "s" and the "e" in the word "laser". In either of the above cases it is clear that the probability of a transition is very low unless the applied frequency ω is very close to the Bohr frequency ω_{fi}; that is, there must be resonance.

We now examine these transition probabilities in more detail. We concentrate on absorption, but the treatment for stimulated emission is identical. The probability of an absorption transition $i \to f$ may be rewritten as

$$P_{i \to f} = G_{fi}(r) F(\omega, t) \tag{13.45}$$

where

$$G_{fi}(r) = \frac{1}{\hbar^2} \left|\hat{p}_{fi}\right|^2 F_0^2 \quad \text{and} \quad F(\omega, t) = \left(\frac{t^2}{4}\right) \frac{\sin^2 \left[\frac{1}{2} (\omega_{fi} - \omega) t\right]}{\left[\frac{1}{2} (\omega_{fi} - \omega) t\right]^2}$$

All of the ω dependence is contained in $F(\omega, t)$; a graph of this function versus $(\omega_{fi} - \omega)$ for a fixed time t is shown in FIGURE 13.4.

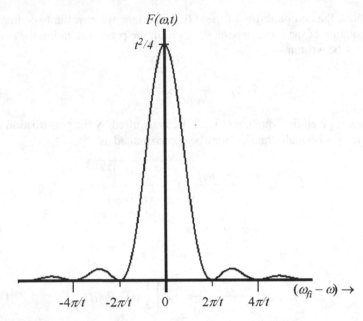

FIGURE 13.4. Graph of the frequency dependence of the probability of a transition from initial to final states.

The positions of the zeros of the function are inversely proportional to time so the width of the central maximum decreases as t increases. Thus, for large t the system becomes less and less forgiving as far as permitting the atom to absorb an off-resonant photon. This is, of course, a manifestation of the energy–time uncertainty principle.

To evaluate $P_{i \to f}$ if there is a range of frequencies of the external field we must replace F_0 by $F_0(\omega)$. That is, different frequency components of the external field have continuously variable amplitudes, so we must sum them by integrating. We are also interested in long times so

$$P_{i \to f} = \lim_{t \to \infty} G_{fi}(r) \, F(\omega, t)$$

$$= \frac{|\hat{p}_{fi}|^2}{\hbar^2} \cdot \int_0^\infty d\omega \, [F_0(\omega)]^2 \lim_{t \to \infty} \left\{ \left(\frac{t^2}{4} \right) \frac{\sin^2\left[\frac{1}{2}(\omega_{fi} - \omega)t \right]}{\left[\frac{1}{2}(\omega_{fi} - \omega)t \right]^2} \right\} \quad (13.46)$$

Now $F_0(\omega)$ represents the electric field amplitude at frequency ω, but $[F_0(\omega)]^2 \, d\omega$ is proportional to the more readily available energy density per frequency interval $\rho(\omega) \, d\omega$. That is,

$$\rho(\omega) = \frac{\varepsilon_0}{2} [F_0(\omega)]^2 \quad (13.47)$$

The standard symbol for energy density $\rho(\omega)$ should not be confused with the standard symbol for charge density $\rho(\mathbf{r}, t)$ used earlier in this chapter.

We have then

$$P_{i \to f} = \frac{2}{\varepsilon_0} \frac{|\hat{p}_{fi}|^2}{\hbar^2} \cdot \int_0^\infty d\omega \, \rho(\omega) \lim_{t \to \infty} \left\{ \left(\frac{t^2}{4} \right) \frac{\sin^2 \left[\frac{1}{2}(\omega_{fi} - \omega) t \right]}{\left[\frac{1}{2}(\omega_{fi} - \omega) t \right]^2} \right\} \quad (13.48)$$

It can be shown that the function in the curly brackets is proportional to a Dirac delta function. Specifically,[2]

$$\delta(x) = \left(\frac{\beta}{2\pi} \right) \lim_{t \to \infty} \left[\frac{\sin^2 \left(\frac{\beta x}{2} \right)}{\left(\frac{\beta x}{2} \right)^2} \right] \quad (13.49)$$

Making the appropriate substitutions we have

$$P_{i \to f} = \frac{2}{\varepsilon_0} \cdot \frac{|\hat{p}_{fi}|^2}{\hbar^2} \cdot \int_0^\infty d\omega \rho(\omega) \delta(\omega - \omega_{fi})$$

$$= t \cdot \left(\frac{\pi}{\varepsilon_0} \right) \frac{|\hat{p}_{fi}|^2}{\hbar^2} \rho(\omega_{if}) \quad (13.50)$$

The transition *rate* W'_{if} is the probability *per unit time* that a transition has taken place. From Equation (13.50),

$$W'_{if} = \frac{dP_{i \to f}}{dt}$$

$$= \frac{\pi}{\varepsilon_0 \hbar^2} |\hat{p}_{fi}|^2 \rho(\omega_{if}) \quad (13.51)$$

A final modification of Equation (13.51) is required because the atoms are randomly oriented with respect to the direction of polarization of the electric field. The matrix element $\hat{p}_{if}$, however, is proportional to $\hat{p} \cdot \hat{n}$ and thus depends upon the orientation of the atom with respect to the direction of polarization $\hat{n}$. It is merely the component of $\hat{p}$ along the direction $\hat{n}$. We must therefore average over all orientations of the atoms, or, equivalently, over all orientations of $\hat{n}$. Because $\hat{p} = -e\mathbf{r}$ this is equivalent to averaging $\hat{n} \cdot \mathbf{r}$ over all possible orientations. For convenience we define the angle between $\mathbf{r}$ and $\hat{n}$ to be θ so we have

$$|\hat{p}_{fi}|^2 = e^2 |\langle u_f | \mathbf{r} \cdot \hat{n} | u_i \rangle|^2$$

$$= e^2 |\langle u_f | \mathbf{r} | u_i \rangle|^2 \cos^2 \theta$$

$$= e^2 |\mathbf{r}_{fi}|^2 \cos^2 \theta \quad (13.52)$$

where

$$|\mathbf{r}_{fi}| \equiv \langle u_i | \mathbf{r} | u_f \rangle \quad (13.53)$$

For unpolarized radiation we must average over the entire solid angle so

$$\frac{1}{4\pi} \int_0^\pi \cos^2 \theta \, (2\pi \sin\theta d\theta) = \frac{1}{2} \cdot \frac{1}{3} \cos^3 \theta \Big|_\pi^0$$

$$= \frac{1}{3} \tag{13.54}$$

Thus, the transition rate W_{if} induced by randomly oriented atoms in the presence of an incoherent external electric field is given by

$$W_{if} = \frac{e^2 \pi}{3\varepsilon_0 \hbar^2} \left| \boldsymbol{r}_{fi} \right|^2 \rho\left(\omega_{if}\right) \tag{13.55}$$

We see then that, given the eigenfunctions for the atom and $\rho\left(\omega_{fi}\right)$, the transition rate may be calculated.

The transition rate seems to imply that we have returned to consideration of a two-level system, i and f. This is because we have examined the transition between the two designated levels. None of the other levels appears in this expression, for example, in the matrix element, because of the strongly peaked function $F(\omega, t)$. Transitions to other "final" levels would have corresponding expressions with, of course, the appropriate matrix element and $\rho(\omega_{fi})$.

It is important to note that W_{if} is proportional to the square of the dipole matrix element connecting the initial and final states. Thus, it is this matrix element that determines the strength of the transition. Indeed, it is the matrix element that determines whether the transition will occur at all. If it vanishes then the transition cannot occur, at least to within the electric dipole approximation. We may therefore derive *selection rules*, that is, rules under which the transition can and cannot occur, by examining the conditions under which the matrix element vanishes. First, however, we consider the process of spontaneous emission which cannot be treated using the perturbation theory of "first quantization" because there is, seemingly, no external perturbation (such as an electromagnetic field) to cause the transition. In fact, this spontaneous transition rate can be derived using second quantization, QED in which the absence of a field is described by the zero-point energy of the field. Stimulated emission and absorption required interactions between the atom and an external field, however, spontaneous emission occurs when the atom interacts with the ground state of the field. Classically this means that spontaneous emission occurs in the absence of a field.

13.6. Spontaneous Emission

Although the derivation of the spontaneous emission rate is properly carried out using the principles of QED, this rate can be derived using an argument first employed by Einstein. Einstein derived a relation between the spontaneous rate and the stimulated rates using a thermodynamic argument. Because we have expressions for the stimulated rates, the spontaneous rate is thus derived. Because of

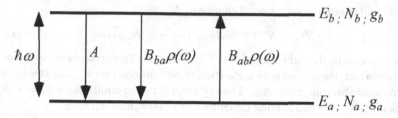

FIGURE 13.5. Schematic energy-level diagram showing the parameters used in Einstein's derivation.

his pioneering work, the probabilities of stimulated emission and absorption and spontaneous emission are often referred to as the Einstein coefficients.

Einstein began using a two-level system. The two levels are presumed to have energies E_a and E_b and are separated by $\hbar\omega$. FIGURE 13.5 shows the relation between A and the Bs; N_a and N_b are the instantaneous populations of the a and b levels, respectively, and g_a and g_b are the degeneracies of each level. The gs merely represent the number of states that have energy E_a or E_b.

The possible transitions are stimulated emission, stimulated absorption, and spontaneous emission. It is assumed that the system is in equilibrium with the surroundings which are at a temperature T that is characterized by a Planck distribution $\rho(\omega)$. The Einstein coefficient A is the spontaneous transition rate in transitions per second or s^{-1}; the quantities $[B_{ab}\rho(\omega)]$ and $[B_{ba}\rho(\omega)]$ also have units of s^{-1}. For the purpose of this two-level derivation we simplify notation by letting $A \equiv A_{ba}$ inasmuch as the term A_{ab} is meaningless because a spontaneous transition from the lower level to the upper level would be quite impossible. Once having derived the form of A, however, subscripts will be necessary to distinguish transitions from a given upper level to different lower levels.

Note that the stimulated transitions require the presence of the external field, but the spontaneous transition does not. Therefore, the probabilities for the stimulated transitions must, necessarily, depend upon the "strength" of the radiation as contained in $\rho(\omega)$. The consequence is that the units of A are different from the units of the Bs; it is A and $[B\rho(\omega)]$ that have the same units, viz. s^{-1}. In terms of the notation of Section 13.5,

$$[B_{ab}\rho(\omega)] = W_{ab} \quad \text{and} \quad [B_{ba}\rho(\omega)] = W_{ba}$$

so it is clear that the B coefficients are contained in the expression for W_{ab} from the last section of this chapter. Comparing with the expression for W_{if}, Equation (13.55) we see that

$$B_{ab} = \frac{e^2\pi}{3\varepsilon_0\hbar^2} |r_{fi}|^2 \tag{13.56}$$

As noted above though, A cannot be obtained from first quantization. We can, however, obtain a relation between A and the Bs using Einstein's treatment without having to resort to QED.

In equilibrium the time rate of change of N_a is

$$\dot{N}_a = N_b A + N_b B_{ba} \rho(\omega) - N_a B_{ab} \rho(\omega) \tag{13.57}$$

The first term on the right-hand side of Equation (13.57) represents the spontaneous emission rate, the second term is the stimulated emission rate W_{ba}, and the third the stimulated absorption rate W_{ab}. The key point is that at equilibrium $\dot{N}_a = -\dot{N}_b = 0$. Setting $\dot{N}_a = 0$ and solving Equation (13.57) for $\rho(\omega)$ we have

$$\rho(\omega) = \frac{A}{\left(N_a / N_b\right) B_{ab} - B_{ba}} \tag{13.58}$$

Now, the ratio of the populations is given by the Boltzmann factor

$$\frac{N_a}{N_b} = \left(\frac{g_a}{g_b}\right) \exp\left(\frac{\hbar\omega}{kT}\right) \tag{13.59}$$

where k is the Boltzmann constant. After replacing N_a/N_b in Equation (13.58) $\rho(\omega)$ may be compared with the Planck distribution for $\rho(\omega)$; that is,

$$\rho(\omega) = \frac{\hbar\omega^3}{\pi^2 c^3} \cdot \frac{1}{\exp\left(\hbar\omega / kT\right) - 1} \tag{13.60}$$

From the comparison we make the following identifications

$$B_{ab} = \left(\frac{g_a}{g_b}\right) B_{ba} \tag{13.61}$$

and

$$A = \frac{\omega^3 \hbar}{\pi^2 c^3} B_{ab} \tag{13.62}$$

Equation (13.62) leads to a value of the spontaneous transition rate A,

$$A = \frac{1}{3\pi \varepsilon_0 \hbar c^3} \omega_{ab}^3 \left| r_{fi} \right|^2$$
$$= \frac{4\alpha}{3e^2 c^2} \omega_{ab}^3 \left| r_{fi} \right|^2 \tag{13.63}$$

where α is the fine-structure constant.

We see that A is proportional to the cube of the Bohr frequency between the levels and to the square of the dipole matrix element connecting the states. It is natural to ask if these dependencies are reasonable. In particular, is the result consistent with classical physics? To investigate this we compare the quantum mechanical expression for A with the classical rate of radiation from an accelerated atomic electron. Of course, this comparison must not be taken too literally, but the dependencies upon the parameters are interesting.

Consider a collection of atoms, N of which are in a given excited state. Transitions to a lower state yield photons of energy $\hbar\omega$. The rate at which energy is lost by spontaneous emission is the spontaneous transition rate A multiplied by the energy per transition, $\hbar\omega$. This may be equated to the Larmor formula for the power radiated by the accelerating charges in the atoms. We have (in SI units)

$$N\,(A\hbar\omega) = N\left(\frac{e^2 a^2}{6\pi\varepsilon_0 c^3}\right) \tag{13.64}$$

where a is the acceleration of the orbiting electrons. For circular orbits of radius R and angular velocity ω the acceleration is

$$a = v^2/R$$
$$= \omega^2 R \tag{13.65}$$

Because, classically, the frequency of the emitted radiation is the same as the frequency of the accelerating charge we use the same value of ω as that for the emitted radiation. Inserting this in the above equation and solving for A we obtain

$$A = \frac{1}{2} \cdot \frac{\omega^3}{3\pi\varepsilon_0 \hbar c^3} \cdot (eR)^2 \tag{13.66}$$

Identifying the quantity $(eR)^2$ with the square of the electric dipole matrix element we see that, except for a factor of two, the derived transition rate is the same as the quantum result. The factor of two is of little consequence because this rough calculation provides only the order of magnitude and the dependencies upon the parameters. Nonetheless, the result is of interest. Note that because of the proportionality on ω^3 high-energy transitions are favored.

Now, how about the lifetime of an excited state? If an atom is in a given excited state how long is it expected to remain there before spontaneous emission causes it to undergo a transition to some lower state? More properly, we should ask how the number of atoms in a given excited state decreases with time. Clearly the higher the value of A, the more rapidly the upper state will be depopulated by spontaneous emission.

We imagine a collection of atoms in a field-free region of space. The collection is not in thermal equilibrium as was assumed for the derivation of the relationship between the Einstein coefficients. Let $N_e\,(t)$ be the number of atoms in an excited state designated by e. Then the time rate of change of this number is given by

$$\frac{dN_e\,(t)}{dt} = -A_{el} N_e\,(t) \tag{13.67}$$

where l designates some lower level. There may, however, be many lower levels to which the excited state can decay by spontaneous emission. The total rate of decay of the excited state is then the sum of these decay rates and is given by

$$\frac{dN_e\,(t)}{dt} = -N_e\,(t) \sum_j A_{ej} \tag{13.68}$$

where the summation is taken over all levels having energies lower than that of the excited state. Integrating this equation and designating the population of the excited state at $t = 0$ as $N_e\,(0)$ we obtain the usual equation for a decay rate of a population that is proportional to the population itself.

$$N_e(t) = N_e(0) e^{-t/t_e}$$

where

$$t_e = \frac{1}{\sum_j A_{ej}}$$

is the lifetime of the state. In many cases a transition to one particular lower state dominates and one of the A_{ej} is much larger than the rest. In this case the lifetime is essentially the reciprocal of this particular spontaneous emission coefficient.

To reiterate, we have used Einstein's method to circumvent the necessity of employing QED to account for spontaneous emission. Thus, the spontaneous emission rate, although it cannot be described in terms of "first quantization", that is, quantizing atomic energy levels, but not the electromagnetic field, can be obtained from thermodynamic considerations that lead to relations between A and B.

To actually compute a spontaneous emission rate and the corresponding lifetime, it is necessary to compute the matrix elements. This can be done for hydrogen because the wave functions are known. We calculate the spontaneous transition probability for the $n = 2 \to 1$ transition in hydrogen. The required matrix elements are

$$\langle u_i | \mathbf{r} | u_f \rangle = \langle 21m | \mathbf{r} | 100 \rangle \quad \text{and} \quad \langle u_i | \mathbf{r} | u_f \rangle = \langle 200 | \mathbf{r} | 100 \rangle$$

Although we have been using SI units throughout this chapter, it is convenient to carry out these integrations using atomic units. The lifetime thus obtained will be in atomic units of time, roughly 2.4×10^{-17} s.

We first perform the angular part of the integral. To integrate over the angular coordinates we express $\mathbf{r} = r\hat{\mathbf{r}}$ in terms of spherical harmonics. This is easily done by replacing the sines and cosines in the unit vector

$$\hat{\mathbf{r}} = \sin\theta \cos\phi \hat{\mathbf{i}} + \sin\theta \sin\phi \hat{\mathbf{j}} + \cos\theta \hat{\mathbf{k}} \tag{13.69}$$

by their equivalent spherical harmonics. We obtain

$$\hat{\mathbf{r}} = \sqrt{\frac{4\pi}{3}} \left[\frac{-\hat{\mathbf{i}} + i\hat{\mathbf{j}}}{\sqrt{2}} Y_{11}(\theta, \phi) + \frac{\hat{\mathbf{i}} + i\hat{\mathbf{j}}}{\sqrt{2}} Y_{1-1}(\theta, \phi) + \hat{\mathbf{k}} Y_{10}(\theta, \phi) \right] \tag{13.70}$$

The angular integrals may be represented as

$$\sqrt{\frac{4\pi}{3}} \int d\Omega Y_{00}^*(\theta, \phi) \left[\frac{-\hat{\mathbf{i}} + i\hat{\mathbf{i}}}{\sqrt{2}} Y_{11}(\theta, \phi) + \frac{\hat{\mathbf{i}} + i\hat{\mathbf{j}}}{\sqrt{2}} Y_{1-1}(\theta, \phi) + \hat{\mathbf{k}} Y_{10}(\theta, \phi) \right]$$
$$\times Y_{\ell m}(\theta, \phi) \tag{13.71}$$

$Y_{00}^*(\theta, \phi)$ is the angular part of the ground state wave function and the $Y_{\ell m}(\theta, \phi)$ in the integrand can have $\ell = 0$ with $m = 0$ or $\ell = 1$ with $m = 0, \pm 1$. We can see immediately that the integral will vanish for $\ell = 0$ because of orthogonality of the spherical harmonics. If we acknowledge that the $Y_{00}^*(\theta, \phi)$ is a constant we can ignore it for the purpose of determining whether the integral vanishes. Because none of the terms in the expression for $\hat{r}$ contain (the remaining) $Y_{00}(\theta, \phi)$, orthogonality demands that the integral vanish. We may thus concentrate on the integral for which $\ell = 1$ and $m = 0, \pm 1$.

Using the orthogonality relation we obtain

$$\sqrt{\frac{4\pi}{3}} \int d\Omega Y_{00}^*(\theta, \phi) \left[\frac{-\hat{i} + i\hat{i}}{\sqrt{2}} Y_{11}(\theta, \phi) + \frac{\hat{i} + i\hat{j}}{\sqrt{2}} Y_{1-1}(\theta, \phi) + \hat{k} Y_{10}(\theta, \phi) \right]$$

$$\times Y_{1m}(\theta, \phi)$$

$$= \frac{1}{\sqrt{3}} \left[\delta_{m,0} \hat{k} - \left(\frac{-\hat{i} + i\hat{j}}{\sqrt{2}} \right) \delta_{m,-1} - \left(\frac{\hat{i} + i\hat{j}}{\sqrt{2}} \right) \delta_{m,1} \right] \tag{13.72}$$

This expression actually represents three matrix elements, one for each allowed value of the quantum number m. Clearly two of the three Kroneker deltas vanish for each value of m. Moreover, the squares of each of these matrix elements are identical because the absolute squares of the coefficients of each of the Kroneker deltas are unity.

Because the angular integral vanishes for $n = 2$, $\ell = 0$, the only radial integral that must be computed is

$$\int_0^\infty dr R_{10}^*(r) r^3 R_{21}(r) = \frac{1}{\sqrt{6}} \int_0^\infty dr\, r^4 \exp\left(-\frac{3r}{2} \right)$$

$$= 4\sqrt{6} \left(\frac{2}{3} \right)^5 \tag{13.73}$$

The square of the matrix element is then

$$|\langle 21m| er |100\rangle|^2 = \frac{2^{15}}{3^{10}} \cdot (\delta_{m,0} + \delta_{m,1} + \delta_{m,-1}) \tag{13.74}$$

If the p-state is not "polarized" then all m-states of $\ell = 1$ are equally populated and we may replace the sum of the Kroneker deltas by its average, viz. $(1/3)(1 + 1 + 1) = 1$ and the spontaneous transition probability is given by

$$A_{2p \to 1s} = \frac{4}{3} \cdot \left(\frac{\omega^3}{c^3} \right) \left(\frac{2^{15}}{3^{10}} \right)$$

$$= \frac{2^{17}}{3^{11}} \cdot \left(\frac{\omega^3}{c^3} \right) \tag{13.75}$$

where all quantities are in atomic units for which $\omega = 3/8 = 3/2^3$ and $c = 137$. The final result is then

$$A_{2p \to 1s} \approx 1.5 \times 10^{-8} \text{ (a.u. of time)}^{-1}$$

$$= 6 \times 10^8 \text{ s}^{-1} \tag{13.76}$$

and the lifetime is

$$\tau_{n=2} = 7 \times 10^7 \text{a.u. of time}$$
$$= 1.7 \times 10^{-9} \text{ sec} \qquad (13.77)$$

Now, the period of the electron in the second Bohr orbit is

$$T_{n=2} = 2\pi \frac{2^2 a_0}{(v_0/2)}$$
$$= 16\pi \text{ a.u} \qquad (13.78)$$

so that

$$\frac{T_{n=2}}{\tau_{n=2}} \approx 10^6 \qquad (13.79)$$

and the electron undergoes roughly one million orbits before the atom decays to the $n = 1$ state. Thus, Bohr's characterization of the allowed energy levels as "stationary" states seems entirely justified.

13.7. Angular Momentum Selection Rules

From our discussion of spontaneous emission it is clear that the electric dipole matrix element connecting the two states involved in the presumed transition determines the strength of that transition. Indeed, if this matrix element vanishes, as was found for the case for the $2s \rightarrow 1s$ transition in hydrogen, then the transition cannot proceed via electric dipole radiation. In this case, the transition is said to be "electric dipole forbidden". It is possible that the transition could proceed via a higher-order multipole moment such as electric quadrupole or magnetic dipole, but these rates are smaller than the electric dipole rates by a factor of α, the fine-structure constant.

Consideration of the conditions under which the matrix element

$$r_{fi} = \langle u_f | r | u_i \rangle \qquad (13.80)$$

vanishes provide "selection rules" for the transitions. Let us first use physical arguments to deduce two of the selection rules. In what follows, we assume that the electron that undergoes the transition, sometimes referred to as the "jumping electron" is subjected to a spherically symmetric potential energy function so the angular parts of the wave functions of the initial and final states are spherical harmonics. Thus, ℓ and m (or L and M) are good quantum numbers.

The simplest of the electric dipole selection rules states that the jumping electron cannot change spin. This can be understood when it is remembered that electronic spin is a magnetic moment, a quantity that cannot be affected by an electric field. Thus, for example, a transition between triplet and singlet states in helium is forbidden because it would require one of the electronic spins to flip.

Another rule that may be deduced is based on conservation of angular momentum. If a single photon is emitted or absorbed in the transition then the total angular momentum of the atom must change by $\pm\hbar$. This is because the photon has spin one and thus carries one unit of angular momentum, $\hbar$. Because angular momentum is a vector quantity the total angular momentum of the atom must change by $\pm\hbar$ and we must have a change in angular momentum quantum number $\Delta\ell = \pm1$. Because ℓ is also the parity of the spherical harmonic the parity of the two wave functions must be different. That is, the parity of the state must change in an electric dipole allowed transition. This is called the Laporte rule.

The selection rules discussed above are quite specific to orbital and spin angular momenta. We can, however, derive these selection rules, and others, by considering a general angular momentum $\hat{\boldsymbol{J}}$, that is, a vector operator $\hat{\boldsymbol{J}}$ that is an angular momentum because its components obey the angular momentum commutation rules. Because the electric dipole operator is proportional to the position vector $\boldsymbol{r}$, a vector operator, we may use the general properties of the matrix elements of such operators derived in Section 2.4. We then specialize to the hydrogen atom and multielectron atoms.

It was found in Section 2.4 that the matrix element $\langle jm|\hat{V}|j'm'\rangle$ vanishes unless

$$j - j' = 0, \pm1 \quad \text{but } j = j' = 0 \text{ is not allowed}$$
$$m - m' = 0, \pm1 \tag{13.81}$$

and this will also be true of the matrix elements $\langle jm|\boldsymbol{r}|j'm'\rangle$. In terms of radiative transitions, we may regard the unprimed quantum numbers as describing the initial state and the primed the final state. The relationships between the primed and unprimed quantum numbers given in Equation (13.81) then constitute selection rules. That is, they specify conditions under which a given transition can occur because the transition cannot occur by electric dipole radiation if the matrix element $\langle jm|\boldsymbol{r}|j'm'\rangle$ vanishes.

According to Equation (13.81) the z-component of the atomic angular momentum need not change, but if it does it must do so by ±1. This is easily understood in terms of the unit spin of the photon that is either absorbed or emitted. Whether m changes, and how much it changes, is determined by the polarization of the photon (or the direction of polarization of the external electric field). Because the photon is a spin 1 particle, there are three possible z-components.

It should be emphasized that these selection rules are general conditions under which the matrix element $\langle jm|\hat{V}|j'm'\rangle$ vanishes. For specific systems, for example, the hydrogen atom, some of these rules may not apply.

13.8. Selection Rules for Hydrogen Atoms

The rules discussed in this section, of course, apply to any one-electron atom such as singly ionized helium and doubly ionized lithium atoms. They also apply to alkali atoms because, as we show, it is the spherical harmonics that determine the selection rules and not the radial part of the eigenfunctions. Because the single valence electron in alkali atoms is subject to the inherently spherically symmetric potential of the inert atom shell the angular parts of their wave functions are also spherical harmonics.

For electric dipole radiation, the spin selection rule discussed above, that no change in spin can occur during a transition, applies to all atoms because spin represents the intrinsic *magnetic* moment of the electron. For hydrogen atoms this rule is simply $\Delta m_s = 0$. The selection rules for hydrogen atoms must, of course, be consistent with those already derived from angular momentum considerations because r is a vector operator. We might therefore presume that these selection rules are $\Delta \ell = 0, \pm 1$ and $\Delta m = 0, \pm 1$, but, as we show below, certain exclusions must be made.

Because the eigenfunctions for the hydrogen atom are known, we can derive the selection rules directly. We use spherical coordinates so the conditions of the previous section are met; in particular, the eigenfunctions must be eigenfunctions of the magnitude of the angular momentum and its z-component. Inasmuch as spin is not a factor, we drop the subscript on the quantum number m_ℓ. The matrix element to be investigated is

$$\langle n\ell m | r | n'\ell'm' \rangle = \int_{all\ space} R_{n'\ell'}(r) R_{n\ell}^*(r) r Y_{\ell'm'}(\theta, \phi) Y_{\ell m}^*(\theta, \phi) dV \quad (13.82)$$

which is actually three matrix elements corresponding to the three components of the vector operator r. The radial integrals are difficult to evaluate,[3] but the important conclusion is that there are no selection rules on the quantum number n. We therefore eliminate this quantum number from the matrix element designations. We have then

$$\langle \ell m | r | \ell'm' \rangle = \langle \ell m | (x\hat{i} + y\hat{j} + z\hat{k}) | \ell'm' \rangle \quad (13.83)$$

Because our goal is to determine when these integrals vanish we concentrate on the integrals themselves and not their actual values.

It is convenient to convert the three integrals to spherical coordinates for which

$$x = r \sin\theta \cos\phi \quad ; \quad y = r \sin\theta \sin\phi \quad z = r \cos\theta \quad (13.84)$$

and to examine linear combinations of the x- and y-integral. In particular, we consider

$$I_z = \langle \ell m | z | \ell'm' \rangle = r \langle \ell m | \cos\theta | \ell'm' \rangle$$
$$I_\pm = \langle \ell m | x | \ell'm' \rangle \pm i \langle \ell m | y | \ell'm' \rangle$$
$$= r \langle \ell m | e^{\pm i\phi} \sin\theta | \ell'm' \rangle \quad (13.85)$$

The integrand of I_z is proportional to $Y_{\ell'm'}(\theta, \phi) \cos\theta Y_{\ell m}(\theta, \phi)$. But, from Equation (2.116) we substitute for $\cos\theta Y_{\ell m}(\theta, \phi)$ and obtain

$$I_z = C_1 \int [Y_{\ell'm'}(\theta, \phi)]^* Y_{\ell+1,m}(\theta, \phi) d\Omega$$
$$+ C_2 \int [Y_{\ell'm'}(\theta, \phi)]^* Y_{\ell-1,m}(\theta, \phi) d\Omega \quad (13.86)$$

where C_1 and C_2 are constants that depend upon ℓ and m. Examining first the restrictions on m, we see from Equation (13.86) that, because the spherical harmonics are orthogonal, I_z vanishes unless $m = m'$.

The integrands of $I_\pm$ are proportional to $Y_{\ell'm'}(\theta, \phi) e^{\pm i\phi} \sin\theta Y_{\ell m}(\theta, \phi)$. Using Equation (2.117) we substitute for $e^{\pm i\phi} \sin\theta Y_{\ell m}(\theta, \phi)$ and obtain

$$I_\pm = C_3 \int [Y_{\ell'm'}(\theta, \phi)]^* Y_{\ell+1, m\pm 1}(\theta, \phi) d\Omega$$

$$+ C_4 \int [Y_{\ell'm'}(\theta, \phi)]^* Y_{\ell-1, m\pm 1}(\theta, \phi) d\Omega \qquad (13.87)$$

where C_3 and C_4 are constants that depend upon ℓ and m. It is clear from Equation (13.87) that these matrix elements vanish unless $m' = m \pm 1$.

Using wave functions and integrals, we have thus retrieved the selection rules on m that were derived using angular momentum methods. Moreover, we have verified that they are valid for transitions in one-electron atoms. We now examine the selection rules on ℓ using Equations (13.86) and (13.87). Again using the orthogonality condition for the spherical harmonics we see that transitions will be forbidden unless $\ell' = \ell \pm 1$. We do not, however, obtain the selection rule $\ell' = \ell$ that resulted from general angular momentum considerations; $\ell' = \ell$ transitions are forbidden because of parity. The parity of the spherical harmonics is determined by ℓ. Thus, the product of two spherical harmonics of the same ℓ in the integrand of the matrix element is necessarily even. Because the rest of the integrand, the operator r, is odd, the integral vanishes identically for such transitions. This is the Laporte rule that states:[4] even terms can combine only with odd, and odd only with even.

FIGURE 13.6 shows some of the allowed transitions of the Lyman, Balmer, and Paschen series for atomic hydrogen states up to $n = 5$. Also included are the wavelengths of these transitions in nm; for visible and infrared transitions the wavelengths listed are in air. In spectroscopy, the state of ionization of an atom is frequently indicated by a Roman numeral following the chemical symbol. Neutral atoms are I, once-ionized atoms, II, and so on. Thus, the diagram is labeled H I. Also included in the diagram are the common designations for a few of the lines of the Lyman series (Lyα and Lyβ) and the Balmer series (Hα and Hβ). Hα and Hβ were designated the C- and F-lines by Fraunhofer in the solar spectrum.

The total angular momentum j also has a selection rule. According to the general selection rules derived in the last section Δj can be 0, ± 1. Indeed, these are applicable because the selection rule on ℓ can be obeyed with $\Delta j = 0$. It is the selection rules on j that determine the total number of lines per transition that appear in the spectrum. For example, the Lyman α transition, $2p \rightarrow 1s$, consists of two lines, the $2p_{3/2} \rightarrow 1s_{1/2}$ and the $2p_{1/2} \rightarrow 1s_{1/2}$ transitions. The wavelengths of these emissions differ by roughly 1 nm.[5] Of course, all lines of the Lyman series are doublets.

The details of the Balmer series (and other series) are more complicated because there are more j-levels involved. FIGURE 13.7 shows the allowed transitions that produce the Hα transition, that is, transitions between $n = 3$ and $n = 2$ levels (ignoring the Lamb shift). This diagram is not to scale because the splitting of the lower level is considerably larger than that of the upper level.

It was seen in Chapter 7 that, ignoring the Lamb shift, the fine-structure energies depend only upon the j quantum number. Therefore, if the Lamb shift is ignored,

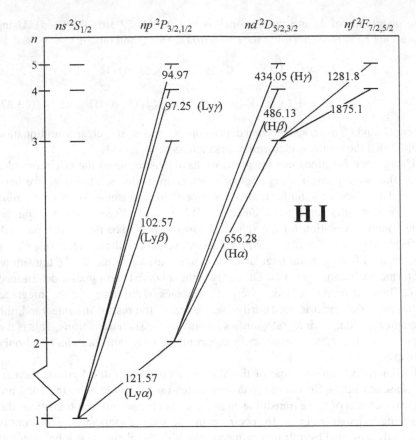

FIGURE 13.6. Partial energy-level diagram for the hydrogen atom showing some of the transitions of the Lyman, Balmer, and Paschen series. The wavelengths are listed in nm.

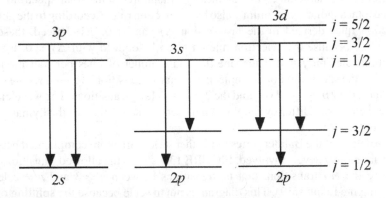

FIGURE 13.7. Allowed transitions between the $n = 3$ and $n = 2$ fine-structure levels of atomic hydrogen ignoring the Lamb shift and hyperfine interactions.

TABLE 13.1. Wavelengths and energies of the Hα emissions.

Transition	Energy (cm^{-1})	Wavelength (nm)
$3d_{5/2} \rightarrow 2p_{3/2}$	15233.07058993(10)	656.4664649
$3d_{3/2} \rightarrow 2p_{3/2}$	15233.03445369(10)	656.4680222
$3d_{3/2} \rightarrow 2p_{1/2}$	15233.40034154(10)	656.4522546
$3p_{3/2} \rightarrow 2s_{1/2}$	15233.365233485(24)	656.4537676
$3p_{1/2} \rightarrow 2s_{1/2}$	15233.256822174(24)	656.4584394
$3s_{1/2} \rightarrow 2p_{3/2}$	15232.93672343(13)	656.4722339
$3s_{1/2} \rightarrow 2p_{1/2}$	15233.302 61127(13)	656.4564662

the seven allowed transitions would yield only five distinct lines. The precise energies of these transitions[6] are, however, available (including the Lamb shift). They are listed in Table 13.1 in cm^{-1}. There are thus seven different energies; the wavelengths for these transitions are shown in the third column of the table.

FIGURE 13.8 shows the locations of the Hα wavelengths in TABLE 1. It can be seen that there are two main groupings of lines. This is a consequence of the fine-structure splitting of the $n = 2$ level being considerably larger than that for $n = 3$. This is the reason that the original work on the Balmer series regarded the lines as being "doubled" rather than broken down into seven distinct components. Of course, the lines are doublets in the sense that $2S + 1 = 1/2$.

It should be noted that the actual energies in TABLE 13.1 are not derivable from observed wavelengths because, customarily, the wavelengths given are as measured in air. Of course, it is the vacuum wavelength that accurately reflects the energy difference between the two states involved in the transition. To illustrate the difference between the air and vacuum measurement, TABLE 13.2 lists the wavelengths in air and in vacuum[7] for the Hα "doublet".

We see that, although the difference, $\sim$0.2 nm, seems small, it is not. It is worth remarking that wavelengths in the visible region of the spectrum are usually reported in Angström units ($1\text{Å} = 0.1$ nm) by spectroscopists. Thus, the difference between the air and vacuum measurements for Hα is about 2Å, which is easily resolvable in the laboratory. Again, note that conversion of the air wavelengths to

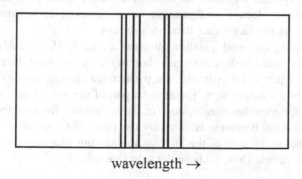

wavelength $\rightarrow$

FIGURE 13.8. The locations of the Hα lines showing that they are grouped to appear to be a doublet.

TABLE 13.2. Wavelengths of Balmer
emissions in air and in vacuum.

Component ($j \to j'$)	Air	Vacuum
$5/2 \to 3/2$	656.2849	656.4662
$3/2 \to 1/2$	656.2725	656.4538

wave numbers (accomplished by multiplying the reciprocal of the wavelength in nm by 10^7) does not produce a number that is close to those listed in the table of energies. Using the vacuum wavelength does, however.

13.9. Transitions in Multielectron Atoms

LS-*Coupling*

It was remarked in Chapter 10 that atomic states are named according to the coupling scheme of the inherent angular momenta that best describe them. We discuss the extreme coupling schemes, LS-coupling and *jj*-coupling. It is important to understand that a given atomic state may be designated according to any scheme. The utility of the designations is that the selection rules for the scheme used accurately describe transitions between states. We begin with LS-coupling, a scheme that is used for a majority of the atoms.

The state designations in LS-coupling have already been discussed in Chapter 10. The selection rules when LS-coupling is pertinent are similar to those for one-electron atoms (with capital letters replacing lowercase letters) with a few important differences. Although the selection rules for one-electron atoms are based on the mathematics of the dipole moment matrix element, selection rules for many electrons are based on both mathematics and experimental evidence. For example, one empirical selection rule for all multielectron atoms is that only one electron "jumps" during the transition. This means that in the electron configuration only one electron changes its orbital designation from initial to final state. Of course, this rule is only valid for atoms for which the wave functions are adequately described as a single-electron configuration. Often, however, it is violated. Moreover, the violation can produce rather strong transitions.

In LS-coupling the good quantum numbers are L, S, M_L, and M_S. Again, the spin is not contained in the perturbing Hamiltonian so we have the selection rule $\Delta S = 0$. Also, just as for hydrogen, the parity must change. To see this, consider a given electron configuration. The angular part of the wave function consists of products of spherical harmonics, one for each electron. Because the parity of a particular spherical harmonic is given by the value of ℓ, odd or even, the parity of the product is the sum of the angular momentum quantum numbers for the individual electrons. Thus, we have the selection rule

$$\Delta\left(\sum_i \ell_i\right) = \pm 1 \qquad (13.88)$$

where $\Delta \left(\sum_i \ell_i \right)$ is the change in the sum of the individual angular momentum quantum numbers for the electrons. Even if there is configuration interaction so that more than one configuration is present in the actual atomic wave function, the parity must change because the constituents of the wave function must each have the same parity.

For total orbital angular momentum, in addition to the selection rule $\Delta L = \pm 1$, we must retain $\Delta L = 0$, but $0 \rightarrow 0$ is forbidden, obtained from general angular momentum considerations. Clearly $\Delta L = 0$ cannot occur in one-electron atoms. In fact, it rarely occurs in atoms for which the ground state electronic configuration contains only s-valence electrons. It can, however, occur in other atoms. A simple example of this is the carbon atom. The ground state configuration is $1s^2 2s^2 2p^2$ which has even parity. From Chapter 10 we know that the possible LS states for a p^2 configuration are ^3P, ^1D, ^1S, the triplet being the ground state. There is also a low-lying excited ^1D state of configuration $1s^2 2s^2 2p 3d$. This state has odd parity. The parity of odd states in multielectron atoms is frequently explicitly noted in the state designation by a following superscript "o". Thus, the ^1D state of $1s^2 2s^2 2p 3d$ configuration is designated $3d\,^1D^\circ$. Clearly a transition between the $2p^2\,^1D$ state and the $3d\,^1D^\circ$ is allowed. It changes parity, one electron jumps, and $\Delta L = 0$. Indeed, in the spectrum of carbon, this transition of wavelength 148.2 nm is rather strong. To illustrate, FIGURE 13.9 shows a partial energy level diagram for singlet states of carbon. Only a few transitions are shown. The wavelengths are given in nm, and, for those in the visible region of the spectrum, the wavelengths are in air. Because only singlet states are included the J-values are omitted.

The two heavy lines are the *raies ultimes*[8] or "sensitive lines". These are so designated[9] because they are the last to be extinguished in an arc discharge as

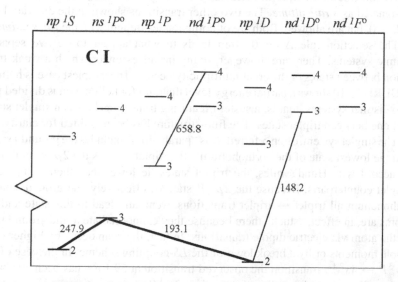

FIGURE 13.9. Partial energy-level diagram for the carbon atom showing a few relevant transitions. The wavelengths are listed in nm.

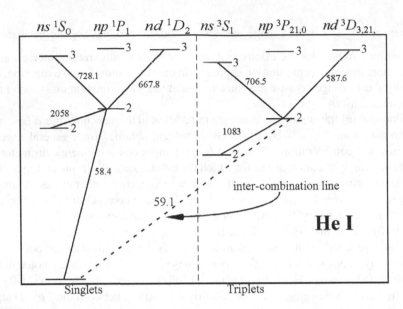

FIGURE 13.10. Partial energy-level diagram for the helium atom showing a few relevant transitions. The wavelengths are listed in nm.

the discharge is made progressively weaker. Early experimental spectroscopic studies employed arc discharges to produce the light that was analyzed. Even today, discharges are used. Often, but not always, the *raies ultimes* are the "resonance lines", that is, lines for which the lower state is the ground state. Note that the 247.9 nm line of CI does not terminate in the ground state, but is nevertheless designated as a *raie ultime*. The two other transitions shown in the diagram have $\Delta L = 0$, but are allowed. Both are of comparable strength with the *raies ultimes*.[5]

The selection rule $\Delta S = 0$ often leads to what appear to be two separate atomic systems. There are, however, many instances under which a weak transition between states of different multiplicity occurs. The simplest case is helium. FIGURE 13.10 shows a partial energy level diagram for helium that is divided into two distinct systems, that is, a system featuring transitions between singlet states and one between triplet states. The fine-structure has been omitted for clarity.

The singlet system, often referred to as "parahelium" includes the ground $1s^2 \, {}^1S_0$ and the lowest state of the "orthohelium", the triplet state, is the $2p \, {}^3P$ multiplet. In accord with Hund's rules, the triplet states lie lower than their comparable singlet counterparts. Because the $2p \, {}^3P$ state is, effectively, the ground state of orthohelium all triplet $\rightarrow$ triplet transitions eventually lead to this state and the atoms are, in effect, "stuck" there because they cannot radiate to the ground state of the atom via electric dipole transitions. In fact, they can decay by higher multipole moments or by a breakdown of the LS-coupling scheme. In the case of the $2p \, {}^3P \rightarrow 1s^2 \, {}^1S$ transition the observed transition at 59.1 nm has been attributed to a small amount of $2p \, {}^1P$ in the nominal $2p \, {}^3P$ wave function.

It is interesting that for many years it was believed that orthohelium and para-helium were two different atoms because there seemed to be two distinct sets of emission lines. Another complication is that the spacing between states of the triplets was so small that orthohelium multiplets were observed to be doublets rather than triplets. For example, the $3s\,^3S_1 \to 2p\,^3P_{2,1,0}$ transition at 706.5 nm was finally resolved, but, even today there are only two wavelengths listed in most compilations.[5,8] This conundrum persisted until the 1920s when, using very high resolution instruments, the triplet nature of the transitions was resolved.

jj-Coupling

The selection rules when *jj*-coupling pertains may be deduced from general angular momentum considerations. Although L and S are no longer good quantum numbers, J, M_J, and the individual electronic js are good. Thus, the selection rules must be $\Delta J = 0, \pm 1$, $\Delta M_J = 0, \pm 1$, and $\Delta j = 0, \pm 1$. All three of these selection rules include the usual rule that $0 \to 0$ is prohibited. In addition, as for *LS*-coupling, it is assumed that only one electron "jumps".

We have noted that the states of an atom may be designated according to any scheme we please because the designations are merely names. The utility of these names lies in whether they obey any selection rules that permit the atomic spectrum to be analyzed. A good example of this is the Hg atom for which the electron configuration is $[\text{Xe}]\,4f^{14}5d^{10}6s^2$. Thus, if *LS*-coupling is appropriate for this atom the spectrum is expected to be heliumlike and divide into two systems, singlet and triplet. FIGURE 13.11 is a partial energy-level diagram that shows some of

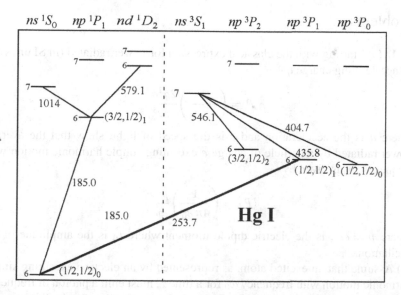

FIGURE 13.11. Partial energy-level diagram for the mercury atom showing a few relevant transitions. The wavelengths are listed in nm.

TABLE 13.3. Summary of selection rules.

Hydrogen atom	$\Delta m_\ell = 0, \pm 1; \Delta \ell = \pm 1; \Delta j = 0, \pm 1; \Delta s = 0$
LS-coupling	$\Delta M_L = 0, \pm 1; \Delta L = 0, \pm 1; \Delta J = 0, \pm 1; \Delta M_J = 0, \pm 1$ but
	$0 \leftrightarrow 0$ is forbidden for $\Delta L = 0; \Delta \ell_i = 0, \pm 1$
jj-coupling	$\Delta J = 0, \pm 1; \Delta M_J = 0, \pm 1$ but $0 \leftrightarrow 0$ is forbidden for
	$\Delta J = 0; \Delta J_i = 0, \pm 1$

the HgI emissions. The states are, as is customary, labeled in *LS*-notation, but the $6p\,^3P_1 \rightarrow 6s\,^1S_0$ ultraviolet line at 253.7 nm, a forbidden transition according to *LS*-coupling, is quite strong. It is, in fact, designated a *raie ultime*. (This emission is the primary cause of fluorescence of the phosphor that coats the inside of a standard fluorescent bulb.) The analogous transition in He, $2p\,^3P_1 \rightarrow 1s^2\,^1S_0$, does occur, however, it is quite weak.

To understand the strength of this apparently forbidden transition in HgI it is necessary to consider the states from the standpoint of *jj*-coupling. To this end the states involved are labeled in *jj*-coupling notation in the diagram. We see that although the $6p\,^3P_1 \rightarrow 6s\,^1S_0$ transition is forbidden in the *LS*-coupling scheme, the same transition, labeled $(1/2, 1/2)_1 \rightarrow (1/2, 1/2)_0$ is allowed according to *jj*-coupling. It is a resonance line in Hg! On the other hand, transitions from the $6p\,^3P_0$ and $6p\,^3P_2$ to the ground state, which are doubly forbidden in *LS*-coupling, are also forbidden in *jj*-coupling.

For convenience, the selection rules in the two coupling schemes are summarized in TABLE 13.3.

Problems

13.1. (a) Starting with the classical expression for power radiated (in SI units) by an accelerating charge q,

$$P = \left(\frac{1}{4\pi\varepsilon_0} \right) \frac{2q^2a^2}{3c^3}$$

where a is the acceleration and c is the speed of light, show that the average power radiated by a particle of charge e executing simple harmonic motion with frequency ω_0 is given by

$$\langle P \rangle = \left(\frac{1}{4\pi\varepsilon_0} \right) \frac{p^2\omega_0^4}{3c^3}$$

where $p = er_0$, is the electric dipole moment where r_0 is the amplitude of the oscillation.

(b) Assume that an excited atom, as represented by an electron executing simple harmonic motion with frequency ω_0 for a time τ, must emit a photon of frequency $\hbar\omega_0$. Using the expression for $\langle P \rangle$ above, find an expression for the lifetime τ of the excited state and for the spontaneous transition rate.

(c) Point out salient features of your answer to (b). For example, the relation between τ and ω_0, τ and the electric dipole moment, relation to the quantum mechanical result, and so on.

13.2. (a) Solve Equations (13.22) that result from the rotating wave approximation for the probabilities $P_{1\to2}$ and $P_{2\to1}$.

b) Show that the sum $P_{1\to2} + P_{2\to1} \equiv 1$.

13.3. A hydrogen atom is placed in a time-dependent homogeneous electric field

$$F(t) = F_0 e^{-t/\tau} \hat{k}$$

where $\hat{k}$ is the unit vector in the z-direction and F_0 and τ are constants. At $t = 0$ the atom is in its ground state. Calculate the probability that it will be in a $2p$ state as $t \to \infty$. Use atomic units.

13.4. Consider a spin-1/2 particle with magnetic moment μ. At time $t = 0$, the particle is in the state $|\alpha\rangle$, that is, spin up with respect to the operator $\hat{S}_z$.

(a) If $\hat{S}_x$ is measured at $t = 0$, what is the probability of measuring a value $+\hbar/2$? Let $|\alpha\rangle_x$ be the spin up eigenfunction of $\hat{S}_x$ and use analogous notation.

(b) Suppose instead of performing the above measurement, the system is allowed to evolve in a magnetic field $B = B_0 \hat{j}$ **B**. Using the $\hat{S}_z$ basis, calculate the state of the system $|\psi(t)\rangle$ at time t.

(c) At time t suppose we measure $\hat{S}_x$; what is the probability that a value $+\hbar/2$ will be found?

13.5. Look up the initial and final states of the Cd^+ ion (perhaps listed as Cd II) that are involved in the transition that yields $\lambda = 4416\,\text{Å}$. What is unusual about the initial and final electron configurations? Rationalize the necessity of this transition proceeding as it does given that the spin of a photon is 1.

References

1. G.B. Arfken and H.J. Weber, *Mathematical Methods for Physicists* (Harcourt, New York, 2001).
2. B.H. Bransden and C.J. Joachain, *Physics of Atoms and Molecules* (John Wiley, New York, 1983).
3. H.A. Bethe and E.E. Salpeter, *Quantum Mechanics of One- and Two-Electron Atoms* (Springer-Verlag, Berlin, 1957).
4. G. Herzberg, *Atomic Spectra and Atomic Structure* (Dover, New York, 1957).
5. A.R. Striganov and N.S. Sventitskii, *Tables of Spectral Lines of Neutral and Ionized Atoms* (Plenum, New York, 1968).
6. S.A. Kotochigova, P. Mohr, and B.N. Taylor, (National Institute of Standards and Technology, Gaithersburg, MD, 2004). http://physics.nist.gov/HDEL
7. F. Phelps, *M.I.T. Wavelength Tables. Volume 2: Wavelengths by Element* (M.I.T. Press, Cambridge, MA, 1982).
8. G.R. Harrison, *M.I.T. Wavelength Tables* (M.I.T. Press, Cambridge, MA, 1969).
9. C. Candler, *Atomic Spectra* (D. Van Nostrand, Princeton, 1964).

Answers to Selected Problems

Chapter 1: Background

1.2 -6.8 eV.

1.3 $E_n = \dfrac{1}{2m} \dfrac{n^2 \pi^2 \hbar^2}{L^2}$

1.4 54.4 eV

1.5 $A \approx 0.2\, a_0$

Chapter 2: Angular Momentum

2.1 Two quantum numbers with degeneracy. The degeneracy suggests that the problem will be separable in some other coordinate system. Because of the cylindrical symmetry the obvious choice is polar (cylindrical coordinates).

2.2 $\Psi(\rho, \phi) = const \cdot J_m\left(\sqrt{\frac{2\mu E}{\hbar^2}}\rho\right) \cdot e^{im\phi}$ where $m = 0, \pm 1, \pm 2 \ldots$
Since the Bessel functions wiggle there are many zeros of the argument. The quantum number m determines the order of Bessel function in the wave function. The energies are given by the condition $J_m\left(\sqrt{\frac{2\mu E}{\hbar^2}}a\right) = 0$ so that the energies are

$E_{mn} = \dfrac{\hbar^2}{2\mu a^2} r_{mn}$ where r_{mn} is the nth root of the mth Bessel function.

2.3 The enegies would be the same as in Problem 2 with the "particle-in-a-box" energies added. The wave functions would be multiplied by $\psi_q(z) = \sqrt{2/L} \sin(q\pi z/L)$ where q is the "extra" quantum number.

2.6 $\dfrac{2}{17}$

2.7 $\dfrac{\hbar}{\sqrt{2}}\left(A^*B + B^*C + B^*A + C^*B\right)$

2.8 a) $|\alpha\rangle_y = \dfrac{1}{\sqrt{2}} \begin{pmatrix} 1 \\ i \end{pmatrix}$; $|\beta\rangle_y = \dfrac{1}{\sqrt{2}} \begin{pmatrix} 1 \\ -i \end{pmatrix}$; b) $\dfrac{1}{2}$

2.10 $\dfrac{1}{4}$

Chapter 3: Angular Momentum—Two Sources

3.1 Probability of finding the electron with $j = 3/2 = |A|^2 = 1/3$
Probability of finding the electron with $j = 1/2 = |B|^2 = 1 - |A|^2 = 2/3$
Probability of finding the electron with $j = 5/2 = 0$

3.2 Probability of $m_\ell = 0$: $\left(\dfrac{1}{\sqrt{3}}\right)^2 = \dfrac{1}{3}$
Probability of $m_\ell = +1$: zero

Probability of $m_\ell = -1$: $\left(-\dfrac{2}{\sqrt{3}}\right)^2 = \dfrac{2}{3}$

3.4 $-3/2$

3.5 $j = 3/2$; $m = 1/2$

Chapter 4: The Quantum Mechanical Hydrogen Atom

4.1 a) $n = 3$; $\ell = 2$; $m = 0$; b) $3^2 a_0$ It is the Bohr radius for $n = 3$.

4.2 It *is* an eigenfunction $(n = 5)$. $E_5 = -0.544\,eV = -\dfrac{1}{5^2}Ry = -\dfrac{1}{2}\cdot$
$\dfrac{1}{5^2}\,au = -4389.492\,\mathrm{cm}^{-1}$

4.3 0.70

4.4 a) $-21/72$ a.u.: b) $(10/9)\hbar^2$ c) $(-1/36)\hbar$

4.5 $\left\{\left[\ell(\ell+1) - m_\ell^2\right] \cdot \hbar^2\right\}$

4.7 a) $-(1/8)$ a.u.; b) $\{n_1 n_2 m\} = \{100\}$ or $\{n_1 n_2 m\} = \{010\}$

4.8 a) $\psi_{par}(r, \theta, \phi) = \dfrac{1}{\sqrt{2}} R_{20}(r) Y_{00}(\theta, \phi) - \dfrac{1}{\sqrt{2}} R_{21}(r) Y_{10}(\theta, \phi)$ where the
subscript par refers to the parabolic quantum numbers given $n = 2$, $n_1 = 1$, $n_2 = 0 = m$.
b) zero; c) 1; d) $1/2$; e) $1/2$; f) 1; g) 0; h) It is the same in spherical and parabolic coordinates.

4.9 a) $-(1/8)$ a.u.; b) $\ell = 0$ and 1 are equally probable; c) $m = 0$ with probability 1

4.10 Zero for spherical coordinates, not necessarily zero for parabolic coordinates.

Chapter 5: The Classical Hydrogen Atom

5.1 $\left[(2\pi n^3) \sqrt{\dfrac{(4\pi \varepsilon_0) m_e}{e^2}} \right] a_0^{3/2}$ or $\dfrac{4\pi \varepsilon_0}{e^2} \cdot 2\pi n^3 \hbar a_0$

5.   d) $\ell = n$, circular orbits ; e) $r_{max} = n^2 a_o$, circular states

5.4   a) $A = r p^2 - p \, (p \bullet r) - \hat{r}$

Chapter 6: The Lenz Vector and the Accidental Degeneracy

6.1   1

Chapter 7: Breaking the Accidental Degeneracy

7.2 $10^{-10} E_0^{(0)}$

7.3 $E_n^{(1)} = - \left(E_n^{(0)} \right)^2 / 2 m_0 c^2$; the result might be questionable for a proton in a nucleus. It is, however, reasonable for an electron in an atom.

7.4 $E_0^{(1)} = - (3/32) \, (\hbar \omega)^2 / \left(m_0 c^2 \right)$ The correction for a diatomic molecule will be $\sim 10^{-8}$ eV.

7.5 $\Delta E_{SO} = \langle \xi \, (r) \rangle \cdot \dfrac{\hbar^2}{2} \, (2\ell + 1)$

7.6   Ground state correction: none

First   excited   state: $j = 3/2 \;\Rightarrow\; (5/2) \hbar \omega + (1/2) \zeta$; $\; j = 1/2 \;\Rightarrow\;$ $(5/2) \hbar \omega - (1/2) \zeta$

Second   excited   state: $j = 5/2 \;\Rightarrow\; (7/2) \hbar \omega + \zeta$; $\; j = 1/2 \;\Rightarrow\;$ $(5/2) \hbar \omega - (3/2) \zeta$

Chapter 8: The Hydrogen Atom in External Fields

8.2 $E \, (n = 2; \ell = 1; m_\ell = 1) = \dfrac{7}{48} \alpha^2 E_n^{(0)} + \dfrac{\mu_B B}{\hbar}$

$E \, (n = 2; \ell = 1; m_\ell = 0) = \dfrac{7}{48} \alpha^2 E_n^{(0)}$

$E \, (n = 2; \ell = 1; m_\ell = 1) = \dfrac{7}{48} \alpha^2 E_n^{(0)} - \dfrac{\mu_B B}{\hbar}$

$E \, (n = 2; \ell = 0; m_\ell = 0) = \dfrac{5}{16} \alpha^2 E_n^{(0)}$

The only states that are affected by the magnetic field are the two states for which $m_\ell \neq 0$.

8.3 $E \left(n, \ell, m_\ell, m_s \pm \dfrac{1}{2} \right) = -\dfrac{1}{2n^2} + \mu_B B \, (m_\ell \pm 1)$

8.5 b) $E_{11} = (W + \xi)$; $E_{1-1} = (W - \xi)$; $E_{10} = \left(-W + \sqrt{4W^2 + \xi^2}\right)$; $E_{00} =$ $\left(-W - \sqrt{4W^2 + \xi^2}\right)$

Chapter 9: The Helium Atom

9.1 a) Ground state energy: $E_0 = E(00) = \hbar\omega$. Ground state ket: $|n_1 n_2\rangle |SM_s\rangle = |00\rangle |00\rangle$.

First excited state energy: $E_1 = E(10) = E(01) = 2\hbar\omega$ with 4 possible kets.

$\left\{\dfrac{1}{\sqrt{2}} |10\rangle + \dfrac{1}{\sqrt{2}} |01\rangle\right\} |00\rangle$ (singlet)   and $\left\{\dfrac{1}{\sqrt{2}} |10\rangle - \dfrac{1}{\sqrt{2}} |01\rangle\right\} |S =$ $1 M_s = 0 \pm 1\rangle$ (triplet)

b)) Total spin-0 state will have the energy lowered more than that of the total spin-1 state.

9.3 a) -14.5 eV below the $\text{He}^{++} + e + e$ energy. $\lambda = 192\,\text{Å}$; **b)** $v = 3.8 \times 10^6\,\text{m/s}$

9.4 a) $-1/2$; **b)** $-(1/2)(8/3\pi)$; **c)** The wave function in a) because it is exact.

9.5 For hydrogen $\left|\dfrac{E_0^{(1)}}{E_0^{(0)}}\right|_H = \dfrac{3/8}{1} = 0.375$; For helium $\left|\dfrac{E_0^{(1)}}{E_0^{(0)}}\right|_H = \dfrac{5/8}{4} = 0.156$.

9.6 b) $(4\pi\varepsilon_0)\left(2^{19}/3^{11}\right) a_0^3$

Chapter 10: Multielectron Atoms

10.1 a) $^3P_{2,1,0}$ & $^1P_1$

b) $^1F_3$; $^1F_{4,3,2,1}$; $^1D_2$; $^3D_{3,2,1}$; $^1P_1$; $^3P_{2,1,0}$

c) $^2P_{1/2,3/2}$ and $^4P_{1/2.3/2.5/2}$; $^1D + s \rightarrow {}^2P_{3/2,5/2}$; $^1S + s \rightarrow {}^2S_{1/2}$

10.4 a) $^3D_{3,2,1}$; $^3P_{2,1,0}$; $^3S_1$; $^1D_2$; $^1P_1$; $^1S_0$

$$\left(\frac{1\,1}{2\,2}\right)_0 ; \left(\frac{1\,1}{2\,2}\right)_1 ; \left(\frac{1\,3}{2\,2}\right)_2 ; \left(\frac{1\,3}{2\,2}\right)_1 ; \left(\frac{3\,1}{2\,2}\right)_2 ; \left(\frac{3\,1}{2\,2}\right)_1 ; \left(\frac{3\,3}{2\,2}\right)_3 ;$$

$$\left(\frac{3\,3}{2\,2}\right)_2 ; \left(\frac{3\,3}{2\,2}\right)_1 ; \left(\frac{3\,3}{2\,2}\right)_0$$

10.5 a) $^3P_{2,1,0}$ $^1D_2$ $^1S_0$ and $\left(\frac{1\,1}{2\,2}\right)_0 ; \left(\frac{3\,1}{2\,2}\right)_2 ; \left(\frac{3\,1}{2\,2}\right)_1 ; \left(\frac{3\,3}{2\,2}\right)_2 ; \left(\frac{3\,3}{2\,2}\right)_0$

10.6 $J = 4$ The multiplet is ^3F.

Chapter 11: The Quantum Defect

11.1 $IP(\text{Li}) = 5.31\,\text{eV};\quad IP(\text{Na}) = 5.00\,\text{eV}$

11.2   a) $F = -eE_{dipole}\,\hat{k} = -\left(\dfrac{e}{4\pi\varepsilon_0}\right)^2 2\alpha_d \dfrac{1}{z^5}\hat{k}$

11.3 $\delta_\ell = b/\ell$

11.4 $\delta_\ell = \left(\ell + \tfrac{1}{2}\right) - \sqrt{\left(\ell + \tfrac{1}{2}\right)^2 - 2b} \approx b/\left(\ell + \tfrac{1}{2}\right)$; The $(\ell + 1/2)$ comes from the $\ell\,(\ell+1)$ term in the solution of the radial Schrödinger equation.

$$E_{n\ell} = -\frac{1}{2n^2}\left[1 + \frac{2b}{n\left(\ell + \tfrac{1}{2}\right)}\right]$$

Chapter 12: Multielectron Atoms in External Fields

12.2 $\alpha_{nm} = \dfrac{2m_e e^2 n^8 a_0^4}{\hbar^2}\left(\dfrac{1}{4m^2 - 1}\right)$ where n and m are the Bohr and rigid rotor quantum numbers.

12.4   b) There will be 19 lines observed.

Chapter 13: Interaction of Atoms with Radiation

13.1   b) $\dfrac{\hbar\omega_0}{\tau} = \left(\dfrac{1}{4\pi\varepsilon_0}\right)\dfrac{D^2\omega_0^4}{3c^3}$ and   transition rate $= \dfrac{1}{\text{lifetime}} =$
$\left(\dfrac{1}{4\pi\varepsilon_0}\right)\dfrac{p^2\omega_0^3}{3\hbar c^3}$
c) The transition probability, which is the same as the Einstein A coefficient, is proportional to ω_0^3 and the square of the electric dipole moment, p.

13.2   a) $P_{1\to 2} = |c_2(t)|^2 = \left(\dfrac{V}{2\hbar\omega_R}\right)^2 \sin^2\omega_R t$ and $P_{2\to 1} = |c_1(t)|^2 =$
$\left(\cos^2\omega_R t + \dfrac{\delta^2}{4\omega_R^2}\sin^2\omega_R t\right)$

13.3 $P_{2p}(\infty) = \dfrac{2^{15}}{3^{10}}F_0^2 \cdot \dfrac{\tau^2}{1 + \left(\dfrac{3}{8}\right)^2\tau^2}$

13.4   a) 1/2

b) $|\psi(t)\rangle = \begin{pmatrix} \cos(\mu Bt/\hbar) \\ \sin(\mu Bt/\hbar) \end{pmatrix}$ on the $\hat{S}_z$ basis.

c) $\tfrac{1}{2}[1 - \sin(2\mu Bt/\hbar)]$

13.5   Upper state: $[\text{Kr}]\,4d^9 5s^2\,{}^2D_{5/2}$ & lower state: $[\text{Kr}]\,4d^{10} 5p\,{}^2P_{3/2}$
The transition involves a "two-electron jump".

Index

A

Absorption, 2, 24, 51, 246, 249, 253, 254–262

Accidental degeneracy, 77–82, 92, 95, 96, 105–124, 126, 162, 163, 214, 221, 230

Angular momentum, 4, 6, 14–41, 46–61, 105–24, 198–208, 214–220, 266

See also Orbital angular momentum; Spin angular momentum; Generalized angular momentum

Apocenter, 95, 102, 221

Apside, 97

Associated Laguerre polynomials, 79–81, 87, 90, 133

See also Laguerre polynomials

Atomic dimensions, 8–10

Atomic units, 11, 12

B

Balmer, 3, 269–272

Bohr
energy, 5, 6, 77,
frequency, 251
magneton, 34, 146
model, 1–10
radius, 4, 95

Boltzmann factor, 262

Bosons, 180

C

Central potential, 18, 29, 73

Centrifugal term, 74

Classical radius of the electron, 11

Clebsch–Gordan coefficients, 55–72, 123, 149, 155, 169, 170

Commutation relations, 18–20, 25, 28, 116, 117

Commutator, 18–21, 25, 26, 48, 106, 108, 111, 117

Completeness relation, 161

Compton
effect, 10
wavelength, 10, 11, 130

Configuration interaction, 197, 273

Core polarization, 217

Correspondence principle, 2, 4, 6, 174, 175, 220, 247

Coulomb potential, 74

Coupled kets, 55, 56, 59, 68, 69, 154, 155

D

D–line, 51

Darwin term, 130, 133–135, 143

deBroglie wavelength, 1, 10, 13, 194

Degeneracy, 17, 78, 79, 82, 87, 92, 95–97, 105–125, 126, 136–138, 145, 148, 153, 154, 214, 221, 230

Delta function, 127, 133, 139, 259

Dipole moment
electric, 88, 101–103, 159, 162, 163, 167, 171, 175, 221, 233–236, 250, 272
magnetic, 33–36, 61, 139, 239, 240

Dirac equation, 126, 128, 130, 136, 137, 142

Doubly excited states, 192–194

E

Eccentricity, 92, 95, 100, 101, 116, 219
Effective potential, 74, 75, 78, 79
Eigenfunctions
    parabolic, 88, 121, 122, 125, 160, 163,
        164, 166, 167–170
    spherical, 82, 101, 113, 114, 122, 123,
        163, 167–170, 246
Einstein coefficients, 261, 263
Electric dipole radiation, 251, 266–268
Electrodynamics
    classical, 2, 4
    quantum, 138, 260, 264
    See also QED
Electric quantum number, 166, 176
Electromagnetic waves, 249, 250
Electron, 2
Electron affinity, 195
Electron configuration, 181, 189, 191, 192,
    196–213, 272, 275, 277
Elliptical orbits, 94, 124
Emission, 2, 24, 51, 137, 152, 242,
    246–277
Energy eigenvalues, 75–77, 85–87,
    118–120
Equivalent electrons, 200, 201, 208
    See also Nonequivalent electrons
Exchange energy, 191, 198
Exchange force, 191, 198, 204
Exchange operator, 178

F

Fermions, 178, 180
Fine structure
    constant, 7
    correction, 126–137
    interval, 136, 141, 156, 159, 160
    splitting, 62, 135–137, 156, 158, 271
    terms, 127
First quantization, 1, 10, 260, 264
Flopping. *See* Rabi flopping frequency
Fraunhofer, J., 51, 269

G

Gauss' trick, 49, 87
Generalized angular momentum, 114
    See also Angular momentum; Orbital
        angular momentum; Spin angular
        momentum

Good quantum numbers, 46–48, 148, 198,
    209, 266, 272, 275
Gyromagnetic ratio, 36, 140

H

Hamiltonian, 14
Harmonic oscillator, 16, 137
    See also Isotropic harmonic oscillator
Hartree, 12
Helium, 178–195, 196–198, 202, 266, 267,
    274, 275
Highly excited atoms, 233
    See also Rydberg atoms
Hund's rules, 192, 202, 204, 210, 211,
    274
Hydrogenic energy, 77, 109, 218
Hydrogenic manifold, 231–233, 237
Hyperfine
    Hamiltonian, 68
    interaction, 61, 62, 70, 140, 141, 270
    levels, 70
    splitting, 61, 68, 137
    structure, 61, 68, 126, 139, 141, 143

I

Indistinguishability, 178, 191, 202, 204
Integral
    Coulomb, 190–192
    exchange, 191, 192
Ionization potential, 12, 75, 183, 194
Isotropic harmonic oscillator, 17, 78, 123
    See also Harmonic oscillator

J

jj–coupling, 207–213, 275, 276

K

Kepler problem, 92, 124, 171
Keplerian
    ellipse, 97, 100, 220, 221, 226
    orbit, 2, 99, 101, 102, 116, 172, 173,
        177, 218–221, 233, 237
Kepler's laws, 96, 102, 103, 218, 221, 223

L

Ladder operators, 20, 23, 109, 110, 114,
    118, 122
    See also Raising and lowering operators
Lagrangian, 96

Laguerre polynomials, 79–81
  See also Associated Laguerre
    polynomial
Lamb shift, 126, 137–139, 141, 143, 151,
  269–271
Landé
  formula, 30
  interval rule, 205, 206
  g–factor, 150, 152, 239, 241–243
Laporte rule, 267, 269
Larmor
  formula, 263
  frequency, 239
Latus rectum, 92
Legendre polynomials, 184
Lenz vector
  classical, 97–102, 105
  operator, 105–124
Levi–Cevita symbol, 18
Lifetime, 263, 264, 266
LS–coupling, 197, 198, 202, 205, 207,
  210–212, 272–276
Lyman, 269, 270

M

Magnetic moment
  orbital, 34, 35, 50, 61, 62, 129,
    146
  spin, 35, 129, 146
Magnetic quantum number, 73, 84,
  145–148, 160, 174
Multiplet, 199, 204–206, 274, 275
Multiplicity, 198–200, 204, 207,
  274

N

Negative polarizability, 230, 238
Nonequivalent electrons, 198, 200, 208,
  209
  See also Equivalent electrons

O

Orbital angular momentum, 29–34, 50, 70,
  73, 109, 116, 130, 138, 182, 197,
  198, 200, 202, 204, 207, 273
  See also Angular momentum; Spin
    angular momentum; Generalized
    angular momentum
Orthohelium, 274, 275

P

Parabolic coordinates, 82–90, 101, 110,
  120, 164–174
Parabolic quantum numbers, 90, 120, 121,
  167, 172
Parahelium, 274, 275
Parity, 160, 267, 269, 272, 273
Paschen, 269, 270
Paschen–Bach effect, 175
Pauli
  principle, 180, 189, 200, 202, 204, 208
  spin matrices, 39
Pauli, W., 105
Pericenter, 93, 95, 100, 102, 104,
  221
Polarizability
  dipole, 160, 220, 223
  quadrupole, 218
Precession, 100, 220–225, 233–240
Principle quantum number, 95

Q

QED, 126, 137, 260, 261
  See also Electrodynamics, quantum
Quantum defect, 214–229, 230–238

R

Rabi flopping frequency, 253
Radial matrix element, 114–116
Radial wave function, 80, 81, 133,
  225–227
Raies ultimes, 273, 274
Raising and lowering operators, 20, 21, 63,
  64, 109, 148, 168
  See also Ladder operators
Recursion relation, 76
Relativistic correction, 127, 128, 130, 134,
  147, 148, 207, 248
Resonance line, 274, 276
Rigid rotor, 14–16, 245
Rotating wave approximation, 251–253,
  257
Russell Saunders coupling, 197
  See also LS–coupling
Rydberg
  atoms, 217, 224
    See also Highly excited atoms
  constant, 4–6, 8
Rydberg, R., 214

S

Secular equation, 65, 153, 156, 179
Selection rule, 26, 28, 152, 153, 210, 242,
    260, 266–276
SI units, 3, 7, 11, 83, 147, 187, 264
Singlet, 55, 56
Spectroscopic notation, 182, 190,
    192
Spherical harmonics, 31–33
Spherical symmetry, 74, 82, 95, 96, 145,
    147, 148, 161
Spin angular momentum, 24, 33–44, 50,
    62, 129, 197, 198, 207, 267
    See also Angular momentum; Orbital
        angular momentum; Generalized
        angular momentum
Spin–orbit correction, 128, 129, 132–135,
    144, 147
Spinor, 37, 62, 66, 134
Spontaneous emission, 137,
    260–266
Stark effect, 159–175, 230–238
Stationary states, 2–5, 10, 13, 218, 246,
    248, 249
Stimulated
    absorption, 249, 253–262
    emission, 253–262
Stern–Gerlach, 41–44
Superposition, 247, 252

T

Thomas correction factor, 129
Transition probability, 249, 250, 253, 256,
    257, 264, 265
Triplet, 56, 57

U

Uncertainty principle, 8–11, 13, 258
Uncoupled kets, 55, 56, 59, 60, 67, 68, 155

V

Valence electron, 201
Variational method (principle), 186, 187
Vector operator, 25–27, 106, 114, 116, 267,
    268
Virial theorem, 128, 131, 187

W

Wigner–Eckart theorem, 114, 116
WKB approximation, 225, 226

Z

Zeeman
    effect
        anomalous, 145–158, 242–244
        normal, 241–243
    energy, 147–150, 15, 241
    splitting, 150, 152, 175, 249
Zitterbewegung, 130, 137, 138